Parthenium Weed

Biology, Ecology and Management

_____

FSC
www.fsc.org
MIX
Paper from
responsible sources
FSC® C013604

CABI INVASIVES SERIES

Invasive species are plants, animals or microorganisms not native to an ecosystem, whose introduction has threatened biodiversity, food security, health or economic development. Many ecosystems are affected by invasive species and they pose one of the biggest threats to biodiversity worldwide. Globalization through increased trade, transport, travel and tourism will inevitably increase the intentional or accidental introduction of organisms to new environments, and it is widely predicted that climate change will further increase the threat posed by invasive species. To help control and mitigate the effects of invasive species, scientists need access to information that not only provides an overview of and background to the field, but also keeps them up to date with the latest research findings.

This series addresses all topics relating to invasive species, including biosecurity surveillance, mapping and modelling, economics of invasive species and species interactions in plant invasions. Aimed at researchers, upper-level students and policy makers, titles in the series provide international coverage of topics related to invasive species, including both a synthesis of facts and discussions of future research perspectives and possible solutions.

Titles Available

1. *Invasive Alien Plants: An Ecological Appraisal for the Indian Subcontinent*
   Edited by J.R. Bhatt, J.S. Singh, R.S. Tripathi, S.P. Singh and R.K. Kohli

2. *Invasive Plant Ecology and Management: Linking Processes to Practice*
   Edited by T.A. Monaco and R.L. Sheley

3. *Potential Invasive Pests of Agricultural Crops*
   Edited by J.E. Peña

4. *Invasive Species and Global Climate Change*
   Edited by L.H. Ziska and J.S. Dukes

5. *Bioenergy and Biological Invasions: Ecological, Agronomic and Policy Perspectives on Minimizing Risk*
   Edited by L.D. Quinn, D.P. Matlaga and J.N. Barney

6. *Biosecurity Surveillance: Quantitative Approaches*
   Edited by F. Jarrad, S.L. Choy and K. Mengersen

7. *Pest Risk Modeling and Mapping for Invasive Alien Species*
   Edited by Robert C. Venette

8. *Invasive Alien Plants: Impacts on Development and Options for Management*
   Edited by C. Ellison, K.V. Sankaran and S. Murphy

9. *Invasion Biology: Hypotheses and Evidence*
   Edited by J.M. Jeschke and T. Heger

10. *Invasive Species and Human Health*
    Edited by G. Mazza and E. Tricarico

11. *Parthenium Weed: Biology, Ecology and Management*
    Edited by S.W. Adkins, A. Shabbir and K. Dhileepan

Parthenium Weed

Biology, Ecology and Management

_____

Edited by

Steve Adkins,[1] Asad Shabbir[2] and Kunjithapatham Dhileepan[3]

[1]*The University of Queensland, Australia*

[2]*University of the Punjab, Pakistan and the University of Sydney, Australia*

[3]*Biosecurity Queensland, Department of Agriculture and Fisheries, Australia*

CABI is a trading name of CAB International

CABI	CABI
Nosworthy Way	745 Atlantic Avenue
Wallingford	8th Floor
Oxfordshire OX10 8DE	Boston, MA 02111
UK	USA

Tel: +44 (0)1491 832111	T: +1 (617)682-9015
Fax: +44 (0)1491 833508	E-mail: cabi-nao@cabi.org
E-mail: info@cabi.org	
Website: www.cabi.org	

A catalogue record for this book is available from the British Library, London, UK.

Library of Congress Cataloging-in-Publication Data

Names: Adkins, Steve W., editor. | Shabbir, Asad, editor. | Dhileepan, Kunjithapatham, editor.
Title: Parthenium weed : biology, ecology and management / edited by Steve Adkins, Asad Shabbir and Kunjithapatham Dhileepan.
Description: Boston, MA : CABI, 2018. | Series: CABI invasives series ; 11 | Includes bibliographical references and index.
Identifiers: LCCN 2018027226 (print) | LCCN 2018036169 (ebook) | ISBN 9781780645261 (ePDF) | ISBN 9781786392060 (ePub) | ISBN 9781780645254 (hbk : alk. paper)
Subjects: LCSH: Parthenium. | Parthenium--Control.
Classification: LCC QK495.C74 (ebook) | LCC QK495.C74 P383 2018 (print) | DDC 632/.52--dc23
LC record available at https://lccn.loc.gov/2018027226

ISBN-13: 978 1 78064 525 4

Commissioning editor: David Hemming
Editorial assistant: Emma McCann
Production editor: Marta Patino

Typeset by AMA DataSet Ltd, Preston, UK
Printed and bound in the UK by Antony Rowe, CPI Group (UK) Ltd

Contents

————————————

Contributors

Steve W. Adkins, School of Agriculture and Food Sciences, the University of Queensland, Gatton, Queensland 4343, Australia. E-mail: s.adkins@uq.edu.au

Muhammad H. Ali, First Capital University of Bangladesh, Chuadanga. E-mail: hazratali11@yahoo.com

Sally Allan, School of Agriculture and Food Sciences, the University of Queensland, Gatton, Queensland 4343, Australia. E-mail: sallyallan01@gmail.com

Ali Ahsan Bajwa, School of Agriculture and Food Sciences, the University of Queensland, Gatton, Queensland 4343, Australia. E-mail: a.bajwa@uq.edu.au

Amalia Belgeri, AGROTERRA S.A., Cuareim 1929, C.P. 11.800, Montevideo, Uruguay. E-mail: abelgeri84@gmail.com

Shane D. Campbell, Biosecurity Queensland, Department of Agriculture and Fisheries, Queensland 4820, Australia. Current affiliation: School of Agriculture and Food Science, the University of Queensland, Gatton, Queensland 4343, Australia. E-mail: shane.campbell@uq.edu.au

Nimal Chandrasena, GHD Water Sciences, 6/20 Smith Street, Parramatta, New South Wales 2150, Australia. E-mail: nimal.chandrasena@gmail.com

Michael Day, Biosecurity Queensland, Department of Agriculture and Fisheries, Ecosciences Precinct, Boggo Road, Dutton Park, Brisbane, Queensland 4102, Australia. E-mail: Michael.Day@daf.qld.gov.au

Kunjithapatham Dhileepan, Biosecurity Queensland, Department of Agriculture and Fisheries, Ecosciences Precinct, Boggo Road, Dutton Park, Brisbane, Queensland 4102, Australia. E-mail: K.Dhileepan@daf.qld.gov.au

Naeem Khan, Department of Weed Science, the University of Agriculture, Peshawar 25130, Pakistan. E-mail: nkhan@aup.edu.pk

Ian A. W. Macdonald, International Environmental Consultant, PO Box 29, Hluhluwe 3960, KwaZulu-Natal, South Africa. E-mail: macdonaldky18@planetcoms.co.za

Alec McClay, McClay Ecoscience, 15 Greenbriar Crescent, Sherwood Park, Alberta T8H 1H8, Canada. E-mail: alec.mcclay@shaw.ca

Andrew McConnachie, Weed Research Unit, Department of Primary Industries, Biosecurity and Food Safety, Orange, New South Wales 2800, Australia. E-mail: Andrew.mcconnachie@dpi.nsw.gov.au

Rachel McFadyen, PO Box 88, Mt Ommaney, Queensland 4074, Australia. E-mail: rachel.mcfadyen@live.com.au

Adusumilli Narayana Rao, ICRISAT Development Center (IDC) and International Rice Research Institute (IRRI), International Crops Research Institute for Semi-Arid Tropics, Building 303, ICRISAT, Patancheru 502324, Hyderabad, India. E-mail: anraojaya1@gmail.com

Asad Shabbir, Department of Botany, University of the Punjab, Quaid-e-Azam Campus, Lahore 54590, Pakistan. Current affiliation: Plant Breeding Institute, the University of Sydney, Narrabri, New South Wales 2390, Australia. E-mail: asad.shabbir@sydney.edu.au

Boyang Shi, Biosecurity Queensland, Department of Agriculture and Fisheries, Ecosciences Precinct, Boggo Road, Dutton Park, Brisbane, Queensland 4102, Australia. E-mail: Boyang.Shi@daf.qld.gov.au

Bharat B. Shrestha, Central Department of Botany, Tribhuvan University, Kathmandu, Nepal. E-mail: shresthabb@gmail.com

Lorraine Strathie, Agricultural Research Council – Plant Protection Research Institute, Hilton 3245, South Africa. E-mail: StrathieL@arc.agric.za

Sushilkumar, ICAR, Directorate of Weed Research Adhartal, Jabalpur-482 004, Madhya Pradesh, India. E-mail: sknrcws@gmail.com

Tamado Tana, School of Plant Sciences, College of Agriculture and Environmental Sciences, Haramaya University, PO Box 138, Dire Dawa, Ethiopia. E-mail: tamado63@yahoo.com

Saichun Tang, Guangxi Institute of Botany, Chinese Academy of Sciences, Guilin, China. E-mail: tangs0448@sina.com

Colette Terblanche, Colterra Environmental Consultants, Vredehof Farm, Vryheid, KwaZulu-Natal, South Africa. E-mail: rudolph.colette@gmail.com

Nguyen Thi Lan Thi, Department of Ecology and Evolutionary Biology, Faculty of Biology, University of Sciences, Ho Chi Minh City, Vietnam. E-mail: thi.mimosa@gmail.com

Wayne D. Vogler, Biosecurity Queensland, Department of Agriculture and Fisheries, Tropical Weeds Research Centre, Charters Towers, Queensland 4820, Australia. E-mail: wayne.vogler@daff.qld.gov.au

Arne Witt, CABI, PO Box 633-00621, Nairobi, Kenya. E-mail: a.witt@cabi.org

1 An Introduction to the 'Demon Plant' Parthenium Weed

Steve W. Adkins,[1]* Asad Shabbir[2] and Kunjithapatham Dhileepan[3]

[1]The University of Queensland, Gatton, Queensland, Australia; [2]University of the Punjab, Lahore, Pakistan; current affiliation: The University of Sydney, Narrabri, New South Wales, Australia; [3]Biosecurity Queensland, Department of Agriculture and Fisheries, Brisbane, Queensland, Australia

1.1 Introduction

In this book we ask the question, in parthenium weed do we have the 'worst weed the world has ever encountered'? The conclusion we have reached is, if not yet, then we soon will have! As this phenomenal 'demon plant' spreads around the world at a remarkable rate, causing such devastating outcomes to all aspects of agriculture, horticulture, forestry and the natural environment, as well as being a significant health concern, it is coming under unparalleled scientific and public scrutiny.

Parthenium Weed: Biology, Ecology and Management has been a collective effort by 26 members of the International Parthenium Weed Network. The book builds on a fundamental understanding of invasive plant biology and weed science that can be acquired from the many good texts available. Given such grounding, our broad aim in this book is to emphasize the practical relevance of understanding biology and ecology to enable effective and sustainable management, hence our subtitle – biology, ecology and management.

Parthenium Weed: Biology, Ecology and Management has a conspicuously world focus, drawing on examples from 48 countries which have found themselves with the misfortune of being invaded by this phenomenal weed. The journey through the biology and ecology reveals the very special nature of this quite amazing plant, as well as the general principles that apply universally within invasive weed science.

Our narrative is to build on credible observations and experiments and is abundantly illustrated with original data and well-selected images. Numerous summary sections provide a clear background to the new knowledge that is readily accessible and structured for easy reading. Within the book, one key theme that has been used to impart coherence through the specialized contributions from the 26 authors has been the integrated thought process to initially understand the weed and then to manage it. This theme refers to a constant interplay between internal (genetics) and external (environment) factors that drive every facet of the weed's existence.

Knowing which traits confer weedy status, and which stages of the life cycle are best to target to achieve meaningful management, continue to be major challenges for weed science – *Parthenium Weed: Biology, Ecology and Management* rises to these challenges.

* s.adkins@uq.edu.au

1.2 Know Your Enemy – Global Impacts and Losses

Originally regarded as a major weed in Australia and India, parthenium weed is now widespread in about 48 countries in Africa, Asia and the South Pacific, and has the potential to spread to new countries in Africa, Asia and parts of Europe. Until the end of the 20th century, most information on parthenium weed came from Australia and India. However, with the emergence of parthenium weed in Asia and Africa, there has been considerable interest and research effort in other countries as well. Despite the weed affecting the livelihoods of millions of people in Asia and Africa, by causing significant economic, health and environmental loss, information available on the weed is scattered, mostly as research publications. Except for conference proceedings focusing on parthenium weed research on a regional scale and general reviews, currently there is no book available on parthenium weed reflecting its global weed status, despite the considerable research achievements over the last five decades in many countries.

Research on parthenium weed is in progress in many countries, including Australia, India, South Africa, Ethiopia, Pakistan, Bangladesh, China, Sri Lanka and Nepal. This book, with contributions from expert researchers with extensive involvement in parthenium weed research from these countries, has collected and synthesized existing knowledge on parthenium weed in 16 chapters covering aspects of: (i) biology; (ii) ecology; (iii) genetics; (iv) introduction histories; (v) geographic distribution; (vi) the impact on agriculture, natural forests and the environment of protected areas; (vii) allelopathy; (viii) impacts on human and animal health; (ix) potential uses; and (x) management strategies, including chemical, cultural and biological control methods.

The book brings in experts from 13 countries (Australia, Bangladesh, Canada, China, Ethiopia, India, Kenya, Nepal, Pakistan, South Africa, Sri Lanka, Uruguay and Vietnam) and is the first book exclusively on parthenium weed. The book also provides current distribution records/status of the weed, along with future risks of spread based on climate change. There are dedicated chapters on the current status of parthenium weed problems in Australia and the Pacific, Southern Asia, East and South-east Asia, North Africa and the Middle East, and southern Africa. All chapters have relevant photos and figures included to make the reading interesting.

This book will be a comprehensive reference book for researchers, students, professionals involved in weed management, government officials and policy makers and anyone interested in parthenium weed. As the research needs and knowledge gaps vary widely across different countries, the book has also identified these gaps, and provides direction for future research for various countries. This is the first book on parthenium weed under the CABI invasive species series. The book will be of immense value to all countries with a parthenium problem, which will benefit by sharing knowledge and experience.

1.3 Biology, Ecology and Spread

Chapter 2 provides comprehensive information on the overall biology and ecology of the weed. It provides specific details on the taxonomy, plant distinguishing characteristics, its likely centre of origin, genetics and intraspecific diversity, growth, reproduction and phenology, seed biology (including seed production, dispersal, germination, longevity and seed bank), population dynamics and preferred climatic requirements for growth. The chapter also highlights how various biological and ecological characteristics of parthenium weed, such as its morphological attributes, biological plasticity, intermediate photosynthesis mechanism, allelopathy, stress tolerance, competitive ability and long-lived seeds, make it one of the most successful global invasive species.

Chapter 3 deals with how parthenium weed, with a humble origin in the neotropics, has ended up as a global weed with a pantropical distribution spreading across about 48 countries in the last five to six decades (Fig. 1.1). The chapter specifically deals with the spread pathways adopted by the weed and the present distribution in the invaded continents. The chapter also highlights that parthenium weed is likely to expand its geographical distribution range even more, particularly into the Mediterranean and eastern European, Western Africa and South and South-east Asian regions, with anticipated future climate change.

1.4 Impacts

Chapter 4 provides comprehensive details on how parthenium weed interferes with and negatively affects: (i) crop production, including grain crops, horticultural crops, vegetables, fibre and field crops, and agroforestry; (ii) farm animal production, including pastures, fodder crops and meat production; and (iii) the socio-economic aspects of agriculture (Fig. 1.2).

Chapter 5 deals with the impact of parthenium weed on the environment, more specifically, the negative impact of the weed on soil properties, and the above- and below-ground community biodiversity, including fauna, flora and microorganisms. Also, the chapter comprehensively examines the negative impact of parthenium weed on insect pollinators and shows how the weed can: (i) alter nutrient cycling; (ii) reduce plant species diversity and abundance; (iii) change vegetation structure; and (iv) alter the assemblage of other organisms such as invertebrates, amphibians, reptiles, birds and mammals.

Chapter 6 deals with various human and animal health impacts of the weed. Diseases caused by the weed include dermatitis, rhinitis, asthma and atopic dermatitis. The detrimental health effects of parthenium weed are attributed to the sesquiterpene lactones and in particular parthenin in the plant which are toxic to farm animals and responsible for allergic diseases in humans (Fig. 1.3).

Fig. 1.1. The dramatic and rapid spread of parthenium weed around an abandoned homestead in South Africa. (Ezemvelo KZN Wildlife and Department of Environmental Affairs, South Africa.)

Fig. 1.2. Heavy infestation of parthenium weed in a maize field in Bangladesh. (Ilias Hossian, Bangladesh.)

1.5 Management

Although parthenium weed has now invaded more than 48 countries around the world, and threatens to invade more in the future, development of various tools to manage the weed is mainly based on availability of resources, level of awareness and socio-economic status of the affected regions.

Chapter 7 focuses on biological control, one of the main strategies of parthenium weed management. The chapter covers key biological control agents available for management of parthenium weed, which are mainly based on Australian initiatives. However, this chapter also encompasses introduction history and status of parthenium weed biological control agents in other parts of the world, for example Eastern and South Africa, South Asia and the Pacific Islands.

Chapter 8 considers the importance of other management strategies and discusses them in the context of integrated weed management. A detailed account is presented of cultural (legislative measures, hygiene practices, crop rotation, cover crops, competition and suppression), physical (manual and mechanical removal, fire and heat) and chemical (synthetic and natural products) approaches used to manage parthenium weed in different parts of the world (Fig. 1.4).

In Chapter 9, the critically important role that coordination and awareness can play in managing parthenium weed is addressed. Of all the invaded countries, only Australia and South Africa have put in place well-developed national coordinated strategies for the management of parthenium weed. This chapter describes in detail the major components of Australian and South African coordination strategies and

Fig. 1.3. A worker showing severe skin allergy symptoms due to parthenium weed. He was employed by the Punjab Forest Department, Lahore to slash and remove the weed from the understorey of Changa Manga Forest Reserve. (Asad Shabbir, Pakistan.)

discusses their role in containment of weed spread. The role of public awareness campaigns in highlighting the issue of parthenium weed is also discussed, followed by a discussion on the potential role of societies, action groups and international networks in creating awareness and linkages on parthenium weed.

There has been a debate about the potential role of utilization in management of parthenium weed. A large volume of published material is available on potential uses of this weed and opportunities for further exploitation. Chapter 10 critically reviews this published information on the actual uses of the plant in different countries. It also weighs the benefits and potential problems associated with utilization of this weed in countries where existing infestations are large and threaten to spread further.

1.6 History and Regional Management

Chapters 11–15 outline the history, background, spread and management tools used for parthenium weed in Australia and the Pacific, Southern Asia, East and South-east Asia, southern Africa and the Western Indian Ocean islands, East and North Africa, and the Middle East. Parthenium weed has become one of the biggest threats to subsistence farming and future food security of some poor nations in Africa and Asia. The negative effects of parthenium weed on the environment in developing countries are extensive, yet they are least documented due to the limited knowledge and resources in these countries. Recently the level of awareness about the weed in East and southern African countries has increased, thanks to international donors who funded research programmes on management, especially biological control in Ethiopia, Tanzania and Uganda. Parthenium weed is a relatively new weed in some parts of South-east Asia, for example Malaysia and Thailand, and local authorities should take notice of the situation and start implementing eradication/containment programmes to stop its further spread.

Chapter 16 draws conclusions and makes future recommendations on the management of parthenium weed, based on extensive information reviewed in different chapters of the book. There are some research gaps, particularly in quantification of economic losses and health effects of parthenium weed. Integrated management options are the way forward in the management of the weed under changing environmental conditions. Finally, this chapter looks at the gaps in our knowledge and how we might close these with collaborative programmes of research around the globe.

1.7 Final Note

If effective management programmes are not put in place, parthenium weed is expected to spread further in all regions

Fig. 1.4. Eradication of parthenium weed from a vegetable- and maize-growing area in Lampoon Province, northern Thailand using physical and chemical management. Thailand is one of the countries most recently invaded by parthenium weed. (Siriporn Zungsontiporn, Thailand.)

mentioned above. The future changing climate would further help in its spread to more countries and exacerbate the problem. It is alarming that just during the process of writing this book, we witnessed three confirmed reports of weed invasion in the Kingdom of Saudi Arabia, Thailand and the United Arab Emirates. Well-coordinated and effective national strategies are required to combat the issue of parthenium weed.

Acknowledgements

We would like to thank all the contributors to this book, for their time and commitment, valuable inputs and patience during the process of writing and editing of the various chapters. We would also like to express our special thanks to all collaborators and colleagues, members of the International Parthenium Weed Network and family members who have helped us in any way. Arne Witt, Ian W. MacDonald, Lorraine Strathie, Sushil Kumar, Andrew McConnachie, Kelli Pukallus, K. Verma, Zahid Ata Cheema, Bharat B. Shrestha, Thi Nguyen, Boyang Shi, Saichun Tang, S.K. Garu, Siriporn Zungsontiporn, Ilias Hossain, Ram B. Khadka, Ezemvelo KZN Wildlife and the Department of Environmental Affairs, South Africa, are greatly acknowledged for supplying some of the photos used in this book.

2 Biology and Ecology

Steve W. Adkins,[1]* Alec McClay[2] and Ali Ahsan Bajwa[1]

[1]*The University of Queensland, Gatton, Queensland, Australia;*
[2]*McClay Ecoscience, Alberta, Canada*

2.1 Introduction

Parthenium weed (*Parthenium hysterophorus* L.) is now recognized as a major invasive weed worldwide. Yet back in the 1950s, when it first came to the attention of land managers in Australia, it was a virtually unknown plant. International focus wasn't drawn to its weed potential until the mid-1970s, after reports of dense infestations forming in central India associated with increasing health problems, and its rapid spread in Australia. Understanding the biology and ecology of this unique weed is essential for determining its impact in the natural and agricultural environment, and also in helping design new and improved, cost-effective management strategies. This chapter provides details of the biology, ecology and origins of this weed and a discussion of why this plant has become such a successful invader of many new landscapes in over 40 countries around the world.

2.2 Parthenium Weed: Biology

2.2.1 Nomenclature

Parthenium is derived from the Greek (parthenos), meaning 'virgin', possibly in reference to the white flowers produced by the majority of species of this genus (Strother, 2006). It may also be derived from the Greek (partheniki) and Latin (parthenice) names for the plant now known as feverfew (*Tanacetum parthenium* (L.) Schultz-Bip.), as feverfew and parthenium weed are both known as a treatment for fever (McFadyen, 1985; Parsons and Cuthbertson, 1992). The species name *hysterophorus* was coined by Vaillant (1720) and is derived from the Greek hystera (womb) and phoros (bearing), from a supposed resemblance of the roughly triangular achene complex, with its two attached sterile ray florets, to female genitalia. The earliest recognizable illustration of parthenium weed was published by Nissole (1711), with the name *Partheniastrum*.

Parthenium hysterophorus L. is most often referred to as parthenium or parthenium weed, but there is an array of alternative common names in use around the world. Some of the more commonly used names in other parts of the world include: (i) bitter weed, carrot weed, carrot grass, gajar grass, broom-brush, congree grass, congress grass and congress weed (India); (ii) gajar booti (Pakistan); (iii) whitetop, escoba amarga, feverfew and false chamomile (Caribbean); (iv) Demoina weed (Zimbabwe); (v) Santa Maria feverfew, false ragweed and ragweed parthenium (USA); and (vi) famine weed (South Africa). Common names used for parthenium weed in Mexico include falsa altamisa, altamisa del campo, altamisa cimarrona, altamisilla, cola de ardilla, manzanilla del campo, romerillo, yerba de la oveja, arrocillo, confitillo, chaile, hierba del burro,

* s.adkins@uq.edu.au

hierba del gusano, huachochole, jihuite amargo, zacate amargo, cicutilla and hierba ama-
rgosa (Martínez, 1979; Villarreal Quintanilla,
1983; Calderón de Rzedowski and Rzedowski,
2004). Several of these names refer to the
plant's bitter ('amargo' in Spanish) taste. In
Brazil the Portuguese name losna-branca is
used (Gazziero *et al.*, 2006).

2.2.2 Taxonomy

The genus *Parthenium* belongs to the tribe
Heliantheae of the family *Asteraceae* (Table
2.1). The most distinctive feature of the
Asteraceae is the configuration of the inflorescence into a head, or capitulum, consisting of numerous florets surrounded by
bracts. The fruit of this family, which is
achene-like and derived from an inferior
ovary, is termed a cypsela (Parsons and
Cuthbertson, 1992). Stuessy (1973) removed
the genus *Parthenium* from the subtribe
Melampodiinae and placed it in the subtribe
Ambrosiinae, which contains genera such as
Ambrosia, *Xanthium*, *Iva* and *Parthenice*.
Members of the *Ambrosiinae* are mainly
characterized by possession of sterile disc
florets, adaptations for wind pollination,
and the presence of distinctive sesquiterpene chemicals (Robinson *et al.*, 1981; Miao
et al., 1995). However, it has also been suggested that *Parthenium* is sister to the genus
Dugesia and may in future be transferred,
with *Parthenice*, to the subtribe *Dugesiinae*.

Table 2.1. A detailed taxonomic hierarchy of
parthenium weed.

Taxon	Parthenium weed
Kingdom	*Plantae*
Subkingdom	*Tracheobionta*
Division	*Magnoliophyta*
Class	*Magnoliopsida*
Subclass	*Asteridae*
Order	*Asterales*
Family	*Asteraceae*
Tribe	*Heliantheae*
Subtribe	*Ambrosiinae*
Genus	*Parthenium*
Species	*Hysterophorus*

Further sampling is needed to clarify the
relationships of these genera (Panero, 2005).

Molecular examination of chloroplast
DNA has led to the identification of two
groupings within the subtribe *Ambrosiinae*, a
basal lineage containing *Parthenium* and the
other comprising the rest of the subtribe
(Miao *et al.*, 1995). A morphological analysis
by Karis (1995) also showed *Parthenium* as a
basal group to the rest of *Ambrosiinae*.

The genus *Parthenium* is regarded as the
evolutionary precursor for this subtribe and
contains 16 species native to the North and
South American continents, with the highest concentration of taxa found in Mexico
(Stuessy, 1975). It includes bitter aromatic
herbs and shrubs that have a fruit that is
achene-like (Rollins, 1950; Mears, 1973;
Stuessy, 1975).

The genus *Parthenium* has a very uniform and distinctive floral structure, with
five fertile female ray florets, each of which
has two attached sterile disc florets and a
subtending bract or phyllary, the whole
being shed as a unit and termed an achene
complex by Rollins (1950). However, the
species of *Parthenium* have a very wide range
of growth forms, from annual and perennial
herbs to cushion plants, shrubs and small
trees. The genus is in need of revision, with
the only comprehensive treatment being
that of Rollins (1950), and there has been no
phylogenetic analysis of the genus using
molecular methods. Rollins recognized four
sections within the genus *Parthenium*: (i)
Parthenichaeta, with seven species, all shrubs
or small trees, occurring mainly in Mexico
but with one species in Bolivia; (ii) *Argyro-
chaeta*, with five herbaceous species, of
which four, including parthenium weed,
occur in Mexico and the southern USA and
one species in Bolivia; (iii) *Partheniastrum*,
with two herbaceous species in the USA; and
(iv) *Bolophytum*, with two alpine cushion-
plant species in found in Utah, Colorado and
Wyoming, USA. However, the distinction of
these sections was mainly based on habit
and Miao *et al.* (1995) have questioned their
validity.

Because of the lack of a modern phylogenetic analysis, the relationships of parthenium weed to other members of the genus

are not clear. Rollins (1950) thought that the most closely related species to parthenium weed were the perennial Gray's feverfew (*Parthenium confertum* Gray) and the annual *Parthenium bipinnatifidum* (Ortega) Rollins, both native to Mexico and a possibly synonymous form, *Parthenium glomeratum* Rollins, native to Bolivia and Argentina. Later studies suggested that *P. glomeratum* may not be distinct from parthenium weed (de la Fuente *et al.*, 1997; Piazzano *et al.*, 1998).

2.2.3 Description

Parthenium weed is an annual, or short-lived perennial, herbaceous plant with a deep tap root and an erect stem system, reaching a height of more than 2.0 m in good soil and moisture conditions (Haseler, 1976; Navie *et al.*, 1996; Adkins and Shabbir, 2014). Flowering can occur just 4–6 weeks after seedling emergence, which, in appropriate soil moisture and temperature conditions, can occur at any time of the year (Navie *et al.*, 1996). Its aerial parts do not tolerate frost and die in winter (Shabbir, 2012), although after mild frost, the plant can regrow from stem bases and therefore is considered by some to show perennial characteristics. The plant is capable of flowering even when still in the rosette stage, 4 weeks after germination. A single plant in the field has been estimated to produce 39,192 flowers (or c.156,768 seeds; Dhileepan, 2012), while glasshouse studies report fewer filled seed produced per plant (25,000; Nguyen *et al.*, 2017b) with each flower head (about 0.5 cm in diameter) containing up to five seeds which may be distributed by vehicles, machinery, fodder, pasture seed, stock feed, wind and water (Auld *et al.*, 1982). Several aspects of the plant's biology and ecology contribute to its invasiveness. These include: (i) the large seed production; (ii) large persistent soil seed banks; (iii) the longevity of its seeds when buried; (iv) its fast germination rate; (v) its quick flowering; (vi) flowering over a long period of time; (vii) its allelopathic capacity; and (viii) its ability to adapt to many different and stressful environments (Bajwa *et al.*, 2016).

2.2.4 Distinguishing characters

Parthenium weed may be confused with several other species, principally ragweed species (i.e. *Ambrosia artemisiifolia* L., *Ambrosia tenuifolia* Sprengel., *Ambrosia psilostachya* DC. and *Ambrosia confertiflora* DC.), especially in the vegetative growth stage (Navie *et al.*, 1996). However, these *Ambrosia* species can be distinguished by their oppositely arranged leaves in the early stages of growth and by the lack of a distinctly grooved stem. They can be more clearly recognized from parthenium weed when flowering. Then the small white flower heads or capitula of parthenium weed can be observed in much branched terminal panicles and are quite distinct from the spike-like racemes of the *Ambrosia* spp. which are predominantly green in colour.

2.2.5 Intraspecific variation

In most invaded countries there seems to be little morphological variation in populations of parthenium weed, probably as a result of the limited size of the initial introductions. However, variation in leaf morphology can be observed in the field, with some differences in reproductive biology (Hanif, 2015) and biochemical profile (Shi, 2016; Ahmad, 2017) also detectable. Of the two separate introductions of parthenium weed into Australia, plants from the second introduction (in Clermont, central Queensland) seem to be much more invasive than those from the first introduction (in Toogoolawah, southeast Queensland). Detailed comparisons between plants from the two introductions have indicated distinct morphological and genetic differences between the two populations.

In the Americas, there are two distinct forms or races of parthenium weed, termed the 'South American' and 'North American' races (Dale, 1981). The 'North American' race is more common and is the one that has been introduced into all parts of the world. The 'South American' race has cream to yellow flowers and differs from the 'North American' race in leaf morphology (S.W. Adkins, Brisbane, 2017, personal communication), pollen colour, capitula size, development of axillary branches, size of disc florets and size of ray corollas (Dale, 1981). Hymenin is often the dominant sesquiterpene lactone found in the plants of the 'South American' race, whereas parthenin is the dominant sesquiterpene lactone in the 'North American' race (Picman and Towers, 1982). These large differences in the chemistry and morphology of the two races of parthenium weed indicate that they may represent subspecies (Picman and Towers, 1982); however, Hanif (2015) suggests that they are genetically close based on chloroplast DNA (cpDNA) analysis and therefore are unlikely to represent separate species. Among the 'North American' race, Parker (1989) has reported two distinct biotypes from different locations in Mexico. The first of these biotypes produces a plant with a rosette of leaves that does not begin stem elongation until flowering has been initiated and the second biotype has no rosette stage (McClay, 1983), with leaves that are more hirsute. However, this might simply be an artefact of the time of the year, as McClay (1983) has observed autumn emerging seedlings to form rosettes while spring emerging seedlings form no rosettes or very loose ones.

Rollins (1950) reports plants from Mexico that appeared to be hybrids of parthenium weed with *P. confertum* var. *lyratum*, and also notes that artificial hybrids of parthenium weed with *Parthenium argentatum* A. Gray and *Parthenium incanum* Kunth have been produced. Rollins (1950) and Picman and Towers (1982) suggested that many of the morphological and chemical characters of the highly variable South American race of parthenium weed appear to originate from both Gray's feverfew (*P. confertum* Gray) and bitter yerba (*P. bipinnatifidum* (Ortega) Rollins). However, they concluded that such hybrids were unlikely to occur in South America because these species are only known to grow in the presence of parthenium weed in Mexico, where such extensive diversity in parthenium weed has not been found. The chromosome number of parthenium weed is $2n = 34$ (Towers *et al.*, 1977).

2.3 Growth and Reproduction

2.3.1 Vegetative morphology

Parthenium weed is a rapidly growing, erect and much-branched annual, reaching up to 2.0 m in height, though most individuals do not exceed 1.0 m (Haseler, 1976). The cotyledons of the seedling are hairless, up to 3 × 6 mm in size, and possess only a short petiole (Fig. 2.1a). The young plant often forms basal rosettes of leaves that are up to 15 cm in length and 2–4 cm wide (Fig. 2.1b). These rosette leaves are pale green, pubescent, scissored into narrow pointed lobes and are arranged alternately onto the main stem (Parsons and Cuthbertson, 1992; Adkins *et al.*, 1997). These leaves are also strongly dissected into narrow lobes and spread horizontally, very close to the ground in a rosette (Fig. 2.1c) and can cover a considerable area, thereby interfering with the emergence of other seedlings (Jayachandra, 1971; Everist, 1976). Upon stem elongation, the upper leaves (which are smaller, narrower and less dissected than the basal leaves) are produced alternately on the stem (Figs 2.2 and 2.3; Everist, 1976). The lower leaves can be as much as 20 cm long and up to 4–8 cm wide, while the upper leaves are shorter (Haseler, 1976; Parsons and Cuthbertson, 1992; Navie *et al.*, 1996). The stem is grooved and succulent at the early stages of growth but becomes inflexible at the later stages of growth (McFadyen, 1985; Parsons and Cuthbertson, 1992). Both the leaves and the stem are covered with short, fine trichomes (Haseler, 1976; Williams and Groves, 1980; Navie *et al.*, 1996). The weed produces a deep tap root with many small root hairs (Parsons and Cuthbertson, 1992) which aid the absorption of moisture and nutrients from the deeper layers of the soil

Fig. 2.1.   The early growth stages of parthenium weed showing the cotyledonary stage (a), the early seedling stage (b) and the rosette stage (c). Scale bars are 1.0 cm. (a, b: R. Mao; c: S. Navie.)

Fig. 2.2. A flowering parthenium weed plant approximately 50 days old. (A. Shabbir.)

Fig. 2.3. A small clump of field-growing parthenium weed plants approximately 80 days old. (S. Adkins.)

profile, as well as helping the plant in withstanding drought (Navie *et al.*, 1996). The root system is capable of storing large quantities of nutrients and water, thus supporting plant regrowth when the plant is cut (Haseler, 1976).

2.3.2 Phenology

Parthenium weed is able to germinate, grow and flower over a wide range of temperatures and photoperiods, hence it can be present in the field at any time of the year (Haseler, 1976). In Australia the main season of growth is during the summer months when rainfall is usually more regular and abundant. In India, Pakistan and Nepal the best growth is observed in the wet season. In either location, four or more successive cohorts of seedlings may emerge on the same site during a good growing season (Everist, 1976; Pandey and Dubey, 1989). Plants that emerge in the spring or early in the growing season seem to attain a greater size and have a longer lifespan than those that emerge in the summer or later in the

growing season (Adkins and Shabbir, 2014). Soil moisture seems to be the major limiting factor to the duration of flowering. Those plants that emerge after early spring rains can have a lifespan of 6–8 months if soil moisture remains adequate (Doley, 1977; McFadyen, 1992), whereas the plants emerging in summer may only live half as long. This may be a consequence of the soil drying more quickly during the hot summer months. Plant growth increases with a rise in temperature up to an optimum of 33°C (Williams and Groves, 1980). Temperature is also a factor controlling the duration of vegetative growth, while day length seems to have little effect on this aspect of the weed's phenology (Navie, 2002).

Flowering occurs most rapidly at a warm 27/22°C day/night temperature regime and less rapidly at cooler regimes (Fig. 2.2). Under ideal conditions, parthenium weed has the ability to reach maturity and set seeds quickly, and hence has the potential to become a serious weed throughout the warm, humid and subhumid regions of most countries. The main limitation to this weed's distribution was initially thought to be in areas where there were extremes of temperature occurring all year round, or in areas where rainfall was limited (Dale, 1981). However, though it shows increased performance in its introduced range, parthenium weed now seems to be able to grow and reproduce under a wider range of temperature regimes and in more arid environments than in its native range. Doley (1977) suggested that the distribution of this weed may also be limited by heavy shading or prolonged drought. Plants emerging just before winter in Mexico produce large, well-defined rosettes, with leaves closely appressed to the ground, and remain in this stage until spring before bolting and flowering occurs. Field observations from north-eastern Mexico also indicate that plants in the rosette stage of growth were only seen in the winter months from October to March. This winter rosette form is not simply a response to water stress, as it was also shown by well-watered plants in the greenhouse; it may be a response to temperature or photoperiod (McClay, 1983). Plants emerging in spring only formed loose rosettes which soon developed into bolting plants flowering in summer and scenescing in autumn (Fig. 2.3).

2.3.3 Floral biology

Flower initiation can start as early as 28 days after seedling emergence from the soil (Jayachandra, 1971), while others report that flower initiation will take up to 42–63 days (Navie et al., 1996). The flower head or capitulum consists of a conical receptacle surrounded by an outer involucre of five persistent bracts, five (some times six to eight) peripheral fertile ray florets, and 12–20 central cylindrical staminate disc florets, each bearing four connate anthers (Fig. 2.4a) (Navie et al., 1998). The appearance of reddish-brown spots on the stigmas of the ray florets indicates successful pollination has taken place. Pollen grains are mostly spheroidal, 15–20 µm in size, and

have short to medium-length spines often permeated with micropores (Lewis et al., 1991). An average of 150,000–350,000 pollen grains are produced in each capitulum, and as thousands of flower heads can be present on each plant, pollen production by an average plant is extremely large, c. 850 million pollen grains per plant (Kanchan and Jayachandra, 1980; Lewis et al., 1988) or over 10 billion/m² in a typical stand of the weed. Large amounts of airborne pollen from parthenium weed have been detected both in America and in India at a variety of altitudes (2–915 m above sea level) and at considerable distances from parthenium weed populations (Lewis et al., 1991).

There are conflicting reports as to whether parthenium weed is self- or cross-pollinated, and also what is the actual mechanism of pollination. Esau (1946) reported that apomixis did not occur in parthenium weed and that the species was only known to reproduce after pollination. Lewis et al. (1988), working on plants from the native range, considered the species to exhibit a high degree of self-pollination (95% of achenes produced in this way were viable) with little or no insect pollination. They concluded that wind or self-pollination was to account for the majority of seed produced. In later studies, Lewis et al. (1991) advanced the notion that the mechanism of wind pollination in parthenium weed was less developed than that seen in many other wind-pollinated species, indicating that self-pollination was probably the most common form of pollination in this species. However, Gupta and Chanda (1991), in the invaded range of India, noted that parthenium weed plants appeared to be insect pollinated or at least with pollen dispersed mainly by insects. The main pollinating agents were thought to be bees, ants, flies and other dipterans that frequently visited its flowers. They concluded that parthenium weed is not normally self-pollinating, but ants may occasionally induce the process of self-pollination after visiting flowers from the same plant. More recently, Hanif (2015) has raised the very interesting idea that in certain parts of the invaded range, where the plants are extremely invasive, they are mainly cross-pollinating plants while less invasive plants are self-pollinated.

2.3.4 Seed production

Parthenium weed is a very prolific seed producer and will continue to flower and fruit profusely until fully senesced (Haseler, 1976). Seed (retained within the achene complex) is shed gradually throughout the latter stages of growth, while other seed is retained in the flowers until after senescence (Fig. 2.4c; Parsons and Cuthbertson, 1992). Each flower-head produces up to five blackish seeds (Fig. 2.5b), occasionally six to eight (Fig. 2.4b), of uniform size (1.0–1.5 mm in length) and weighing from 0.4 mg to 0.8 mg (Auld *et al.*, 1982; Lewis *et al.*, 1988) enclosed in a straw-coloured achene complex with two lateral attached sterile florets (Fig. 2.5a; Navie *et al.*, 1996). Other studies have reported only four obovate and flat-shaped seeds to be produced in each capitulum (Jayachandra, 1971; Williams and Groves, 1980; Auld *et al.*, 1982). While there have been a range of estimations of parthenium weed seed production per plant in the field (*c.*15,000 according to Haseler, 1976; to *c.*156,000 according to Dhileepan, 2012), the most recent and accurate counts for glasshouse-grown plants is between *c.*18,000 and 26,000 filled seeds per plant (Fig. 2.5C; Nguyen *et al.*, 2017b). In India, Kanchan and Jayachandra (1980) found that there was an average of 15 plants/m^2 in a typical stand of parthenium weed; however, this can go as high as 315/m^2 in eastern Ethiopia when the plants are young (Tamado and Milberg, 2004). In a field trial at Mt Panorama, central Queensland in 1996/97, it was reported that parthenium weed populations of over 800 plants/m^2 existed, but when such densities did occur the average number of flowers per plant was only around 250 (Dhileepan, 2003). Further studies have shown that there is a significant negative correlation between

Fig. 2.4. Flowers of parthenium weed showing the typical five-ray floret capitulum (a), a rare seven-ray floret capitulum (b) and a five-ray floret capitulum preparing to release its five cypsela fruits each consisting of a seed-containing floret and two subtending infertile ray florets that aid fruit dispersal (c). Scale bars are 2 mm. (a, b: S. Adkins; c: R. Mao.)

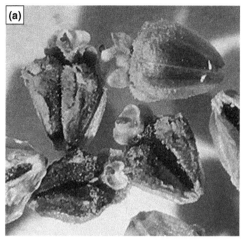

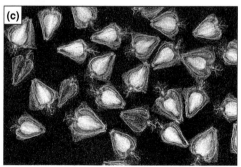

Fig. 2.5. Fruit and seed of parthenium weed showing cypsela (a), the black seed contained within the cypsela (b) and an X-ray taken of 25 cypsela showing 88% seed fill (c). Scale bars are 2 mm. (R. Mao.)

parthenium weed plant density and the number of flowers produced per plant (Dhileepan, 2012).

Such data indicate that in excess of 300,000 seeds/m^2 could be produced in many field circumstances. These figures for seed production are only applicable when sufficient moisture and warm temperatures are available to produce vigorously growing stands of plants, since Nguyen *et al.* (2017b) have shown that filled seed production will be reduced (from *c.*20,000 to *c.*9000) when plants are grown under cool/dry conditions as compared with warm/wet environmental conditions. Pandey and Dubey (1988) reported parthenium weed to produce polymorphic seed that vary in size and weight. They placed seeds into six different categories, based on size and weight and suggested that the variation in seed morphology may be due to differences in the maturation time of the capitula produced at different positions on the parent plant. They also found that small seeds were more commonly produced at lower latitudes (i.e. in southern India) as compared with the larger seeds produced at higher latitudes (i.e. in northern India). Therefore, it seems that the climatic conditions have a bearing on both seed production and size (Dubey and Pandey, 1988).

2.3.5 Seed dispersal

Dispersal of parthenium weed seeds (see Shabbir *et al.*, Chapter 3, this volume) can occur locally by wind and water (Maheshwari and Pandey, 1973). Wind transport is usually only in the order of a few metres, but whirlwinds can carry large numbers of the light cypsela fruit for considerable distances (Haseler, 1976). Short-distance dispersal of seed by water is important, as indicated by the large populations of the weed observed spreading along the edges of waterways and irrigation channels (Auld *et al.*, 1982; Adkins and Shabbir, 2014). Native animals, livestock and feral animals are also believed to be involved in the dispersal of parthenium weed seeds over short distances (Holman and Dale, 1981; Parsons and Cuthbertson, 1992). The spread of parthenium weed seed by cattle from infested to uninfested land has been observed in southern Queensland (D. Chandler, Queensland, 2010, personal communication) and in dung in south-east Queensland (S.W. Adkins, Brisbane, 2017, personal communication). The human spread of seed is mainly by vehicles as well as upon agricultural machinery (Blackmore and Johnson, 2010). These pathways for parthenium weed seed spread can be over very long distances and are thought to be the most important pathways in most countries. Blackmore and Johnson (2010) reported that 73% of all the parthenium weed populations appearing in New South Wales arrived as seed carried on vehicles from Queensland. Parthenium weed seed can also be spread within fodder or seed lots (Gupta and Sharma, 1977). All of these various means of seed dispersal play a role in the overall spread of the weed within invaded countries, making management more difficult. Other mechanisms of spread of parthenium weed are also known, including when the weed is used as an ornamental plant in floral bouquets, when using its vegetative and reproductive parts as packaging of items in crates, and in its use as a green manure (see Shabbir *et al.*, Chapter 3, this volume).

2.3.6 Seed banks

Variation in the size of a plant's soil seed bank may depend on several factors including the rainfall pattern of the region, the time of year the bank is sampled and the presence or absence of seed predators (McIvor *et al.*, 2004; Navie *et al.*, 2004). In Australia, Navie *et al.* (2004) determined the size of the germinable parthenium weed soil seed bank at two infested pasture sites and found the total seed bank (all species) to range from 3200 to 5100 seed/m^2 in a black, cracking clay soil with low ground-cover level, to 20,500 to 44,700 seed/m^2 in a sandy loam soil close to a creek. At these two sites the parthenium weed seed bank accounted

for 47–73% and 65–87%, respectively, of the total seed bank present. Nguyen *et al.* (2017a) have recently reported that the parthenium weed soil seed bank at the same two sites 10 years later is still large at around 6000–8000/m^2. Do (2009) has shown the germinable soil seed bank under another pasture in south-east Queensland to vary between 11,500 seeds/m^2 and 23,250 seeds/m^2 in a gully and between 10,100 seeds/m^2 and 12,800 seeds/m^2 at the top of a ridge. In other locations, much larger seed banks have been recorded, for example Joshi (1991) estimated the parthenium weed soil seed bank in India, in a series of abandoned fields invaded by parthenium, to be 200,000 seeds/m^2. However, the methods used in this study were not robust and sample sizes were not large. A series of studies undertaken at a central Queensland site, before and after a major river flood event, showed that following the flood the native seed-bank abundance increased but its species richness and diversity decreased. The presence of a large parthenium weed seed bank that was no more affected by the flood than the native species was a major factor intensifying the effect of the flood on the riparian community (Osunkoya *et al.*, 2014).

2.3.7 Seed dormancy

It has been assumed, in Australia at least, that parthenium weed seeds will germinate readily when shed and possess no primary dormancy mechanism(s). McFadyen (1994), using ripe seeds that were collected directly from the plant, reported that nearly 100% germination could be achieved within 21 days. In this case it was concluded that no physical or physiological dormancy mechanism(s) were present when the seed was first shed (McFadyen, 1994). However, research conducted overseas has demonstrated that water-soluble germination inhibitors (i.e. parthenin and certain phenolic acids) are present in the fruit layers surrounding the seed and that these inhibitors need to be leached out before maximum germination is attained (Picman and Picman,

1984). Parthenium weed seeds may also be induced into a state of conditional physiological dormancy by the ambient environmental conditions, as is the case with many other weed species that develop a light requirement for germination following burial (Baskin and Baskin, 1989). It could be expected that parthenium weed seeds, when buried, will exhibit a form of imposed dormancy that leads to the formation of persistent seed banks (McFadyen, 1994). White (1994) reported significant emergence of parthenium weed seedlings from soil that had been disturbed after a period of 4–6 years without disturbance, showing that parthenium weed seeds may in fact possess an induced or secondary dormancy, as do seeds of many other weeds of disturbed environments. Recently, Nguyen *et al.* (2017b) have shown that the proportion of dormant seed produced by parthenium weed plants is dermined by the maternal environment in which the seeds mature, with a warm/wet environment producing up to *c.*2000 dormant seeds per plant, and a cool/dry environment producing only *c.*100 dormant seeds per plant.

2.3.8 Seed longevity

As is the case with seed dormancy, only a little is known about the longevity of parthenium weed seeds, either when on the soil surface or when in the soil seed bank. In one study looking at the survival of buried seeds in the field it was found that, of the seeds recovered, their germination declined from 66% after 1 week of burial to 29% after burial for 2 years (Butler, 1984). Depth of burial did not seem to affect the subsequent germination percentage (Butler, 1984). These post-exhumation germination tests were conducted in the dark, and at a constant incubation temperature, so conditions may not have been adequate for breaking any induced dormancy that may have been present. It is possible, therefore, that the percentage viability of these seeds may have been much higher than was reported (McFadyen, 1994). As has already been

noted, there is some field evidence that parthenium weed seeds can remain viable after being buried for at least 4–6 years (White, 1994). More recently, Navie *et al.* (1998) have demonstrated in a field study that more than 70% of seed buried 5 cm below the soil surface can live for at least 2 years with a half-life of 7 years, and Tamado *et al.* (2002) have reported that more than 50% of buried seed remained viable for up to 2.5 years, with an anticipated lifespan of 3–4 years. One further study (Nguyen, 2011) has shown that the quality of the seed at the time of being buried also affects seed bank persistence, with high-quality seed living longer in the soil seed bank than poor-quality seed. The longevity of surface-lying seeds seems to be quite short. Research shows that most unburied parthenium weed seeds either germinate, are harvested by insects, or lose viability within 2 years (Butler, 1984), while a further study (Navie *et al.*, 1998) indicated that seed on the soil surface will die within 6 months.

2.3.9 Seed germination

Many authors have noted that parthenium weed fruit have a very high rate of seed fill and those seeds have a viability of 85% or higher, when collected directly from the adult plant (Haseler, 1976; Williams and Groves, 1980; Dubey and Pandey, 1988; Pandey and Dubey, 1988; McFadyen, 1994). Williams and Groves (1980), working with parthenium weed seeds from Queensland, reported that the maximum germination (88%) could be achieved in the dark, under a diurnal temperature regime of 21/16°C. They also noted that the germination percentage would decrease if the diurnal temperature differential was increased to more than 5°C. In another study using seed from Queensland, Navie *et al.* (1996) found that the optimum single temperature for germination of two populations was between 22°C and 25°C, but these populations had a wide range of temperatures over which they would germinate (9–36°C). According to Tamado *et al.* (2002), temperature regimes ranging from 12/2°C to 35/20°C (day/night) were all suitable for the germination of Ethiopian seed, while a recent study has showed that seed can also germinate in extremely hot summers (38.7°C) or in extremely cold winters (2.6°C). In the field it is possible for the weed to have more than four cohorts of seed germination during a single summer. This will occur when there has been good rainfall, and when the soil seed banks are large (Everist, 1976; Pandey and Dubey, 1989). According to Tamado *et al.* (2002), the depth of seed burial can have a significant impact upon seed germination and/or seedling emergence. A depth of 0.5 cm seems to be ideal for the weed's germination, while 5.0 cm or more will result in slow or no germination. Pandey and Dubey (1988), using seeds of Indian origin, found that there was a high percentage of germination of parthenium weed seeds in continuous light or continuous dark, and suggested that this species does not have a strict light requirement for germination. However, they observed germination to be enhanced under the influence of a diurnal photoperiod and/or under an alternating temperature regime. They found that exposing seeds to a light pretreatment led to an increase in subsequent germination in the dark, and this effect increased as the period of the light pretreatment was increased from 6 h to 48 h (Pandey and Dubey, 1988). They also concluded that a 14 h photoperiod and a 25/20°C day/night temperature regime were optimal for the germination of parthenium weed seeds. Parthenium weed seeds from Australian populations were shown to exhibit more than 20% germination in regimes when the night temperature was as low as 10°C or when the day temperature was as high as 36°C (Williams and Groves, 1980). However, germination was noted to decrease from 91% when the soil was at field capacity, to just 50% when the soil moisture was reduced to −0.07 MPa, and to 0% when the soil moisture was reduced to −0.90 MPa. This demonstrates that parthenium weed seeds are very dependent upon there being a high moisture availability for germination (Williams and Groves, 1980). In one further germination study conducted in India, the percentage of seeds germinating gradually increased as their seed size increased (Dubey and Pandey, 1988; Pandey and Dubey, 1988).

2.3.10 Population dynamics

Pandey and Dubey (1989) observed in the field in India that seedlings of parthenium weed were recruited in three successive cohorts after the first rains of the new season. They found that seedling density and survivorship to maturity declined in successive cohorts. In particular, the first cohort showed a slightly better growth and survival rate than the second cohort, but the third cohort was adversely affected to a much greater extent, and very few plants from this cohort survived the seedling stage. In the same study, the pattern of recruitment and population density was remarkably similar in two successive years. The average recruitment for the 2 years of the study was 110 plants/m², of which 14 plants/m² (13%) attained maturity. The results also indicated that the first established cohort adversely affected the growth, and probably the survivorship, of the latter cohorts through resource competition (Pandey and Dubey, 1989) and possibly allelopathy (Adkins and Sowerby, 1996). The reported plant density in this mature stand of parthenium weed plant was compatible with an earlier report of 15 plants/m² (Kanchan and Jayachandra, 1980), but was less than the 25 plants/m² reported by Joshi (1991). As parthenium weed often grows in pure stands, only a few studies have been conducted on its population dynamics in relation to competition with other species. Khan *et al.* (2014, 2015) have shown that a number of introduced pasture plants are quite competitive, while other native species were less so.

2.3.11 Microbial associations

There are few reports of microbial associations with parthenium weed. Sixteen species of endophytic fungi (*Alternaria helianthi*

(Hansf.) Tubaki & Nishih., *Alternaria alternata* (Fr.) Keissl., *Fusarium* sp., *Nigrospora oryzae* (Berk. & Broome) Petch, *Penicillium funiculosum* Thom, *Periconia* sp., *Curvularia brachyspora* Boedijn, *Cylindrocarpon* sp. and eight unidentified species) were isolated from leaves of parthenium weed in the state of Veracruz, Mexico (Romero *et al.*, 2001). Eleven species of pathogenic fungi (*Sphaerotheca fuliginea* (Schlecht.) Poll., *Alternaria* spp., *Cercospora partheniphila* Chupp & Greene, *Colletotrichum capsici* (Syd.) Butler & Bisby, *Colletotrichum gloeosporioides* (Penz.) Sacc., *Curvularia lunata* (Walker) Boedijn, *Fusarium* spp., *Myrothecium roridum* Tode ex Fr., *Oidium parthenii* Satyaprasad & Usharani, *Rhizoctonia solani* Kuhn and *Sclerotium rolfsii* Sacc.) were isolated from parthenium weed in India (Evans, 1997).

In plots in Yucatán, Mexico, 5.2% of plants were colonized by arbuscular mycorrhizal fungi (Ramos-Zapata *et al.*, 2012). It has been observed that the microorganism population in the weed's rhizosphere is quite different to non-infested soils (Jeyalakshmi *et al.*, 2011).

In another study, parthenium weed has been shown to express an antifungal activity against *Fusarium solani* (Mart.) Sacc., *Alternaria alternata* (Fries) Keissler, *Bipolaris oryzae* (Breda de Haan) and several other species (Zaheer *et al.*, 2012; Bezuneh, 2015). In addition, the plant has an antibacterial activity against *Escherichia coli* Migula, *Bacillus subtilis* (Ehrenberg) Cohn and several other species (Fazal *et al.*, 2011; Bezuneh, 2015). The root leachates from parthenium weed have been shown to impair biological nitrogen fixation and to inhibit the growth of the nitrogen-fixing (*Rhizobium phaseoli* and *Azotobacter vinelandii*) and nitrifying (*Nitrosomonas*) bacteria (Kanchan and Jayachandra, 1981). It was established that the growth inhibition of these bacteria was in part due to parthenium weed allelochemicals such as parthenin, caffeic acid, vanillic acid and anisic acid as these chemicals also caused partial growth inhibition when applied separately. Parthenium weed has the ability to manipulate invaded communities directly (through phytotoxic effects on neighbouring plants' roots) or indirectly (through inhibiting the growth of beneficial microorganisms), which could make its introduction and establishment possible in diverse edaphic regimes.

2.4 Native Range

2.4.1 Distribution

Parthenium weed is recorded from virtually all countries in the western hemisphere (Fig. 2.6; and see Shabbir *et al.*, Chapter 3, this volume). However, most of its occurrences are concentrated in two disjunct ranges, one centred in Mexico and the other in northern Argentina, Paraguay, Bolivia, Chile, Peru and Uruguay (Dale, 1981). It is generally considered to be native in both of these ranges. However, there are morphological and biochemical differences, with most plants in the South American range having cream or yellow flowers and containing hymenin as the dominant sesquiterpene lactone (Picman and Towers, 1982), while those in Mexico, the adjacent countries of the Caribbean and southern USA have white flowers and contain the sesquiterpene lactone parthenin (Dale, 1981). Interestingly, the white-flowered form is also found in Brazil and may have been introduced to this location in more recent times. It is the white-flowered form that has become invasive in 40 or more countries around the world. Studies have confirmed that this white-flowered form is genetically distinct from the cream/yellow-flowered form, and that the genotypes invading Australia, India, Mozambique and South Africa are all likely to have originated from the white-flowered populations of southern Texas, USA (Graham and Lang, 1998). It is the white-flowered form that has spread to other parts of the USA, including Alabama, Arkansas, Connecticut, Delaware, Florida, Illinois, Kansas, Louisiana, Maryland, Massachusetts, Michigan, Mississippi, Missouri, New Jersey, New York, Ohio, Oklahoma, Pennsylvania, Texas and Virginia, including some in temperate areas (USDA-NRCS, 2003).

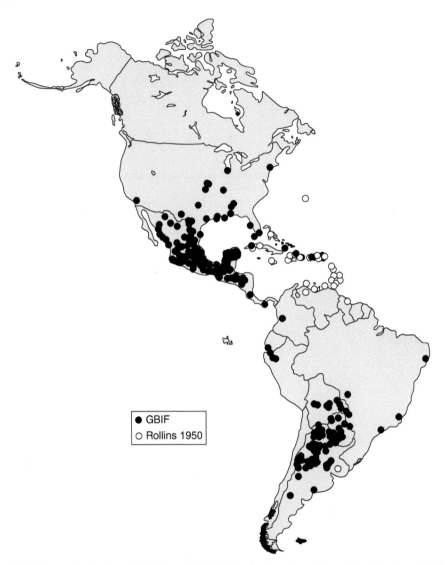

Fig. 2.6. Western hemisphere distribution of *Parthenium hysterophorus* from Global Biodiversity Information Facility (●) (2014) with selected additional locations (○) from Rollins (1950).

2.4.2 Region of origin

According to Rollins (1950), the original distributional area of parthenium weed is difficult to establish, but the best supposition is that it is native to the West Indies and to the North American continent adjacent to the Gulf of Mexico (Fig. 2.6). His reasons for supposing it to be native to the West Indies are not clear, as no other species of *Parthenium* are native to this region and many of

the specialist insects that feed on parthenium weed are not to be found there (K. Dhileepan, Brisbane, 2017, personal communication). Given that the centre of diversity of the genus is in central Mexico (Fig. 2.7), where the closely related *P. bipinnatifidum* and *P. confertum* are also found, it seems more likely that this is the area of origin of parthenium weed. In addition, from the natural enemy surveys conducted while searching for biological control agents

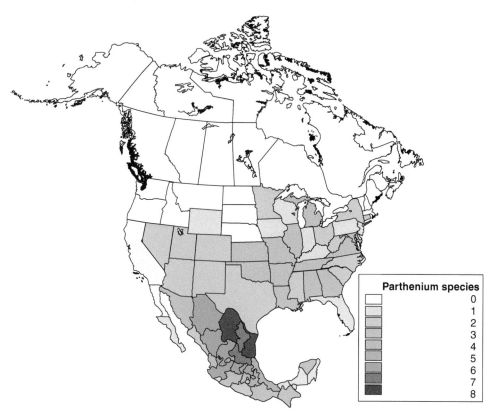

Fig. 2.7. Numbers of *Parthenium* species recorded per state or province in North America. (Data from USDA-NRCS (2016) and Villaseñor (2016).)

for parthenium weed (Bennett, 1977; McFadyen, 1976; McClay *et al.*, 1995), Mexico was found to have the highest diversity of natural enemies associated with parthenium weed, and therefore is the most likely centre of origin (A. McClay, Canada, 2017, personal communication).

2.4.3 Distribution in Mexico

Location records for parthenium weed in Mexico (Fig. 2.8) obtained from the Red Mundial de Información Sobre Biodiversidad (REMIB) database (CONABIO, 2015b), show that according to Mexico's biogeographic province classification (CONABIO, 1997) the majority (85%) are in the Golfo de México, Yucatán, Costa del Pacífico, Tamaulipeca, Sierra Madre Oriental, Petén and Eje

Volcánico provinces. There are smaller numbers of reports from the Altiplano Norte, Altiplano Sur and Sonorense provinces, few in Depresión del Balsas, Altos de Chiapas, Sierra Madre del Sur, Soconusco and Sierra Madre Occidental provinces, but no records from Baja California, California, Oaxaca or Del Cabo provinces. In general terms, this means that parthenium weed is most abundant in lower elevation areas along the Gulf Coast and in the Yucatán peninsula, and to a lesser extent, the temperate parts of central Mexico. It is scarcer or absent in the colder or drier areas such as the southern Altiplano and the Sierra Madre Occidental, and in the largely desertic Baja California peninsula (Mainali *et al.*, 2015). In more recent years it appears to have become much more frequent around the Federal District of Mexico City (H. Vibrans, Colegio de Postgraduados en Ciencias Agrícolas, Montecillo, 2014,

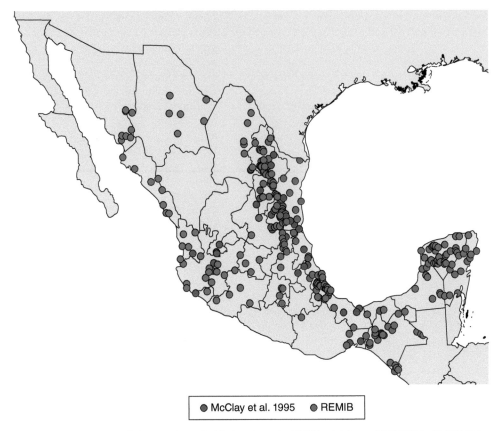

| ● McClay et al. 1995 | ● REMIB |

Fig. 2.8. Distribution of *P. hysterophorus* in Mexico, from the REMIB database (CONABIO, 2015b) (●) and the Queensland project (●) (McClay, 1983; McClay *et al.*, 1995).

personal communication). All records for parthenium weed in the REMIB database for the Federal District and the nearby states of Guanajuato, Querétaro and Morelos date from 1981 or later, again suggesting that parthenium weed has been increasing in abundance in central Mexico in recent decades.

2.4.4 Preferred habitat

The habitat associations of parthenium weed in Mexico were analysed using 253 georeferenced records from the REMIB database in relation to temperature, precipitation, elevation and dominant soil types, using data from the Geoinformation Portal of the National Biodiversity Information System (CONABIO, 2015a). This revealed

that almost 90% of occurrences were at elevations below 1500 m, around 90% were in areas with mean annual minimum temperatures above 4°C, and 85% were in areas with mean annual precipitation above 600 mm. In terms of soil types, occurrences of parthenium weed were more frequent on eutric vertisols (base-rich cracking clays) and calcic leptosols, which are shallow rocky or gravelly soils, widespread in the Yucatán peninsula. There are few occurrences in areas with eutric regosols, haplic calcisols or haplic arenosols (primarily mountain and desert soil types). Dale (1981) also describes the preferred soil type in Mexico as 'predominantly dark, alkaline, self-mulching, cracking, clay soils'. Habitats frequently mentioned in the Mexican herbarium records for parthenium weed occurrence include: (i) riparian sites; (ii) cultivated or agricultural areas, including field edges and abandoned cultivation; (iii) roadsides and railway tracks; (iv) waste areas and vacant lots; (v) disturbed sites; (vi) secondary vegetation; (vii) pastures; and (viii) sunny or open areas. Herbarium labels also frequently refer to the species occurring in forest, forest edges or clearings. These probably refer to relatively open forest or thorn scrub (matorral) as the plant does not usually occur under a dense forest canopy. In the drier areas where parthenium weed was observed, it was often found in a strip along roadsides, presumably where water runoff from paved road surface increases the availability of moisture (Dale, 1981).

In Mexico, parthenium weed behaves as a pioneer species, and dense infestations do not usually persist for more than a year in undisturbed sites. As an example, a site on the southern outskirts of Monterrey, an area which had been cleared of scrub for use as a dump, was densely covered with an almost pure stand of vigorous parthenium weed in 1979. However, by 1980 this infestation began to be replaced by various *Malvaceae* and *Solanaceae* weeds, and by 1981 it was mostly covered by perennial grasses, with virtually no parthenium weed to be found (A. McClay, Canada, 2017, personal communication).

2.4.5 Species associated with parthenium weed

Species associated with parthenium weed in Mexico and recorded in the REMIB database include:

- herbs such as hornbeam copperleaf (*Acalypha caroliniana* Elliott), burr ragweed (*Ambrosia confertiflora* DC.), cobblers pegs (*Bidens pilosa* L.), nettle-leafed goosefoot (*Chenopodium murale* (L.) S. Fuentes, Uotila & Borsch), yellow weed (*Reseda luteola* L.), white twinevine (*Sarcostemma clausum* (Jacq.) Schult.), London rocket (*Sisymbrium irio* L.), sowthistle (*Sonchus oleraceus* L.) and species of *Amaranthus*, *Asclepias*, *Dasylirion*, *Euphorbia*, *Gomphrena*, *Heliotropium*, *Lippia*, *Ludwigia*, *Melochia*, *Parthenium*, *Phlox*, *Simsia*, *Tinantia* and *Ungnadia*;
- grasses and sedges including Bermuda grass (*Cynodon dactylon* (L.) Pers.), desert saltgrass (*Distichlis spicata* (L.) Greene), kikuyu grass (*Pennisetum clandestinum* Hochst. ex Chiov.) and species of *Bouteloua*, *Cyperus* and *Heteropogon*; and
- trees and shrubs including achuchil (*Astianthus viminalis* (Kunth) Baill.), celtis (*Celtis ehrenbergiana* (Klotzsch) Liebm.), logwood (*Haematoxylum campechianum* L.), tree morning glory (*Ipomoea arborescens* (Humb. & Bonpl. ex Willd.) G. Don), sycamore (*Platanus glabrata* Fernald), poplar (*Populus mexicana* Wesm.), castorbean (*Ricinus communis* L.) and species of *Acacia*, *Ceiba*, *Cordia*, *Esenbeckia*, *Ficus*, *Garrya*, *Helietta*, *Inga*, *Karwinskia*, *Larrea*, *Leucophyllum*, *Liquidambar*, *Mimosa*, *Opuntia*, *Pachira*, *Sabal*, *Pinus*, *Pithecellobium*, *Prosopis*, *Quercus*, *Salix* and *Yucca*.

2.4.6 Climatic associations

In the Monterrey area, around Apodaca, Nuevo León in north-east Mexico, growth and flowering of parthenium weed can occur at any time of year provided that sufficient soil water is available. The climate in this area is characterized by cool, dry winters and hot, wet summers (Fig. 2.9a). The most vigorous growth occurs in the summer wet season, approximately from June to September. The amount of growth occurring over the winter months depends on winter rainfall; in dry winters the plant may be almost impossible to find in January and February, while if some rain occurs growth may continue uninterrupted. In the more tropical, coastal areas along the Gulf of Mexico including Cárdenas in the state of Tabasco there is much greater precipitation, more evenly distributed through the year, and relatively little seasonal fluctuation in temperature (Fig. 2.9b). In these areas growth and flowering of parthenium weed can occur at any time of year. Areas of the Altiplano

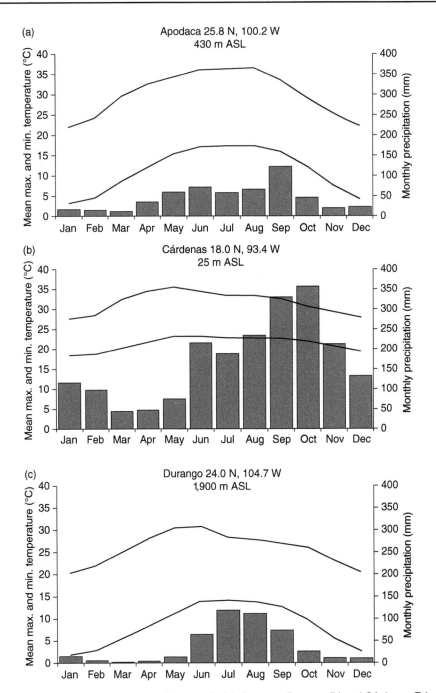

Fig. 2.9. Climate diagram for Apodaca, Nuevo León (a), Durango, Durango (b) and Cárdenas, Tabasco (c), where the *P. hysterophorus* study plots were located from 1978 to 1983. ASL, above sea level; filled bars, monthly precipitation (mm); top line, mean maximum daily temperature (°C); lower line, mean minimum daily temperature (°C). (Based on climate normals for 1971–2000, Servicio Meteorológico Nacional, 2016.) (A. McClay.)

Sur and Sierra Madre Occidental biogeographic provinces where parthenium weed is almost absent are typified by Durango (Fig. 2.9c). Although the climatic reasons for its absence from this area are not entirely clear, it is likely that they are related to the generally cooler conditions and long winter dry season.

2.4.7 Natural enemies

The natural enemies of parthenium weed in North America are reported by McClay *et al.* (1995) and in Dhileepan *et al.* (Chapter 7, this volume). Briefly, in the study of McClay *et al.* (1995) 144 arthropod species were found to feed on the plant at some stage of their life cycle. This included seven insect species that were sent to Australia for evaluation as possible biological control agents. Eight fungal pathogens (*Plasmopara halstedii* (Farl.) Berl. & de Toni, *Puccinia melampodii* Diet. & Holw., *Puccinia abrupta* Diet. & Holw. var. *partheniicola* (Jackson) Parmelee, *Entyloma parthenii* Sydow (= *E. compositarum*), *Erysiphe cichoracearum* DC var. *cichoracearum* Braun, *Alternaria protenta* E.G. Simmons, *Alternaria zinniae* M.B. Ellis, *Cercospora partheniphila* Chupp & Greene) were reported on parthenium weed in its native range by Evans (1997), and plants were also found in Mexico wilting due to attack by *Sclerotinia sclerotiorum* (Lib.) de Bary (A. McClay, Canada, 2017, personal communication). The rust fungi *P. abrupta* var. *partheniicola* and *P. melampodii* (now classified as *Puccinia xanthii* var. *parthenii-hysterophorae* Seier, H.C. Evans & A. Romero), from Mexico, were also evaluated as possible biological control agents (Seier *et al.*, 2009; Dhileepan *et al.*, Chapter 7, this volume).

2.4.8 Weed status, impacts and management

There is very little research on the impacts of parthenium weed as a weed in its native range, but it is certainly recognized in Mexico as a weedy species. Villaseñor and Espinosa-García (1998) report it as a weed in alfalfa (*Medicago sativa* L.), cotton (*Gossypium* spp.), rice (*Oryza sativa* L.), coffee (*Coffea arabica* L.), sugarcane (*Saccharum officinarum* L.), safflower (*Carthamus tinctorius* L.), marigolds (*Tagetes erecta* L.), chilli (*Capsicum* spp.), citrus (*Citrus* spp.), loofah (*Luffa aegyptiaca* Mill.), sunflower (*Helianthus annuus* L.), lentils (*Lens culinaris* Medikus), maize (*Zea mays* L.), mango (*Mangifera indica* L.), okra (*Abelmoschus esculentus* (L.) Moench), banana (*Musa* spp.), sorghum (*Sorghum bicolor* (L.) Moench), soybean (*Glycine max* (L.) Merr.), tomato (*Solanum lycopersicum* L.), grapes (*Vitis vinifera* L.) and beans, fruit and vegetables and in nurseries. In a survey of producers in the state of Tamaulipas it was found that parthenium weed was the fifth most prevalent weed in irrigated maize (Fernández Garza, 1993). Parthenium weed was also one of the five most frequent weeds in an experimental study of cropping systems for maize and beans (*Phaseolus vulgaris* L.) in the state of Nuevo León (Leos Moreno, 1988). Parthenium weed was reported as one of the most prevalent weeds of citrus plantations in Nuevo León, occurring particularly in the alleys between rows of trees (Rocha-Peña and Padrón-Chávez, 2009). It was the second most abundant weed in surveys of citrus plantations in the area of Montemorelos (Alanís Flores, 1974).

It is one of the commonest weeds in sorghum in northern Tamaulipas state, where 2,4-D amine and prosulfuron herbicides were widely used for its management, in addition to cultivation and manual weeding (Rosales Robles *et al.*, 2014). It has been reported as a weed of beans in the state of Veracruz (Esquivel *et al.*, 1997). It is considered a principal weed in the cultivation of bananas in the state of Colima, where recommended control methods include manual slashing and application of glyphosate or paraquat (Gobierno del Estado de Colima, 2005). There is little further information on use patterns of herbicides for control of parthenium weed in Mexico. A search of chemical manufacturers' websites indicated that products currently marketed in Mexico with claims of effectiveness against parthenium weed include Sumimax (flumioxazin), Krovar (bromacil and diuron), Callisto (mesotrione), Prensil (oxyfluorfen), Gamit (clomazone), Peak (prosulfuron), glyphosate, Focus (carfentrazone-ethyl and 2,4-D) and Finale (glufosinate-ammonium). In maize plots in which various weed control techniques were evaluated in the Yucatán, parthenium weed was abundant (over 50% cover) in plots where paraquat had been used for 13 years (Ramos-Zapata *et al.*, 2012), suggesting that this active ingredient is no longer effective for control. Resistance to glyphosate has been reported in Colombia, where this product has been used to control parthenium weed in orchards for more than 15 years (Rosario *et al.*, 2013), and in the Dominican Republic (Jiménez *et al.*, 2014). In most areas of the USA parthenium weed is not particularly troublesome; however, it has been reported as a weed of crops (Everist, 1976) and more recently as a new weed in Florida pastures (Abe *et al.*, 2016) and cropping lands.

2.4.9 Health impacts and uses

Despite the well-documented impacts of parthenium weed as a cause of allergic dermatitis and rhinitis in its introduced range (Kohli *et al.*, 2006; Allan *et al.*, Chapter 6, this volume), this does not appear to be recognized as a problem in Mexico (H. Vibrans, Colegio de Postgraduados en Ciencias Agrícolas, Montecillo, 2014, personal communication). In addition to not being toxic, parthenium weed has gained a wide range of traditional medicinal uses in Mexico, including for treatment of digestive upsets, skin infections and rashes, hair loss, menstrual irregularity, fevers, aches, rheumatism, haemorrhoids, and to promote healing of wounds (Anonymous, 2009; Méndez-González *et al.*, 2014; Chandrasena and Narayana Rao, Chapter 10, this volume).

2.5 What Makes It Invasive?

2.5.1 The mechanism of invasion

Parthenium weed has invaded diverse climatic and biogeographic regions in more than 40 countries across five continents. While efforts are under way to minimize the parthenium weed-induced environmental, agricultural, social and economic impacts, there is a need to better understand how and why this weed is so successful. This section will review what we know about the mechanism of parthenium weed invasion. The discussion will focus around: (i) the plant's morphological advantages, its unique reproductive biology, its competitive ability and its escape from natural enemies in the introduced range; (ii) its tolerance to abiotic stresses and ability to grow in a wide range of edaphic conditions; (iii) its allelopathic potential against most plants it comes up against in the introduced range; (iv) its genetic diversity found among different populations and biotypes; and (v) its positive response to climate change including that of rising temperatures and atmospheric carbon dioxide (CO_2) concentrations and changing rainfall patterns. Through an understanding of the key phenomena regulating its invasion biology, better understanding of its spread and improved methods for its management can be developed.

2.5.2 Morphological attributes

Parthenium weed has a unique set of botanical features, which support its invasive nature. Rapid emergence of its seedlings from the seed bank has been demonstrated by Navie (2002). Its distinctive rosette growth habit, an erect and rigid stem, a tall stature, its multiple branching and high rate of leaf production, the presence of trichome hairs on leaves and stems (Fig. 2.10), the formation of a large tap root and an extensive flowering capacity resulting in a massive,

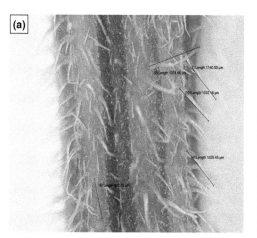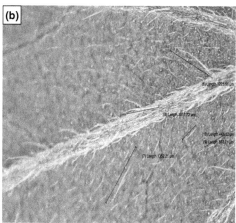

Fig. 2.10. A stem section (a) and abaxial surface of a fully opened leaf (b) taken from a 40-day-old parthenium weed plant (Clermont biotype) showing the trichome hairs. Scale bars represent 1.0 mm and 0.5 mm, respectively. (Khin Zaw Tun.)

viable seed load, are all thought to play an important role in its successful growth and adaptability (Navie *et al.*, 1996; Adkins and Shabbir, 2014; Tanveer *et al.*, 2015; Bajwa *et al.*, 2016). In areas where winter temperature drops to low levels, the rosette stage helps juvenile plants to overwinter. Although all these morphological characteristics enable parthenium weed to be an aggressive invasive species, the relative contribution of each trait is not known. In a recent review, Adkins and Shabbir (2014) proposed that its superior morphology plays a key role in its invasiveness along with other ecological and physiological mechanisms.

2.5.3 Biological life-cycle plasticity

Parthenium weed is an enduring plant species that can grow in a wide range of soils and climatic conditions. In general, it has the characteristics of an annual plant but may behave more like a short-lived perennial in certain locations (Shrestha *et al.*, 2015). It may complete vegetative growth within 28–42 days after germination and may then sustain its reproductive growth phase for up to 90–120 or even 240–300

days, depending upon soil moisture, temperature and relative humidity conditions (Gnanavel, 2013; Tanveer *et al.*, 2015). Although typically it is a warm-season plant, it can prolong or reduce its lifespan, by modifying the duration of any of its growth phases depending upon the environmental conditions (Navie *et al.*, 1996; Adkins and Shabbir, 2014). After germination, it can stay in the rosette stage for variable periods of time and similarly, after entering into the reproductive phase it can prolong or reduce flowering time as well as the time for seed set (Adkins and Shabbir, 2014). Due to such a flexible growth habit and lifespan, it can prevail under diverse environmental conditions, especially within the introduced range (Tanveer *et al.*, 2015). In a recent study from Nepal, Shrestha *et al.* (2015) reported that parthenium weed can be seen throughout the year in the Kathmandu Valley, and at any given time, at any stage of growth.

2.5.4 Seed biology and dispersal

Seed biology plays a fundamental role in the determination of the extent and size of most weed problems (Bajwa *et al.*, 2015). Parthenium weed only reproduces via seeds and has great ability to produce a large number of seeds (up to 156,000 seeds per plant under ideal field conditions; Dhileepan, 2012). The start of seed production can be as early as 30 days but up to *c.*60 days after emergence depending upon temperature, soil moisture and other environmental conditions. Seed germination is also possible under a wide range of temperatures (9–36°C) with an optimum temperature range around 21–25°C (Navie, 2002). Parthenium weed seeds can also germinate at soil moisture levels as low as 40–60% of field capacity (Tanveer *et al.*, 2015). Parthenium weed seeds may also exhibit innate dormancy, which may be removed by leaching of germination inhibitors from the fruit tissues. Navie *et al.* (1998) found that seeds have a predicted half-life in the soil seed bank of between 5 and 7 years. However, the quality of the seed when buried also affects the seed bank persistence (Nguyen, 2011) with high-quality seed living longer in the seed bank than poor-quality seed. This long-term seed bank persistence is an important feature contributing towards invasiveness. The ability of parthenium weed to germinate and grow in a range of soil types is another distinct feature aiding its vigorous invasion (Navie *et al.*, 1996; Timsina *et al.*, 2011; Adkins and Shabbir, 2014). Once produced, parthenium weed seeds can be dispersed by a wide range of biotic and abiotic pathways, including domestic and feral animals, by vehicles and farm machinery, with the transportation of crop seed and fodder, within irrigation and flood water, and by wind (Chippendale and Panetta, 1994; Navie *et al.*, 1996). In a recent study, Nguyen (2011) found considerable amounts of parthenium weed seed in the slurry pit of a number of wash-down facilities used to clean vehicles of weed seeds in central Queensland, indicating that vehicular transport of seed was a significant pathway for the spread of parthenium weed. In addition, Nguyen (2011) has demonstrated that parthenium weed cypselas can remain viable in river water for at least 30 days.

2.5.5 Competitive ability

Parthenium weed is often described as having a strong competitive ability (Gnanavel, 2013; Adkins and Shabbir, 2014). It may exhibit this competitiveness in grasslands, cropping lands and forests, as well as in native ecosystems, depending upon where it is invading (Evans, 1997; Kohli *et al.*, 2006). The invasion of parthenium weed into several cropping systems across the globe is threatening food security and agricultural sustainability (Tanveer *et al.*, 2015). The rapid and early emergence of parthenium weed as compared with crop plants is a major reason for its competitive ability (Tanveer *et al.*, 2015). It has become an important weed in more than 40 major field crops in over 40 different countries. In Australia (central Queensland), parthenium weed has caused substantial yield losses to grain sorghum (*S. bicolor*) and sunflower (*H. annuus*) crops due to its strong competition for resources (Parsons and Cuthbertson, 1992). Similarly, up to 40% losses in pastures in Australia have been reported (McFadyen, 1992). In India, yield losses were 40% and 90% in grain crops and pastures, respectively (Khosla and Sobti, 1979; Nath, 1981). Tamado *et al.* (2002b) reported up to 97% yield loss in grain sorghum due to parthenium weed infestation at a lowland site. Even a very low parthenium weed density of three plants/m^2 caused 69% yield loss, again highlighting its strong competitiveness. In a recent study from Pakistan, Safdar *et al.* (2015) reported that an increase in parthenium weed density from 0 to 20 plants/m^2 in a maize crop resulted in 145% and 433% increase in the relative competition index and parthenium weed biomass, respectively. Such a competitive character shows another possible reason behind the successful invasion mechanism of parthenium weed.

2.5.6 Liberation from natural enemies

According to the popular 'enemy release hypothesis' of biological invasion (Keane and Crawley, 2002), many invasive species establish and thrive in non-native regions because of the absence of their natural enemies and competitors that exist in their native range (e.g. Andonian and Hierro, 2011). Parthenium weed has numerous insect pests and fungal pathogens that keep its population relatively low in its native range (Dhileepan, 2009). In its native range parthenium weed rarely grows over 1 m tall, while it can reach a height of up to 2 m in the introduced range; this may also be a reflection of natural enemy pressure in the native range. Some important insect herbivores of parthenium weed that exist in the native range include the Mexican beetle (*Zygogramma bicolorata* Pallister), the stem-galling moth (*Epiblema strenuana* Walker) and the seed-feeding weevil (*Smicronyx lutulentus* Dietz). A number of pathogenic species have also been reported to be significantly damaging within its native range (Shabbir, 2012). However, none of these natural enemies existed in the invaded range at the time of the weeds first introduction, and as a result it has established well. So, the escape from natural enemies (e.g. insect herbivores and plant pathogens) is likely to be another strong reason for parthenium weed invasion.

2.5.7 Stress tolerance

Parthenium weed can tolerate certain abiotic stresses, which enables it to invade, establish and even flourish under a wide range of stressful conditions (Kohli *et al.*, 2006). Tolerance to heat stress through different morphological and physiological adaptations is a key attribute of many successful invasive species. Parthenium weed has been shown to increase its antioxidant enzyme production upon exposure to high temperatures (Sharma *et al.*, 2014). These higher enzyme activities are associated with a greater rate of reactive oxygen species scavenging and, therefore, a better physiological regulation of plant function as well as protection against structural damage (Sharma *et al.*, 2014). This physiological adaptation against heat stress may partly explain the invasion of parthenium weed in

harsh tropical and subtropical environments. Parthenium weed also tolerates relatively high levels of salt stress, which may enable it to invade salt-affected lands. Parthenium weed seedlings were able to survive up to 17 mM sodium chloride (NaCl) without any reduction in their growth as compared with those growing in water (Upadhyay *et al.*, 2013). At 35–45 mM NaCl, parthenium weed seedlings increased their proline content by up to 68% over the control, which facilitated osmotic adjustment and allowed a proportion of the seedlings to survive. Similarly, mature plants did not show any reduction in their leaf chlorophyll content at high salinity levels and this would also aid its physiological ability to cope with salt stress. Tolerance to water stress has been reported (Nguyen, 2011) and in one study the weed's lifespan was reduced by up to 43% while its growth was enhanced by 20% due to modifications in the plant's vegetative and reproductive biology. Parthenium weed can uptake and accumulate a range of heavy metals, including copper, cobalt, lead, nickel, chromium and zinc from soils contaminated with industrial waste (Malik *et al.*, 2010). Parthenium weed was the most successful species, of 16 tested, for its ability to stabilize then extract nickel and lead from contaminated substrates. The higher bioaccumulation and translocation rates were, in part, made possible due to the vigorous growth of parthenium weed plants. Hadi and Bano (2009) also found parthenium weed to be a successful bioremediation agent for lead due to its efficient extraction mechanism. Parthenium weed also flourishes in roadside communities that become greatly affected by vehicle emissions, and polluted with dust and dirt. Therefore, it is highly likely that the invasion success of the weed is related to its possession of strong mechanism(s) to tolerate environmental pollution.

2.5.8 Allelopathy

Allelopathy is the positive or negative impact of secondary metabolites (allelochemicals) released from a plant which then affect the growth and/or development of other plants growing in its vicinity (Farooq et al., 2013; Bajwa, 2014). Parthenium weed is thought to be a powerfully allelopathic plant species and, as such, is believed to suppress the growth of most neighbouring plants (Aslam et al., 2014); however, the strength of its chemical weapon(s) has been brought into question (Shi, 2016). Parthenium weed produces a variety of secondary metabolites belonging to a number of chemical classes, some of which can act as allelochemicals. Parthenin (a sesquiterpene lactone) produced in the stems and leaves and stored mainly in the trichome hairs (Fig. 2.10), has been recognized as one of the most common and influential allelochemicals released by parthenium weed. In addition to parthenin, parthenium weed produces a range of water-soluble phenolics (caffeic, p-coumaric, vanillic, ferulic, anicic and fumaric acids), which are considered to be another group of important allelochemicals (Kanchan and Jayachandra, 1981a; Das and Das, 1995; Pandey, 1997; Batish et al., 2002; Aslam et al., 2014). These phenolics are released from living roots (Shi, 2016), leaves (Valliappan and Towers, 1988) and seeds (Picman and Picman, 1984), as well as from dead or decaying residues (Kanchan and Jayachandra, 1980; Mersie and Singh, 1988). Recently, volatile chemicals released from parthenium weed plants have been shown to contain high concentrations of the terpenoids myrcene and β-pinene, and these chemicals have also been shown to have allelopathic activities (Chen et al., 2011). A number of other sesquiterpene lactones, flavonoids and tannins have also been reported to be potential allelochemicals produced by parthenium weed (Adkins and Shabbir, 2014). The scientific literature also abounds with laboratory studies reporting the allelopathic potential of parthenium weed against a large number of crops including rice (O. sativa), wheat (Triticum aestivum L.) and lettuce (Lactuca sativa L.) as well as other weed species through the inhibition of a number of different physiological processes (Mersie and Singh, 1987; Tefera, 2002; Singh et al., 2003; Bajwa et al., 2004; Wakija

et al., 2009; Shabbir and Javaid, 2010; Aslam et al., 2014; Bajwa et al., Chapter 4, this volume). In addition, parthenium weed has been shown to suppress the growth of several pasture grass species (Belgeri and Adkins, 2015). It is interesting to note that parthenium weed plants from the native range in Argentina and Bolivia exclusively contained the sesquiterpene lactone, hymenin, which was different from the more common sesquiterpene lactone, parthenin which was present in the native range in North America and the introduced range of India (Towers et al., 1977) and Australia (Shi, 2016). Such differences in allelopathic chemical composition within biogeographically different parthenium weed populations may have implications for its invasion success, with the parthenin types being more invasive than the hymenin types.

2.5.9 Photosynthesis

Parthenium weed is able to contend well with neighbouring vegetation as it has an efficient photosynthetic system. The plant is described as a C3/C4 intermediate and therefore has lower photorespiration losses than typical C3 plants, even under harsh, arid climatic conditions (Tang et al., 2009). Under conditions of increasing atmospheric CO_2 it can benefit from the extra carbon better than the neighbouring C4 plants it is often found growing with (Fig. 2.11). Parthenium weed rosette leaves possess a typical C4 Kranz anatomy and this is in contrast with its close Asteraceae relatives mariola (P. incanum) and guayule (P. argentatum), which utilize the C3 mode of photosynthesis in all stages of growth (Tang et al., 2009). Using C4 photosynthesis in its rosette leaves, the CO_2 compensation concentration is lower in parthenium weed than it is in mariola or guayule, which might help parthenium weed to tolerate heat and strong light conditions, which are often encountered at this early stage of growth (Moore et al., 1987). In addition, the upper portion of the plant has a lower sensitivity to oxygen

Fig. 2.11. Two biotypes (T is Toogoolawah, C is Clermont) of 40-day-old parthenium weed plants either grown under an elevated atmospheric CO_2 concentration (600 ppmv) (left) or under an ambient CO_2 concentration (400 ppmv) (right). (R. Mao.)

concentration, which is also beneficial for photosynthetic efficiency (Moore *et al.*, 1987), and especially under elevated CO_2 concentrations where its growth and reproductive capacity is greatly increased (Navie *et al.*, 1996; Shabbir *et al.*, 2014).

2.5.10 Genetics

The genetic make-up of any plant species will determine its principal morphological and physiological attributes and consequently will have an important role in its adaptation to any new environments encountered. The genetic diversity created by the breeding system employed by a plant species or biotype can play a vital role in creating new variations to improve invasion success (Prentis *et al.*, 2008). Transition from a self- to a cross-pollination mechanism in an invading plant species may help it to establish in a new environment as cross-pollination allows for the production of greater genetic diversity (Coyer *et al.*, 2006). Hanif *et al.* (2011) reported that two Australian biotypes of parthenium weed, differing in their invasiveness, also exhibited differences in their pollination mechanism. When

bagged, flowers of the invasive biotype only produced 28% filled seeds, while the less invasive biotype produced 73% filled seeds under the same conditions. The invasive biotype of parthenium weed may be more successful in new environments it encounters due to its tendency to cross-pollinate, resulting in more vigorous and rapid growth in such locations. Only a small number of other studies have been conducted to evaluate the role of genetic diversity in parthenium weed invasiveness. The use of internal transcribed spacer (ITS) markers showed Australian biotypes to be more closely related to Mexican biotypes but significantly different from Indian biotypes (Adkins *et al.*, 1997). The use of randomly amplified polymorphic DNA (RAPD), inter simple sequence repeats (ISSR), cpDNA and microsatellite markers to assess the genetic diversity of parthenium weed have also been reported in the literature (Adkins *et al.*, 1997; Tang *et al.*, 2009; Qian *et al.*, 2012). Recently, Jabeen *et al.* (2015) studied the genetic structure of 11 populations (a total of 95 individuals) from across Pakistan using ISSR fingerprinting. About 82% genetic diversity existed between populations, while 18% existed within populations.

2.5.11 Climate change response

The changes going on in global climate are having significant impacts upon weed invasions. Certain invasive species have a specific set of morphological and physiological traits that have been recognized as being beneficial to those species under a changing climate (Dukes and Mooney, 1999). The C3 invasive weed species are anticipated to increase their photosynthetic rates, and thus their growth rates and seed production, within a climate of increasing CO_2 concentration and rising temperature (Sheppard and Stanley, 2014). Parthenium weed has shown the ability to grow remarkably well under such a changing climate. Despite having C4 rosette leaves, the main vegetative part of the parthenium weed is thought to be C3, showing significant improvement in growth under a higher CO_2 concentration (480 ppmv) as compared with an ambient concentration (360 ppmv; Navie *et al.*, 2005; see also Fig. 2.11). It not only explains its invasive behaviour under the present climatic conditions but also warns about the severity of its likely success in the future. Parthenium weed has been shown to have an improved photosynthetic rate, water use efficiency and better growth rate under a high CO_2 concentration (700 ppmv) coupled with a warm temperature range of 25–35°C (Pandey *et al.*, 2003). The increase in the photosynthetic rate due to the elevated CO_2 was, however, compromised by the elevated transpiration rates at very high temperatures (e.g. 47°C). Parthenium weed growth was significantly increased at a high CO_2 concentration (550 ppmv) when it was grown alone and in combination with other grass and legume species (Khan *et al.*, 2015). Recently, Shabbir *et al.* (2014) also reported substantial increases of 52%, 55%, 62%, 120%, 94% and 400% in plant height, biomass, branching, leaf area, photosynthesis and water use efficiency of parthenium weed, respectively, under an elevated CO_2 concentration (550 ppmv) when compared with an ambient CO_2 concentration (380 ppmv). A remarkable increase in parthenium weed growth has been recorded under water-deficit conditions together with high

temperature and elevated CO_2 levels (Nguyen, 2011; Belgeri, 2013). So, the predictions are that parthenium weed is going to become more aggressive, grow more vigorously and produce more seed under a future climate comprising an elevation in temperature and CO_2 concentration as well as under an environment of soils that remain dry for longer. Predictive modelling has been conducted to estimate the potential spread of parthenium weed in different parts of the world under a changing climate (McConnachie *et al.*, 2011; Shabbir, 2012; Kriticos *et al.*, 2015; Mainali *et al.*, 2015). All models suggest that the sub-Saharan African countries, the Asia-Pacific region and some European countries are most at risk of parthenium weed invasion in the future (see Shabbir *et al.*, Chapter 3, this volume).

2.6 Conclusions

Parthenium weed is an annual, herbaceous plant of the family *Asteraceae* having a deep tap root and an erect stem system, reaching a height of 2.0 m. The plant is capable of flowering even when still in the rosette stage of growth and a single plant in the field has been estimated to produce more than 150,000 seeds. These seeds can be distributed by vehicles, agricultural and road construction machinery, in animal fodder, pasture seed lots, stock feed, and naturally by wind and water. Parthenium weed is able to germinate, grow and flower over a wide range of temperatures and photoperiods, hence it can be present in the field at any time of the year. Flowering occurs best at a warm 27/22°C day/night temperature regime and less rapidly at cooler regimes. The weed's seed bank can account for as much as 87% of the total seed present and can be as large as 200,000 seeds/m². Parthenium weed seeds can remain viable in the soil seed bank for many years, with a half-life of *c.*7 years, while seeds on the soil surface die within 6 months. Maximum germination is best achieved under a diurnal temperature regime of *c.*21/16°C, but populations do have a wide range of temperatures over which they can germinate (9–36°C). The average recruitment is c.110 plants/m^2, of which only c.14 plants/m^2 attain maturity. Several cohorts of seedlings may be produced each year. The preferred habitat for parthenium weed includes soils that are predominantly dark cracking clay, and found in riparian sites, or cultivated areas, including field edges, abandoned and vacant land, on roadsides and railway tracks, and many sunny and open areas. The centre of diversity of the genus *Parthenium* is in northern Mexico and it seems most likely that this is the area of origin of the species, although a second centre of diversity also exists in central South America. The 'North American' race has white flowers and is the one that has been introduced into all other parts of the world. The 'South American' race has cream to yellow flowers and differs from the 'North American' race in leaf morphology, pollen colour, capitula size, development of axillary branches, size of disc florets and size of ray corollas. Hymenin is often the dominant sesquiterpene lactone found in the plants of the 'South American' race, whereas parthenin is the dominant sesquiterpene lactone in the 'North American' race. In the native range the most vigorous growth occurs in the summer wet season, with a smaller amount of growth occurring over the winter months depending on winter rainfall. Several aspects of the plant's biology and ecology contribute to its invasiveness. These include the large seed production, large persistent soil seed banks, the longevity of its seeds when buried, its fast germination rate and quick flowering, flowering over a long period of time, its allelopathic capacity and its ability to adapt to many different and stressful environments.

Acknowledgements

The authors wish to acknowledge contributions on the current status of parthenium weed in Mexico by Dr H. Vibrans and thank Dr S. Navie for sharing various information and records.

References

Abe, D.G., Sellers, B.A., Ferrell, J.A. Leon, R.G. and Odero, D.C. (2016) Weed control in Florida pastures with the use of aminocyclopyrachlor. *Weed Technology* 30, 271–278.

Adkins, S. and Shabbir, A. (2014) Biology, ecology and management of the invasive parthenium weed (*Parthenium hysterophorus* L.). *Pest Management Science* 70, 1023–1029. doi:10.1002/ps.3708.

Adkins, S.W. and Sowerby, M.S. (1996) Allelopathic potential of the weed, *Parthenium hysterophorus* L., in Australia. *Plant Protection Quarterly* 11, 20–23.

Adkins, S.W., Navie, S.C., Graham, G.C. and McFadyen, R.E. (1997) Parthenium weed in Australia: Research underway at the Cooperative Research Centre for Tropical Pest Management. In: Mahadevappa, M. and Patil, V.C. (eds) *First International Conference on Parthenium Management*. University of Agricultural Sciences Dharwad, Karnataka, pp. 13–17.

Ahmad, S. (2017) Chemical profile of native and introduced populations of *Parthenium hysterophorus* L. MPhil thesis, Department of Botany, University of the Punjab, Lahore, Pakistan.

Alanís Flores, G.J. (1974) Estudio floristico-ecológico de las malezas en la región citrícola de Nuevo León, México. *Publicaciones Biológicas, Universidad Autónoma de Nuevo León* 1, 41–64.

Andonian, K. and Hierro, J.L. (2011) Species interactions contribute to the success of a global plant invader. *Biological Invasions* 13, 2957–2965. doi:10.1007/s10530-011-9978-x.

Anonymous (2009) Biblioteca Digital de la Medicina Tradicional Mexicana. Available at: http://www.medicinatradicionalmexicana.unam.mx/index.php (accessed 27 July 2016).

Aslam, F., Khaliq, A., Matloob, A., Abbas, R.N., Hussain, S. and Rasul, F. (2014) Differential allelopathic activity of *Parthenium hysterophorus* L. against canary grass and wild oat. *Journal of Animal and Plant Science* 24, 234–244.

Auld, B.A., Hosking, J. and McFadyen, R.E. (1982) Analysis of the spread of tiger pear and parthenium weed in Australia. *Australian Weeds* 2, 56–60.

Bajwa, A.A. (2014) Sustainable weed management in conservation agriculture. *Crop Protection* 65, 105–113. doi:10.1016/j.cropro.2014.07.014.

Bajwa, A.A., Mahajan, G. and Chauhan, B.S. (2015) Non-conventional weed management strategies for modern agriculture. *Weed Science* 63, 723–747. doi:10.1614/WS-D-15-00064.1.

Bajwa, A.A., Chauhan, B.S., Farooq, M., Shabbir, A. and Adkins, S.W. (2016) What do we really know about alien plant invasion? A review of the invasion mechanism of one of the world's worst weeds. *Planta* 244, 39–57.

Bajwa, R., Shafique, S., Shafique, S. and Javaid, A. (2004) Effect of foliar spray of aqueous extract of *Parthenium hysterophorus* on growth of sunflower. *International Journal of Agricultural Biology* 6, 474–478.

Baskin, J.M. and Baskin, C.C. (1989) Physiology of dormancy and germination in relation to seed bank ecology. In: Leek, M.A., Parker, V.T. and Simpson, R.L. (eds) *Ecology of Soil Seed Banks*. Academic Press, San Diego, California, pp. 53–66.

Batish, D.R., Singh, H.P., Pandher, J.K., Arora, V. and Kohli, R.K. (2002) Phytotoxic effect of parthenium residues on the selected soil properties and growth of chickpea and radish. *Weed Biology and Management* 2, 73–78. doi:10.1046/j.1445-6664.2002.00050.x.

Belgeri, A. (2013) Assessing the impact of parthenium weed (*Parthenium hysterophorus* L.) invasion and its management upon grassland community composition. PhD thesis, the School of Agriculture and Food Sciences, the University of Queensland, Brisbane, Australia.

Belgeri, A. and Adkins, S.W. (2015) Allelopathic potential of invasive parthenium weed (*Parthenium hysterophorus* L.) seedlings on grassland species in Australia. *Allelopathy Journal* 36, 1–14.

Bennett, F.D. (1977) A preliminary survey of insects and diseases attacking *Parthenium hysterophorus* L. (Compositae) in Mexico and the USA to evaluate the possibilities of its biological control in Australia. Commonwealth Institute of Biological Control, Curepe, Trinidad.

Bezuneh, T.T. (2015) Phytochemistry and antimicrobial activity of *Parthenium hysterophorus* L.: a review. *Scientific Journal of Analytical Chemistry* 3, 30–38. doi:10.11648/j.sjac.20150303.11.

Blackmore, P.J. and Johnson, S.B. (2010) Continuing successful eradication of parthenium weed (*Parthenium hysterophorus*) from New South Wales, Australia. In: Zydenbos, S.M. (ed.) *Proceedings of the 17th Australasian Weeds Conference*. Christchurch, New Zealand, pp. 382–385.

Butler, J.E. (1984) Longevity of *Parthenium hysterophorus* L. seed in the soil. *Australian Weeds* 3, 6.

Calderón de Rzedowski, G. and Rzedowski, J. (2004) Manual de malezas de la región de Salvatierra, Guanajuato. *Flora Del Bajío y Regiones Adyacentes* XX.

Chen, Y., Wang, J., Wu, X., Sun, J. and Yang, N. (2011) Allelopathic effects of *Parthenium hysterophorus* L. volatiles and its chemical components. *Allelopathy Journal* 27, 217–224.

Chippendale, J.F. and Panetta, F.D. (1994) The cost of parthenium weed to the Queensland cattle industry. *Plant Protection Quarterly* 9, 73–76.

CONABIO (1997) Provincias Biogeográficas de México. Comisión Nacional para el Conocimiento y Uso de la Biodiversidad (CONABIO), Mexico. Available at: http://www.conabio.gob.mx/informacion/gis/maps/geo/rbiog4mgw.zip (accessed 4 December 2015).

CONABIO (2015a) Geoportal de Geoinformación: Sistema Nacional de Información Sobre Biodiversidad. Comisión Nacional para el Conocimiento y Uso de la Biodiversidad (CONABIO), Mexico. Available at: http://www.conabio.gob.mx/informacion/gis/ (accessed 8 December 2015).

CONABIO (2015b) REMIB – Red Mundial de Información sobre Biodiversidad. Comisión Nacional para el Conocimiento y Uso de la Biodiversidad (CONABIO), Mexico. Available at: http://www.conabio.gob.mx/remib/doctos/remib_esp.html (accessed 7 December 2015).

Coyer, J.A., Galice, H., Gareth, A.P., Ester, A.S., Wytze, T.S. and Olsen, J.L. (2006) Convergent adaptation to a marginal habitat by homoploid hybrids and polyploid ecads in the seaweed genus *Fucus*. *Biological Letters* 2, 405–408. doi:10.1098/rsbl.2006.0489.

Dale, I.J. (1981) Parthenium weed in the Americas. *Australian Weeds* 1, 8–14.

Das, B. and Das, R. (1995) Chemical investigation in *Parthenium hysterophorus* L. an allelopathic plant. *Allelopathy Journal* 2, 99–104.

de la Fuente, J.R., Novara, L., Alarcon, S.R., Diaz, O.J., Uriburu, M.L. and Sosa, V.E. (1997) Chemotaxonomy of *Parthenium*: *Parthenium hysterophorus* – *P. glomeratum*. *Phytochemistry* 45, 1185–1188.

Dhileepan, K. (2003) Seasonal variation in the effectiveness of the leaf-feeding beetle *Zygogramma bicolorata* (Coleoptera: Chrysomelidae) and stem-galling moth *Epiblema strenuana* (Lepidoptera: Tortricidae) as biocontrol agents on the weed *Parthenium hysterophorus* (Asteraceae). *Bulletin of Entomological Research* 93, 393–401.

Dhileepan, K. (2009) Managing parthenium weed across diverse landscapes: prospects and limitations. In: Inderjit (ed.) *Management of Invasive Weeds*. Springer, Dordrecht, the Netherlands, pp. 227–259.

Dhileepan, K. (2012) Reproductive variation in naturally occurring populations of the weed

Parthenium hysterophorus (Asteraceae) in Australia. *Weed Science* 60, 571–576. doi:10.1614/WS-D-12-00005.1.

Do, A.T.K. (2009) Variation of the germinable soil seed banks in a pasture infested by *Parthenium hysterophorus* L. at Kilcoy, South-eastern Queensland. MSc thesis, School of Land, Crop and Food Sciences, the University of Queensland, Brisbane, Australia.

Doley, D. (1977) Parthenium weed (*Parthenium hysterophorous* L.): Gas exchange characteristics as a basis for prediction of its geographical distribution. *Australian Journal of Agricultural Research* 28, 449–460.

Dubey, S.K. and Pandey, H.N. (1988) Ray achene polymorphism and germination in ragweed parthenium (*Parthenium hysterophorus*). *Weed Science* 36, 566–567.

Dukes, J.S. and Mooney, H.A. (1999) Does global change increase the success of biological invaders? *Trends in Ecology and Evolution* 14, 135–139. doi:10.1016/S0169-5347(98)01554-7.

Esau, K. (1946) Morphology of reproduction in guayule and certain other species of parthenium. *Hilgardia* 17, 61–101.

Esquivel, V.A., Cano, O. and López, E. (1997) Control químico de malezas en fríjol en el estado de Veracruz. *Agronomia Mesoamericana* 8, 53–58.

Evans, H.C. (1997) *Parthenium hysterophorus*: a review of its weed status and the possibilities for biological control. *Biocontrol News Information* 18, 89–98.

Everist, S.L. (1976) Parthenium weed. *Queensland Agricultural Journal* 102, 2.

Farooq, M., Bajwa, A.A., Cheema, S.A. and Cheema, Z.A. (2013) Application of allelopathy in crop production. *International Journal of Agriculture and Biology* 15, 1367–1378.

Fazal, H., Ahmad, N., Ullah, I., Inayat, H., Khan, L. and Abbasi, B.H. (2011) Antibacterial potential in *Parthenium hysterophorus*, *Stevia rebaudiana* and *Ginkgo biloba*. *Pakistan Journal of Botany* 43, 1307–1313.

Fernández Garza, C. (1993) Experiencia regional sobre el control de malezas de hoja ancha en el cultivo del maíz (*Zea mays* L.), en el centro IV. Santa Apolonia, S.A.R.H., delegación Tamaulipas Norte. Undergraduate thesis, Universidad Autónoma de Nuevo León, Marín, N.L., Mexico.

Gazziero, D.L.P., Brighenti, A.M. and Voll, E. (2006) Resistência cruzada da losna-branca (*Parthenium hysterophorus*) aos herbicidas inibidores da enzima acetolactato sintase. *Planta Daninha* 24, 157–162.

Gnanavel, I. (2013) *Parthenium hysterophorus* L.: a major threat to natural and agro eco-systems

in India. *Science International* 1, 124–131. doi:10.5567/sciintl.2013.124.131.

Gobierno del Estado de Colima (2005). Paquete tecnológico para el cultivo del plátano. Available at: http://www.siac.org.mx/tecnos/9001.pdf (accessed 26 July 2016).

Graham, G.C. and Lang, C.L. (1998) Genetic analysis of relationship of parthenium occurrences in Australia and indications of its origins. Cooperative Research Centre for Tropical Pest Management, University of Queensland, Brisbane, Australia.

Gupta, O.P. and Sharma, J.J. (1977) Parthenium menace in India and possible control measures. *FAO Plant Protection Bulletin* 25, 112–117.

Gupta, S. and Chanda, S. (1991) Aerobiology and some chemical parameters of *Parthenium hysterophorus* pollen. *Grana* 30, 497–503.

Hadi, F. and Bano, A. (2009) Utilization of *Parthenium hysterophorus* for the remediation of lead-contaminated soil. *Weed Biology and Management* 9, 307–314. doi:10.1111/j.1445-6664.2009.00355.x.

Hanif, Z. (2015) Characterization of genetic diversity and invasive potential of parthenium weed in Australia. PhD thesis, the School of Agriculture and Food Sciences, the University of Queensland, Brisbane, Australia.

Hanif, Z., Adkins, S.W., Prentis, P.J., Navie, S.C. and O'Donnell, C. (2011) Characterization of the reproductive behaviour and invasive potential of parthenium weed in Australia. Paper presented at the *23rd Asian-Pacific Weed Science Society Conference* 26–29 September 2011, The Sebel, Cairns, Australia.

Haseler, W.H. (1976) *Parthenium hysterophorus* L. in Australia. *PANS* 22, 515–517.

Holman, O.J. and Dale, I.J. (1981) Parthenium weed threatens Bowen Shire. *Queensland Agricultural Journal* 107, 57–60.

Jabeen, R., Prentis, P., Anjum, T. and Adkins, S.W. (2015) Genetic structure of invasive weed *Parthenium hysterophorus* in Australia and Pakistan. *International Journal of Agriculture and Biology* 17, 327–333.

Jayachandra (1971) Parthenium weed in Mysore State and its control. *Current Science* 40, 568–569.

Jeyalakshmi, C., Doraisamy, S. and Valluvaparidasan, V. (2011) Occurrence of soil microbes under parthenium weed in Tamil Nadu. *Indian Journal of Weed Science* 43, 222–223.

Jiménez, F., Fernández, P., Rosario, J., Gonzales, F. and Prado, R.D. (2014) Primer caso de Resistencia a glifosato en la República Dominicana. *Revista Agropecuaria y Forestal APF* 3, 17–22.

Joshi, S. (1991) Biocontrol of *Parthenium hysterophorus* L. *Crop Protection* 10, 429–431. doi:10.1016/S0261-2194(91)80129-4.

Kanchan, S.D. and Jayachandra (1980) Allelopathic effects of *Parthenium hysterophorus* L. *Plant and Soil* 55(1), 67–75. doi:10.1007/BF02149710.

Kanchan, S. and Jayachandra, K.A. (1981) Effect of *Parthenium hysterophorus* on nitrogen-fixing and nitrifying bacteria. *Canadian Journal of Botany* 59, 199–202.

Karis, P.O. (1995) Cladistics of the subtribe Ambrosiinae (Asteraceae, Heliantheae). *Systematic Botany* 20, 40–54.

Keane, R.M. and Crawley, M.J. (2002) Exotic plant invasions and the enemy release hypothesis. *Trends in Ecology and Evolution* 17, 164–170. doi:10.1016/ S0169-5347(02)02499-0.

Khan, N., Shabbir, A., George, D., Hassan, G. and Adkins, S.W. (2014) Suppressive fodder plants as part of an integrated management program for *Parthenium hysterophorus* L. *Field Crops Research* 156, 172–179.

Khan, N., George, D., Shabbir, A., Hanif, Z. and Adkins, S.W. (2015) Rising CO_2 can alter fodder–weed interactions and suppression of *Parthenium hysterophorus*. *Weed Research* 55, 113–117. doi:10.1111/wre.12127.

Khosla, S.N. and Sobti, S.N. (1979) Parthenium – a national health hazard, its control and utility – a review. *Pesticides* 13, 121–127.

Kohli, R., Batish, D., Singh, H. and Dogra, K. (2006) Status, invasiveness and environmental threats of three tropical American invasive weeds (*Parthenium hysterophorus* L., *Ageratum conyzoides* L., *Lantana camara* L.) in India. *Biological Invasions* 8, 1501–1510.

Kriticos, D.J., Brunel, S., Ota, N., Fried, G., Lansink, A.G.O., Panetta, F.D., Prasad, T.V.R., Shabbir, A. and Yaacoby, T. (2015) Downscaling pest risk analyses: identifying current and future potentially suitable habitats for *Parthenium hysterophorus* with particular reference to Europe and North Africa. *PLoS One* 10, e0132807. doi:10.1371/journal.pone.0132807.

Leos Moreno, A. (1988) Evaluacion de diferentes sistemas de siembra en maiz (*Zea mays* L.) como factor de competencia a la maleza en Marin, N.L., 1986. Undergraduate thesis, Universidad Autónoma de Nuevo León, Marin, N.L., Mexico.

Lewis, W.H., Dixit, A.B. and Wedner, H.J. (1988) Reproductive biology of *Parthenium hysterophorus* (Asteraceae). *Journal of Palynology* 23–24, 73–82.

Lewis, W.H., Dixit, A.B. and Wedner, H.J. (1991) Asteraceae aeropollen of the western United States Gulf Coast. *Annals of Allergy* 67, 37–46.

Maheshwari, J.K. and Pandey, R.S. (1973) Parthenium weed in Bihar State. *Current Science* 42, 733.

Mainali, K., Dhileepan, K., Warren, D., McConnachie, A., Strathie, L., Hassan, G., Karki, D., Shrestha, B.B. and Parmessan, C. (2015) Projecting future expansion of invasive species: comparing and improving methodologies. *Global Change Biology* 21, 4464–4480. doi:10.1111/ gcb.13038.

Malik, R.N., Husain, S.Z. and Nazir, I. (2010) Heavy metal contamination and accumulation in soil and wild plant species from industrial area of Islamabad. *Pakistan Journal of Botany* 42, 291–301.

Martínez, M. (1979) *Catálogo de Nombres Vulgares y Científicos de Plantas Mexicanas*. Fondo de Cultura Económica Mexico, Mexico City.

McClay, A.S. (1983) Natural enemies of *Parthenium hysterophorus* L. in Mexico: Final report 1978–1983. Commonwealth Institute of Biological Control, Slough, UK.

McClay, A.S., Palmer, W.A., Bennett, F.D. and Pullen, K.R. (1995) Phytophagous arthropods associated with *Parthenium hysterophorus* (Asteraceae) in North America. *Environmental Entomology* 24, 796–809.

McConnachie, A.J., Strathie, L.W., Mersie, W., Gebrehiwot, L., Zewdie, K., Abdurehim, A. and Tana, T. (2011) Current and potential geographical distribution of the invasive plant *Parthenium hysterophorus* (Asteraceae) in eastern and southern Africa. *Weed Research* 51, 71–84. doi:10.1111/j.1365-3180.2010.00820.x.

McFadyen, P.J. (1976) A survey of insects attacking *Parthenium hysterophorus* L. (F. Compositae) in South America. Department of Lands, State Government of Queensland, Australia.

McFadyen, R.E. (1985) The biological control programme against *Parthenium hysterophorus* in Queensland. In: Delfosse, E.S. (ed.) *Proceedings of the Sixth International Symposium on the Biological Control of Weeds*. Agriculture Canada, Vancouver, Canada, pp. 789–796.

McFadyen, R.E. (1992) Biological control against parthenium weed in Australia. *Crop Protection* 11, 400–407. doi:10.1016/0261-2194(92)90 021-V.

McFadyen, R.E. (1994) Longevity of seed of *Parthenium hysterophorus* in the soil. Report for Queensland Department of Lands. Queensland Department of Lands, Brisbane, Australia.

McIvor, J.G., Saeli, I., Hodgkinson, J.J. and Shelton, H.M. (2004) Germinable soil seedbanks in native pastures near Crows Nest, south-east Queensland. *Rangeland Journal* 26(1), 72–87.

Mears, J.A. (1973) Systematics of parthenium section Bolophytum (Compositae, Heliantheae): a correlation of morphological, biochemical and habitat data. *Proceedings of the Academy of Natural Sciences of Philadelphia* 125, 121–135.

Méndez-González, M.E., Torres-Avilez, W.M., Dorantes-Euán, A. and Durán-García, R. (2014) Jardines medicinales en Yucatán: una alternativa para la conservación de la flora medicinal de los mayas. *Revista Fitotecnia Mexicana* 37, 97–106.

Mersie, W. and Singh, M. (1987) Allelopathic effect of parthenium (*Parthenium hysterophorus* L.) extract and residue on some agronomic crops and weeds. *Journal of Chemical Ecology* 13, 1739–1747. doi:10.1007/BF00980214.

Mersie, W. and Singh, M. (1988) Effect of phenolic acids and ragweed *Parthenium hysterophorus* L. extracts on tomato (*Lycopersicum esculentum*) growth and nutrient and chlorophyll contents. *Weed Science* 36, 278–281.

Miao, B., Turner, B.L. and Mabry, T.J. (1995) Systematic implications of chloroplast DNA variation in the subtribe *Ambrosiinae* (Asteraceae: Heliantheae). *American Journal of Botany* 82, 924–932.

Moore, B.D., Franceschi, V.R., Cheng, S.H., Wu, J. and Ku, M.S. (1987) Photosynthetic characteristics of the C3–C4 intermediate *Parthenium hysterophorus*. *Plant Physiology* 85, 978–983. doi:10.1104/pp. 85.4.978.

Nath, R. (1981) Note on the effect of parthenium extract on seed germination and seedling growth in crops. *Indian Journal of Agricultural Science* 51, 601–603.

Navie, S.C. (2002) The biology of *Parthenium hysterophorus* L. in Australia. PhD thesis, School of Agriculture and Food Sciences, University of Queensland, Brisbane, Australia.

Navie, S.C., McFadyen, R.E., Panetta, F.D. and Adkins, S.W. (1996) The biology of Australian weeds. 27. *Parthenium hysterophorus* L. *Plant Protection Quarterly* 11, 76–88.

Navie, S.C., Panetta, F.D., McFadyen, R.E. and Adkins, S.W. (1998) Behaviour of buried and surface-sown seeds of parthenium weed. *Weed Research* 38, 335–341. doi:10.1046/j.1365-3180.1998.00104.x.

Navie, S.C., Panetta, F.D., McFadyen, R.E. and Adkins, S.W. (2004) Germinable soil seedbanks of central Queensland rangelands invaded by the exotic weed *Parthenium hysterophorus* L. *Weed Biology and Management* 4, 154–167. doi:10.1111/j.1445-6664.2004.00132.x.

Navie, S.C., McFadyen, R.E., Panetta, F.D. and Adkins, S.W. (2005) The effect of CO_2 enrichment on the growth of a C3 weed (*Parthenium hysterophorus* L.) and its competitive interaction with a C4 grass (*Cenchrus ciliaris* L.). *Plant Protection Quarterly* 20, 61–66.

Nguyen, T.L.T. (2011) The invasive potential of parthenium weed (*Parthenium hysterophorus* L.) in Australia. PhD thesis, the School of Agriculture and Food Sciences, University of Queensland, Brisbane, Australia.

Nguyen, T., Bajwa, AA., Navie, S., O'Donnell, C. and Adkins, S. (2017a) The soil seedbank of pasture communities in central Queensland invaded by *Parthenium hysterophorus* L. *Rangeland Ecology and Management* 70, 244–254.

Nguyen, T., Bajwa, A.A., Navie, S., O'Donnell, C.O. and Adkins, S. (2017b) Parthenium weed (*Parthenium hysterophorus* L.) and climate change: the effect of CO_2 concentration, temperature, and water deficit on growth and reproduction of two biotypes. *Environmental Science and Pollution Research* 24, 10727–10739.

Nissole, G. (1711) Établissement de quelques nouveaux genres de plantes. *Mémoires de l'Académie Royale des Sciences [Paris]*, 316–320.

Osunkoya, O.O., Sadiq Ali, S., Nguyen, T., Perrett, C., Shabbir, A., Navie, S., Belgeri, A., Dhileepan, K. and Adkins, S. (2014) Soil seed bank dynamics in response to an extreme flood event in a riparian habitat. *Ecological Research* 29, 1115–1129.

Pandey, D.K. (1997) Inhibition of najas (*Najos gromineo* Del.) by parthenium (*Parthenium hysterophorus* L.). *Allelopathy Journal* 4, 121–126.

Pandey, H.N. and Dubey, S.K. (1988) Achene germination of *Parthenium hysterophorus* L.: effects of light, temperature, provenance and achene size. *Weed Research* 28, 185–190.

Pandey, H.N. and Dubey, S.K. (1989) Growth and population dynamics of an exotic weed *Parthenium hysterophorus* Linn. *Proceedings of the Indian Academy of Science (Plant Sciences)* 99, 51–58.

Pandey, D.K., Palni, L.M.S. and Joshi, S.C. (2003) Growth, reproduction, and photosynthesis of ragweed parthenium (*Parthenium hysterophorus*). *Weed Science* 51, 191–201. doi:10.1614/0043-1745(2003)051[0191:GRAPOR]2.0.CO;2.

Panero, J.L. (2005) New combinations and infrafamilial taxa in the Asteraceae. *Phytologia* 87, 1–14.

Parker, A. (1989) Biological control of parthenium weed using two rust fungi. In: Delfosse, E.S. (ed.) *Proceedings of the Seventh International Symposium on the Biological Control of Weeds*. CSIRO, Melbourne, Australia, pp. 531–537.

Parsons, W.T. and Cuthbertson, E.G. (1992) *Noxious Weeds of Australia*. Inkata, Melbourne, Australia.

Piazzano, M., Bernardello, G., Novara, L., Alarcón, S.R., de la Fuente, J.R. and Hadid, M. (1998) Evaluación de los límites específicos entre *Parthenium hysterophorus* y *P. glomeratum* (Asteraceae-Ambrosiinae): evidencias morfológicas, anatómicas, cromosomáticas y fitoquímicas. *Anales Del Jardin Botanico de Madrid* 56, 65–76.

Picman, J. and Picman, A.K. (1984) Autotoxicity in *Parthenium hysterophorus* and its possible role in control of germination. *Biochemical Systematics and Ecology* 12, 287–292.

Picman, A.K. and Towers, H.N. (1982) Sesquiterpene lactones in various populations of *Parthenium hysterophorus*. *Biochemical Systematics and Ecology* 10, 145–153.

Prentis, P.J., Wilson, J.R.U., Dormontt, E.E., Richardson, D.M. and Lowe, A.J. (2008) Adaptive evolution in invasive species. *Trends in Plant Science* 13, 288–294. doi:10.1016/j.tplants.2008.03.004.

Qian, M., Li, G., Tang, S., Qu, X., Lin, Y., Yin, J., Klinken, R.D.V. and Peng-geng, Y. (2012) Development of microsatellites for the invasive weed parthenium weed (Asteraceae). *American Journal of Botany* 99, 277–279. doi:10.3732/ajb.1100579.

Ramos-Zapata, J.A., Marrufo-Zapata, D., Guadarrama, P., Carrillo-Sánchez, L., Hernández-Cuevas, L. and Caamal-Maldonado, A. (2012) Impact of weed control on arbuscular mycorrhizal fungi in a tropical agroecosystem: a long-term experiment. *Mycorrhiza* 22, 653–661.

Robinson, H., Powell, A.M., King, R.M. and Weedin, J.F. (1981) Chromosome number in Compositae, X11 Heliantheae. *Smithsonian Contribution Botany* 52, 1–28.

Rocha-Peña, M.A. and Padrón-Chávez, J.E. (eds) (2009) *El Cultivo de los Cítricos en el Estado de Nuevo León*. Instituto Nacional de Investigaciones Forestales Agrícolas y Pecuarias, Campo Experimental, General Terán, México.

Rollins, R.C. (1950) The guayule rubber plant and its relatives. *Contributions from the Gray Herbarium of Harvard University* 172, 3–72.

Romero, A., Carrion, G. and Rico-Gray, V. (2001) Fungal latent pathogens and endophytes from leaves of *Parthenium hysterophorus* (Asteraceae). *Fungal Diversity* 7, 81–87.

Rosales Robles, E., Sánchez de la Cruz, R. and Rodríguez del Bosque, L.A. (2014) Tolerancia de sorgo para grano a dos herbicidas. *Revista Fitotecnia Mexicana* 37, 89–94.

Rosario, J., Fuentes, C. and Prado, R.D. (2013) Resistencia de *Parthenium hysterophorus* L. al glifosato: un nuevo biotipo resistente a herbicida en Colombia. *Revista Agropecuaria y Forestal APF* 2, 15–18.

Safdar, M.E., Tanveer, A., Khaliq, A. and Riaz, M.A. (2015) Yield losses in maize (*Zea mays*) infested with parthenium weed (*Parthenium hysterophorus* L.). *Crop Protection* 70, 77–82. doi:10.1016/j.cropro.2015.01.010

Seier, M.K., Morin, L., Evans, H.C. and Romero, Á. (2009) Are the microcyclic rust species *Puccinia melampodii* and *Puccinia xanthii* conspecific? *Mycological Research* 113, 1271–1282.

Servicio Meteorológico Nacional (2016) Normales Climatológicas por Estación. Available at: http://smn1.conagua.gob.mx/index.php?option=com_content&view=article&id=42&Itemid=28 (accessed 10 June 2016).

Shabbir, A. (2012) Towards the improved management of parthenium weed: complementing biological control with plant suppression. PhD thesis, School of Agriculture and Food Sciences, University of Queensland, Brisbane, Australia.

Shabbir, A. and Javaid, A. (2010) Effect of aqueous extracts of alien weed *Parthenium hysterophorus* and two native Asteraceae species on germination and growth of mungbean, *Vigna radiata* L. Wilczek. *Journal of Agricultural Research* 48, 483–488.

Shabbir, A., Dhileepan, K., Khan, N. and Adkins, S.W. (2014) Weed–pathogen interactions and elevated CO_2: growth changes in favour of the biological control agent. *Weed Research* 54, 217–222. doi:10.1111/wre.12078.

Sharma, A.D., Bhullar, A., Rakhra, G. and Mamik, S. (2014) Analysis of hydrophilic antioxidant enzymes in invasive alien species *Parthenium hysterophorus* under high temperature abiotic stress like conditions. *Journal of Stress Physiology and Biochemistry* 10, 228–237.

Sheppard, C.S. and Stanley, M.C. (2014) Does elevated temperature and doubled CO_2 increase growth of three potentially invasive plants? *Invasive Plant Science Management* 7, 237–246. doi:10.1614/IPSM-D-13-00038.1.

Shi, B. (2016) Invasive potential of the weed *Parthenium hysterophorus*: the role of allelopathy. PhD thesis, School of Agriculture and Food Sciences, University of Queensland, Brisbane, Australia.

Shrestha, B.B., Shabbir, A. and Adkins, S.W. (2015) *Parthenium hysterophorus* in Nepal: a review of its weed status and possibilities for management. *Weed Research* 55, 132–144. doi:10.1111/wre.12133

Singh, H.P., Batish, D.R., Pandher, J.K. and Kohli, R.K. (2003) Assessment of allelopathic properties of *Parthenium hysterophorus* residues.

Agriculture, Ecosystems and Environment 95, 537–541. doi:10.1016/S0167-8809(02)00202-5.

Strother, J.L. (2006) 246. *Parthenium* Linnaeus. In: Barkley, T.M., Brouillet, L. and Strother, J.L. (eds) *Flora of North America Vol. 21: Asteraceae.* Flora of North America Editorial Committee, New York.

Stuessy, T.F. (1973) A systematic review of the subtribe Melampodiinae (Compositae, Heliantheae). *Contributions from the Gray Herbarium* 203, 65–80.

Stuessy, T.F. (1975) Flora of Panama. *Annals of the Missouri Botanical Garden* 62, 1091–1096.

Tamado, T. and Milberg, P. (2004) Control of parthenium (*Parthenium hysterophorus*) in grain sorghum (*Sorghum bicolor*) in the smallholder farming system in eastern Ethiopia. *Weed Technology* 18, 100–105. doi:10.1614/WT-03-033R.

Tamado, T., Schutz, W. and Milberg, P. (2002) Germination ecology of the weed *Parthenium hysterophorus* in eastern Ethiopia. *Annals of Applied Biology* 140, 263–270. doi:10.1111/j.1744-7348.2002.tb00180.x.

Tang, S.Q., Wei, F., Zeng, L.Y., Li, X.K., Tang, S.C., Zhong, Y. and Geng, Y.P. (2009) Multiple introductions are responsible for the disjunct distributions of invasive *Parthenium hysterophorus* in China: evidence from nuclear and chloroplast DNA. *Weed Research* 49, 373–380. doi:10.1111/j.1365-3180.2009.00714.x.

Tanveer, A., Khaliq, A., Ali, H.H., Mahajan, G. and Chauhan, B.S. (2015) Interference and management of parthenium: the world's most important invasive weed. *Crop Protection* 68, 49–59. doi:10.1016/j.cropro.2014.11.005.

Tefera, T. (2002) Allelopathic effects of *Parthenium hysterophorus* extracts on seed germination and seedling growth *of Eragrostis tef. Journal of Agronomy and Crop Science* 188, 306–310. doi:10.1046/j.1439-037X.2002.00564.x.

Timsina, B., Shrestha, B.B., Rokaya, M.B. and Münzbergová, Z. (2011) Impact of *Parthenium hysterophorus* L. invasion on plant species composition and soil properties of grassland communities in Nepal. *Flora* 206, 233–240. doi:10.1016/j.flora.2010.09.004.

Towers, G.H.N., Mitchell, J.C., Rodriguez, E., Bennett, F.D. and Rao, S.P.V. (1977) Biology and chemistry of *Parthenium hysterophorus* L: a problem weed in India. *Journal of Scientific and Industrial Research [India]* 36, 672–684.

Upadhyay, S.K., Ahmad, M. and Singh, A. (2013) Ecological impact of weed (*Parthenium hysterophorus* L.) invasion in saline soil. *International Journal of Science Research Publication* 3, 1–4.

USDA-NRCS (2003) Plant profile for *Parthenium hysterophorus* L., The Plants Database, Version 3.5. United States Department of Agriculture (USDA)–Natural Resources Conservation Service (NRCS), National Plant Data Center, Baton Rouge, Louisiana. Available at: https://plants.usda.gov/core/profile?symbol=PAHY (accessed 13 January 2003).

USDA-NRCS (2016) The Plants Database. United States Department of Agriculture (USDA)–Natural Resources Conservation Service (NRCS), National Plant Data Team, Greensboro, North Carolina. Available at: https://plants.usda.gov/java/ (accessed 10 June 2016).

Vaillant, S. (1720) Suite des Corymbifères, ou de la seconde classe des plantes à fleurs composées. *Histoire de l'Académie Royale des Sciences [Paris]*, 277–339.

Valliappan, K. and Towers, G.H.N. (1988) Allelopathic effect of root exudates from the obnoxious weed, *Parthenium hysterophorus* L. *Indian Journal of Weed Science* 20, 18–22.

Villarreal Quintanilla, J.A. (1983) *Malezas de Buenavista, Coahuila.* Universidad Autónoma Agraria Antonio Narro, Buenavista, Saltillo, Coahuila, Mexico.

Villaseñor, J.L. (2016) Checklist of the native vascular plants of Mexico. *Revista Mexicana de Biodiversidad* 87, 559–902.

Villaseñor, J.L. and Espinosa-García, F.J. (1998) *Catálogo de Malezas de México.* UNAM/Fondo de Cultura Economica, Mexico, DF.

Wakija, M., Berecha, G. and Tulu, S. (2009) Allelopathic effects of an invasive alien weed *Parthenium hysterophorus* L. compost on lettuce germination and growth. *African Journal of Agricultural Research* 4, 1325–1330.

White, G.G. (1994) Workshop report: parthenium weed. Report by Cooperative Research Centre for Tropical Pest Management, Brisbane, Australia.

Williams, J.D. and Groves, R.H. (1980) The influence of temperature and photoperiod on growth and development of *Parthenium hysterophorus* L. *Weed Research* 20, 47–52. doi:10.1111/j.1365-3180.1980.Tb00040.x.

Zaheer, Z., Shafique, S., Shafique, S. and Mehmood, T. (2012) Antifungal potential of *Parthenium hysterophorus* L. plant extracts against *Fusarium solani. Science Research Essays* 7, 2049–2054.

3 Spread

Asad Shabbir,[1]* Andrew McConnachie[2] and Steve W. Adkins[3]

[1]University of the Punjab, Lahore, Pakistan; current affiliation: Plant Breeding Institute, the University of Sydney, Narrabri, New South Wales, Australia; [2]Department of Primary Industries, Biosecurity and Food Safety, Orange, New South Wales, Australia; [3]The University of Queensland, Gatton, Queensland, Australia

3.1 Origin and Native Distribution

Parthenium weed (*Parthenium hysterophorus* L.) is an annual herb of Neotropical origin that has developed a pantropical distribution (Evans, 1997). The weed is thought to be native to the tropical and subtropical Americas, possibly around the Gulf of Mexico, including the southern USA and the Caribbean islands, or possibly in northern Argentina and southern Brazil (Dale, 1981). Parthenium weed is a highly invasive species and is now found in 92 countries around the globe, of which only 44 are possibly in its native range (Fig. 3.1; Table 3.1).

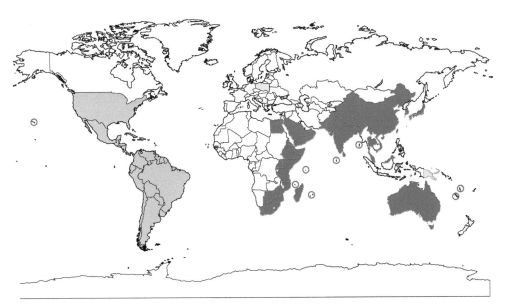

Fig. 3.1. A worldwide map of parthenium weed distribution. Parthenium weed has invaded the countries shaded or circled in red. Blue-shaded countries represent transient populations of the weed. Countries shaded green are those considered to be within its native range. (A. Shabbir.)

* asad.shabbir@sydney.edu.au

Table 3.1. Global records of parthenium weed from its native and introduced ranges.

Range	Country
Native range (44 countries)	South America: Argentina, Paraguay, Uruguay, Venezuela (Rollins, 1950); Brazil (Parsons and Cuthbertson, 1992); Chile, Peru, French Guiana (Dale, 1981); Bolivia (Friend, 1983); Guyana (Ramaswami, 1997); Ecuador, Guadeloupe (EPPO, 2014); Netherlands Antilles, Surinam, Turks and Caicos Islands (USDA-ARS, 2012) North America: USA, Mexico (Rollins, 1950) Central America: Haiti, Cuba, Honduras (Rollins, 1950); Guatemala, Puerto Rico, Trinidad (Friend, 1983); Nicaragua (Lewis *et al.*, 1988); Republic of Panama, Costa Rica (Hammel, 1997); Anguilla, Aruba, Bahamas, Cayman Islands, Saint Lucia (EPPO, 2014); Antigua and Barbuda, Grenada, Saint Kitts and Nevis, Saint Vincent and the Grenadines, United States Virgin Islands (USDA-ARS, 2012); Barbados, Bermuda, Curacao, Dominica, Dominican Republic, Jamaica, Martinique (Dale, 1981); Belize (Khosla and Sobti, 1979)
Introduced range (48 countries)	South Asia: India (Rao, 1956); Pakistan (Razaq *et al.*, 1994); Sri Lanka (Jayasuriya, 2005); Bangladesh (Dhileepan, 2009); Nepal (Mishra, 1991); Bhutan (Parker, 1992) East and South-east Asia: China (Tang *et al.*, 2010); Vietnam (Nguyen, 2011); Japan (Mito and Uesugi, 2004); Korea (Kim, 2013); Taiwan (Peng *et al.*, 1988); Malaysia (Karim, 2014); Thailand (M. Day, Brisbane, 2017, personal communication) Middle East: Israel (Joel and Liston, 1986); Palestine (Dafni and Heller, 1982); Oman, Yemen (Kilian *et al.*, 2002); United Arab Emirates (Mahmoud *et al.*, 2015); Kingdom of Saudi Arabia (Thomas *et al.*, 2015) Australia and Oceania: Australia (Everist, 1976); Vanuatu (Nath, 1981); New Caledonia (MacKee, 1994); Hawaii (Motooka, 2003); Papua New Guinea (Anonymous, 2003); French Polynesia (Aneja *et al.*, 1991); Tahiti (Queensland Government, 2011) East Africa: Ethiopia (Medhin, 1992); Somalia (Tamado *et al.*, 2002); Uganda, Tanzania (Wabuyele *et al.*, 2014); Kenya (Njoroge, 1991); Madagascar (Tamado *et al.*, 2002); Seychelles (Nath, 1988); Mauritius (Parsons and Cuthbertson, 1992); Eritrea (Tadesse, 2004); Rwanda (A. Witt, Kenya, 2014, personal communication); Mayotte (USDA-ARS, 2012); Comoros, Reunion (EPPO, 2014); Djibouti (Etana *et al.*, 2015) Southern Africa: South Africa (Rollins, 1950); Mozambique (Tamado and Milberg, 2000); Zimbabwe (Tamado *et al.*, 2002); Swaziland (McConnachie *et al.*, 2011); Botswana (L. Strathie, South Africa, 2017, personal communication) Northern Africa: Egypt (Zahran and Willis, 1992) Europe: Belgium, Poland (EPPO, 2014)

3.2 Invaded Range Distribution

The spread of parthenium weed around the globe in the last three to four decades has been rapid and has occurred through a number of biotic and abiotic pathways. This rapid invasion suggests that parthenium weed is still capable of spreading to many other countries, especially those in West Africa, South-east Asia, Eastern Europe and other parts of the world with a suitable climate in the near future (McConnachie *et al.*, 2011; Shabbir, 2012; Kriticos *et al.*, 2015; Mainali *et al.*, 2015). The weed has become a major invasive species in Australia, where it is mainly found in the state of Queensland and has invaded by one estimation *c.*600,000 km^2 (Anonymous, 2011a). Parthenium weed is present in all countries of South Asia and the Indian subcontinent is the region worst affected. Parthenium weed is now present in all states of India (Sushilkumar and Varshney, 2010), while in Pakistan it was first reported in the most northern districts of the Punjab Province in the 1990s and since then it has invaded Khyber Pakhtunkhwa, FATA (Federally Administered Tribal Areas), Azad Jammu and Kashmir (Shabbir *et al.*, 2012) and more recently the Sind Province

(A. Shabbir, Lahore, 2012, personal observation).

One of the most severely affected regions outside Australia and South Asia is East Africa where the weed has invaded Ethiopia, Kenya, Tanzania, Uganda, Rwanda, Somalia and Eritrea (Table 3.1; see McConnachie and Witt, Chapter 15, this volume). Among the East African countries, Ethiopia is one of the worst affected and parthenium weed has become a serious problem in grazing and cropping lands. It is thought to have arrived in the country as a food grain contaminant in the mid-1970s. There is little information available on its spread to North Africa; however, this weed is reported in Egypt, where it is thought to have been introduced in contaminated grass seed imported from the USA (Zahran and Willis, 1992; see McConnachie and Witt, Chapter 15, this volume).

In southern Africa, parthenium weed is present in South Africa, Mozambique, Zimbabwe, Swaziland, Mauritius and Madagascar (Table 3.1; McConnachie et al., 2011; see Strathie and McConnachie, Chapter 14, this volume). The weed was first recorded in South Africa in 1880 at Inanda, KwaZulu-Natal province, but was not common up to 1977; however, it became common and invasive after cyclone Domoina caused extensive flooding along the east coast of southern Africa in 1984. The weed is declared invasive in South Africa, where it is present and spreading in the north-eastern parts of the country. Parthenium weed occurs in the savannah biome of South Africa; its distribution extends from the eastern subtropical region of KwaZulu-Natal province, from around Durban, northwards to Mpumalanga province and also into the neighbouring countries of Swaziland and Mozambique (Henderson, 2001; see Strathie and McConnachie, Chapter 14, this volume). It is also present north to north-west of Pretoria, up into Zimbabwe (Strathie et al., 2005).

In East and South-east Asia, the weed is present in China, South Korea, Vietnam and Japan. It has been recently recorded in Thailand and Malaysia, where it is spreading rapidly and concerned authorities are now trying to stop its further spread (Table 3.1;

see Shi et al., Chapter 13, this volume). The mechanism of introduction of parthenium weed into this region is unknown, but it is believed to have been present in south China for c.100 years (S. Tang, Guilin, 2012, personal communication). In the Middle East, parthenium weed has been reported in Israel and Palestine, while in the Gulf States it is found in Yemen and Oman, Kingdom of Saudi Arabia. Recently it has been reported in the United Arab Emirates, where it is thought to have arrived as a contaminant in potting mix or soil associated with imported ornamental plants (Mahmoud et al., 2015; see McConnachie and Witt, Chapter 15, this volume).

Parthenium weed is an up-and-coming weed in Oceania, where it has been reported in Vanuatu, New Caledonia, Hawaii, French Polynesia, Papua New Guinea (PNG) and Christmas Island (Table 3.1; see McFadyen et al., Chapter 11, this volume). By 2009, parthenium weed was considered to be eradicated from PNG after being introduced in 2001. In 2009, a quarantine survey indicated that no parthenium was to be found at the introduction sites (Kawi and Orapa, 2010).

In addition, there have been numerous personal communications about the presence of parthenium weed in Afghanistan, Sudan, South Sudan and Democratic Republic of Congo but as yet no published information is available on these possible invasions (Shabbir, 2012).

3.3 Climatic Suitability

Climatic models for the potential geographic distribution of parthenium weed around the world suggests that the weed has not yet reached its full geographic range and many areas that are ecoclimatically suitable for invasion, remain vulnerable (Fig. 3.2; McConnachie et al., 2011; Shabbir, 2012; Mainali et al., 2015; Kriticos et al., 2015). Parthenium weed has now been reported in a total of 92 countries worldwide with Asia, Africa, and Australia and the Pacific being recipients of major incursions (Table 3.1). With the increasing trend in global trade

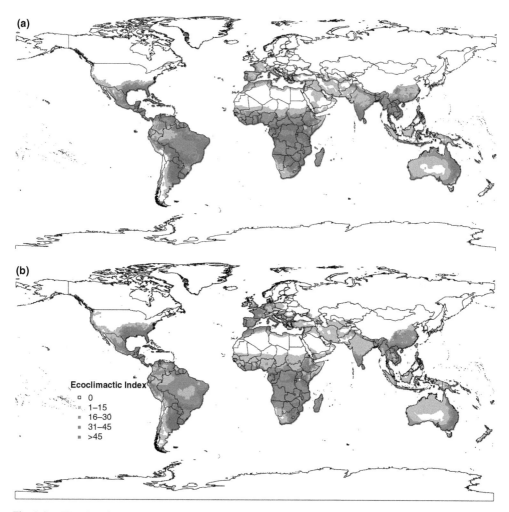

Fig. 3.2. The global ecoclimatic suitability for parthenium weed for the current climate (a) and predicted suitability following a +3°C climate change scenario (b) using the CLIMEX model. The ecoclimatic suitability (Ecoclimatic Index) increases with increasing intensity of colour in (a) and (b). (b: Shabbir, 2012.)

and tourism, more regions are likely to be at risk of invasion. Considering the aggressive nature of this weed, especially in disturbed areas, and the ecoclimatic suitability of large geographical areas, it is quite likely that parthenium weed is already present in some of the poorest African and Asian countries, but has yet to be reported due to lack of awareness, funding and scientific support (Shabbir, 2012).

3.3.1 Africa and Europe

The current African distribution of parthenium weed, which is composed primarily of north-eastern and southern populations, fits well with the projections of models developed by McConnachie *et al.* (2011), Shabbir (2012), Mainali *et al.* (2015) and Kriticos *et al.* (2015). The models show that the current climate of most of Central and

West Africa and some Mediterranean parts of North Africa are highly suitable for parthenium weed invasion (Fig. 3.2). Similarly, the models reveal that south and south-eastern Europe are ecoclimatically suitable for parthenium weed, highlighting its wide climatic plasticity across continents. Under climate change, most of the southern and central-western parts of Europe, and the central parts of China, will become highly suitable for the weed (Fig. 3.2). In a recent weed risk analysis conducted for parthenium weed, it was also demonstrated that within Europe, most Mediterranean countries are at risk (Kriticos *et al.*, 2015).

3.3.2 Australia and Oceania

For Australia, in broad terms, the models suggest that favourable areas for parthenium weed growth occur in all the states (Fig. 3.2). The models also adequately account for the present distribution of parthenium weed in Queensland and suggest that additional uninvaded areas in Australia are ecoclimatically suitable for parthenium weed in Australia, under current and climate change scenarios. Parthenium weed is becoming a problem in the south-west regions of Queensland where the weed is able to outcompete native and introduced pasture plants, in this low input–output country (N. Judd, Queensland, 2012, personal communication). The core infestation of parthenium weed lies in central Queensland; however, the weed is also present in south-east Queensland where satellite populations are being targeted for eradication while biological control is being implemented for the management of larger populations (Callander and Dhileepan, 2016). New South Wales (NSW), being ecoclimatically suitable and adjoining Queensland, is thought to be at greatest threat to parthenium weed invasion and already has the greatest number of reported incursions outside of Queensland (Blackmore and Johnson, 2010). The NSW Government has spent a great deal of money to prevent parthenium weed spread and establishment in the state and every effort is made to eradicate all new infestations that are found (Blackmore and Johnson, 2010; S. Johnson, Orange, 2016, personal communication). The projected range expansion of the distribution of parthenium weed in NSW, under the climate change scenario, is of great concern as large fertile areas of the Murray–Darling Basin will be under threat (Fig. 3.2). In addition, every year new infestations of parthenium weed are reported in other Australian states, and the Northern Territory is on particular alert since the discovery of a new infestation in the Tennant Creek area in 2010 (Anonymous, 2018). Similarly, recent reports of parthenium weed outbreaks in Western Australia (WA) have also been documented (Anonymous, 2011b).

Parthenium weed is present in at least seven Pacific islands including Vanuatu, New Caledonia, Hawaii, PNG, Christmas Island, French Polynesia and Tahiti (Table 3.1). The weed was eradicated in PNG after it was thought to be introduced as a contaminant with old vehicles imported from Australia. In most of the Pacific islands, parthenium weed is becoming problematic and has been targeted for biological control (Day and Winston, 2016). In Vanuatu, parthenium weed is a declared noxious weed. It is suspected that the weed may have come in to the country on imported machinery from Queensland (Day and Bule, 2016; see McFadyen *et al.*, Chapter 11, this volume).

3.3.3 South Asia

The models predict that the whole of South Asia is suitable for parthenium weed growth, with more northern parts of Pakistan, southern India, southern Bhutan and Nepal, most of Sri Lanka and Bangladesh being projected as highly suitable. Parthenium weed has been reported in all mainland countries of South Asia except Afghanistan, although a few parts of Afghanistan were predicted as being ecoclimatically suitable for parthenium weed growth, especially in the future, under a changing climate (Fig. 3.2). Parthenium weed is reported in all Indian states, with more frequent records of occurrence found in the southern states (Dhileepan and Senaratne, 2009). Parthenium weed is now considered to be an emerging weed in Nepal and Bangladesh and the very rapid spread of the weed in Nepal is an alarming development (Akter and Zuberi, 2009; Shrestha *et al.*, 2015). The weed has invaded various habitats that include grasslands, roadsides, wastelands, fallow lands and national parks such as Chitwan National Park, a World Natural Heritage site (Shrestha *et al.*, 2015). A recent survey has revealed that parthenium weed is now present in 54 out of 75 districts of Nepal (B.B. Shrestha, Kathmandu, 2016, personal communication).

In Pakistan, parthenium weed is spreading quickly and invading areas that have been deemed previously to be ecoclimatically unsuitable (Shabbir, 2012). The spread of the weed in these locations suggests that it may be undergoing genetic change due to natural selection of types that are more suited to abiotic conditions, especially in the Khyber Pakhtunkhwa province where it is being found at higher and colder altitudes, and in dryer regions of the southern Punjab and Sind. The other possible reason could be that the weed has not reached the full extent of its niche, since models are not going to factor in all scenarios, especially when it comes to microclimate. Parthenium weed is now spreading quickly into the irrigated south-west districts of the Punjab Province. The mean summer temperature in these districts is *c*.34°C (May–June), with daily maximum readings often up to 50°C, with only 250 mm of annual rainfall. The persistence of parthenium weed in these hotter, dryer areas could be explained through the provision of extra moisture in the form of irrigation (from canals and groundwater pumping). The current spread of parthenium weed from the north-east districts to the south-west districts of the Punjab Province seems to be through the movement of vehicles over a large network of major and minor roads and an extensive water canal system for irrigation (Shabbir *et al.*, 2012). In the central districts of the Punjab, parthenium weed is the most dominant weed growing in wastelands, along roadsides,

water channels, in abandoned fields and crops such as sugarcane (*Saccharum officinarum* L), maize (*Zea mays* L.), sorghum (*Sorghum bicolor* (L.) Moench) and a variety of vegetable and floriculture crops (Riaz *et al.*, 2007; Anwar *et al.*, 2012; Shabbir *et al.*, 2012). Incursions of parthenium weed into the south-west districts of Punjab Province present a very significant threat to the cotton (*Gossypium hirsutum* L.) industry of Pakistan.

The climatic suitability also suggests that parthenium weed can tolerate predicted climate change conditions in most of the Indus Basin in Pakistan due to the availability of the extra moisture in the form of irrigation. Under a climate change scenario, a reduction in the range of the weed is expected in the southern parts of Pakistan. However, irrigation will facilitate the persistence of the weed throughout most of the south-west basin of the Indus River under both current and changing climate scenarios (Shabbir, 2012). This was confirmed when parthenium weed was detected in an irrigated farm in the Khairpur District of Sind Province in 2012.

3.4 Spread Pathways

The long-distance spread of parthenium weed seed is accidental and generally thought to occur through the transport of contaminated food grain and feed and seed lots, seed attached to vehicles and machinery, and in packaging materials (Adkins and Shabbir, 2014). Shorter natural spread vectors include water, wind (Auld *et al.*, 1983; Parsons and Cuthbertson, 1992) and dung (Fig. 3.3; Beyene and Tessema, 2015).

3.4.1 Vehicles, machinery and soil

Parthenium weed seed can also enter new territories as a contaminant, either on machinery (e.g. the headers of grain harvesters) or as seed in soil attached to vehicles (e.g. mud in car wheel arches or on earth-moving equipment). Road and rail transport appear to have been the main means of dispersal of parthenium weed seed over long distances in the Indian sub-continent and Africa (Jayachandra, 1971; McConnachie *et al.*, 2011; Fig. 3.4). Parsons and Cuthbertson (1992) also believed that most of the long-distance spread of parthenium weed seed was as a result of vehicles (cars, trains, trucks) and farm machinery

(especially when entrapped in mud on these vehicles). The first published record of parthenium weed in Australia (Auld *et al.*, 1983) was in 1955 for southern Queensland (however, an earlier Queensland herbarium record shows a plant being collected in southern Queensland in 1950), and it was attributed to the movement of contaminated aircraft and machinery parts into

Fig. 3.3. The possible spread of parthenium weed seeds through animals feeding on mature flowering plants: (a) a group of goats feeding in India and (b) a cow in South Africa. (S.K. Guru and I.A.W. Macdonald.)

Fig. 3.4. Spread pathways of parthenium weed: (a) an infestation along the roadside and (b) a rail track in South Africa. (A. McConnachie and A. Witt.)

Australia during the Second World War, originating from the USA (Parsons and Cuthbertson, 1992). Nguyen (2011) undertook a study examining material washed off vehicles at off-road washdown facilities at five sites in central Queensland, Australia which identified that parthenium weed seed was present in the material removed from vehicles at all five facilities with an average of 1340 seeds/t of dry sludge. One washdown facility was reported to be removing c.2000 viable parthenium weed seeds/week.

To assess the role of small vehicles in the spreading of weed seeds, utility and sedan vehicles that had been in the field in south-east and central Queensland were analysed for their weed seed loads (Khan, 2012). During these studies it was found that such vehicles could collect and transport a large number of viable weed seeds (i.e. from 116 to 397 viable seeds per vehicle) representing 25–59 species. The majority of these species were found to be introduced into Australia, with some being of particular significance. For example, the seeds of parthenium weed were found on vehicles that had been in parthenium-infested areas of central Queensland. Weed seeds were carried on almost all parts of the vehicles, either in mud or directly attached to vehicle parts, with the majority of them found on the underside or on the back and front mudguards and in the cabin.

Of the outbreaks of new parthenium weed infestations on private properties in NSW during the period 1982–2004, over 70% were attributed to the combined movement of contaminated grain harvesters (59%), and vehicles/trucks and other machinery (14%) (Blackmore and Johnson, 2010). Haseler (1976) reported that vehicles used for transportation of stock, trucks, cars and earth-moving equipment were all major vectors for spreading parthenium weed seed over long distances within Australia. In the Shandong Province (China), harvesters were reported to act as a significant vector for parthenium weed seed movement (Li and Gao, 2012). The release of parthenium weed seeds from vehicles along roadsides (and roadside reserves) facilitates its transfer to unintended habitats connected by roads, as

observed in India (Krishnamurthy et al., 1977; Ramachandra Prasad et al., 2010; Sushilkumar and Varshney, 2010).

Parthenium weed has now spread from India into most of its neighbouring countries. It is suspected that it was spread by vehicles or carried as a contaminant of seed lots to Nepal (Mishra, 1991), Bhutan (Parker, 1992), Bangladesh (Anonymous, 2009) and Pakistan (Shabbir et al., 2012). The species was considered to have entered PNG attached to second-hand vehicles imported from Australia. There were two populations recorded, one from a compound where these imported vehicles were parked and the other some 10 km away in a coconut (Cocos nucifera L.) plantation (Kawi and Orapa, 2010).

3.4.2 Seed and grain

Parthenium weed was accidentally introduced into Clermont, Australia as a contaminant of a pasture grass seed lot imported from the USA in 1958 (Haseler, 1976). This introduction led to a relatively rapid spread of this weed, and by 1994 it had invaded about 170,000 km^2 of prime grazing pastures in central Queensland (Chippendale and Panetta, 1994), and more recently it was estimated to have invaded over 600,000 km^2 (Anonymous, 2011a).

Parthenium weed is also believed to have been introduced into some parts of Africa (Ethiopia, Mozambique) and Asia (India, Sri Lanka and Nepal) in cereal or grass seed shipments (Rao, 1956; Bhowmik and Sarkar, 2005; Wise et al., 2007). According to Jayasuriya (2005), one of the possible modes of entry of parthenium weed seed into Sri Lanka was within imported condiment seed lots such as mustard (Brassica nigra L.), chilli (Capsicum annuum L.) or onion (Allium cepa L.) from India. It was suspected to have been introduced to Nepal, on at least one occasion, as a contaminant of seed lots from India (Mishra, 1991). Some of the local spread of parthenium weed from central Queensland to various areas of south-east Queensland and into northern and western NSW is thought to have been within contaminated seed lots (Anonymous, 1994). For instance, a seed analysis of the unused portion of a purple pigeon grass (*Setaria incrassata* (Hochst.) Hack.) seed lot showed that between 300 and 600 parthenium weed seeds were present in every kilogram of that pasture seed (Kelly, 1991). It is thought to have been introduced to Israel in 1980, most likely through the import of contaminated grains from the USA for its use in fish feed (Dafni and Heller, 1982).

Similarly, spread of parthenium weed into Egypt was through a 1960 introduction in contaminated grass seed imported from Texas (Boulos and El-Hadidi, 1984). In Shandong Province in China, the weed is reported to have been imported in 2004 in a soybean (*Glycine max* L. Merr.) seed lot from the USA (Li and Gao, 2012). One possible source of parthenium weed spread into Africa may have been through one of the world food aid programmes, such as Food Aid. These seed lots do not receive the same agricultural scrutiny, from either the exporting or the importing country (A. Witt, Kenya, 2011, personal communication).

In India, parthenium weed was first observed in Poona in 1956 (Rao, 1956) on the rubbish heaps of the local agricultural college where it is reported to have been accidentally introduced with consignments of contaminated wheat (*Triticum aestivum* L.) and other cereals imported from the USA under the PL480 grant to counter a food shortage within the country (Sushilkumar and Varshney, 2010). Through public distribution of these contaminated cereal seed lots and by other vectors, parthenium weed has now spread over 350,000 km^2 of the Indian subcontinent through various spread pathways (except to the Western Ghats and snow-covered areas of the northern and north-eastern parts) (Yaduraju *et al.*, 2005; Sushilkumar and Varshney, 2010; Ramachandra Prasad *et al.*, 2010).

3.4.3 Water

The propagule of parthenium weed is a cypsela, which, with the help of two sterile florets attached on either side, makes the propagule buoyant and aids in the long-distance dispersal by water (Navie *et al.*, 1996). Its seed is adapted to spread in high flows along streams and waterways in central Queensland, Australia (Auld *et al.*, 1983) and in many parts of India (Ramachandra Prasad *et al.*, 2010). Haseler (1976) pointed out that unaided, lateral movement from established parthenium weed plants is slow, although many of the larger parthenium weed populations are known to spread more rapidly along waterways, drainage flats and floodplains, indicating the significant dispersal role of water. The cypselas of parthenium weed vary in size, from 2 mm to 3 mm in length and *c.* 2 mm in width (Auld *et al.*, 1983). When freshly shed, they are buoyant and will float for some time on water and potentially travel over long distances in river or floodwater. In Sri Lanka, viable parthenium weed seeds were discovered in river sand removed from the Mahaweli River and used for building purposes (Jayasuriya, 2005). Auld *et al.* (1983) also thought that short-distance dispersal of seeds by water was important, as indicated by the large populations of parthenium weed spreading along waterways in central Queensland. The spread of parthenium weed in water is thought to be aided by the cypsela's ability to aid in floatation and then to be deposited along the edges of creeks and dams (McFadyen, 1984). In addition, movement in sheet water flow (flooding) is thought to be important, as indicated by large colonies establishing along waterways and on drainage flats and over floodplains (Parsons and Cuthbertson, 1992).

A decline in parthenium weed seed viability following immersion in water was associated with an increase in temperature and with the time of exposure. The viability of seed remained high (*c.* 100%) even after 45 days of immersion in river water, but only when the temperature was low (10°C and 15°C). Moderate temperatures (20°C and 24°C) increased the rate of seed viability loss; however, seed viability remained high for up to 15 days and above 50% for 30 days (Nguyen, 2011). Further, Nguyen found that the seed viability was lowest in the sludge pond water that had been used to wash down vehicles. In these sludge-pond treatments, viability remained high for 5 days, but when exposed to warm temperatures (27–34°C), the viability decreased rapidly to below 20% after 15 days, with all viability being lost after 20 days at 30°C or 34°C and after 30 days at 27°C. This more rapid decline in seed viability in the vehicle washdown water was not associated with its greater acidity, as pH alone had no effect upon seed viability in the range of pH 4.0–9.0.

These findings suggest the long-lived nature of parthenium weed seed in river water will substantially increase the chances of seed dispersal by water bodies, while the life in vehicle washdown ponds indicates that the water should not be reused for agricultural purposes for at least 45 days after the last washdown event has occurred.

Parthenium weed is spreading into southern and south-western Queensland along several rivers and stream systems. These rivers are the headwaters of the Murray–Darling Basin and flow directly into NSW. Prior to floods in early 2011, the infestations were upstream from the St George irrigation area and it is likely that further downstream spread has occurred. Until now, NSW has only needed to manage the spread of parthenium weed through human-mediated vectors. By 2009 it had become evident that established infestations of parthenium weed in the upper reaches of the Maranoa and Balonne Rivers, northern tributaries of the Darling River, had created a pathway for spread to NSW. To date no infestations have reached NSW via this route, but local control authorities remain vigilant in regularly inspecting the waterways (Blackmore, 2011).

In Pakistan, parthenium weed is very common along the watercourses supplying water to agricultural farms. Data on its current distribution shows that parthenium weed has spread from the north-eastern districts of the Punjab Province and Islamabad Capital Territory to the south-western districts. In Pakistan there is an extensive canal network carrying water from the Indus river basin into the cropping regions where it is used to irrigate the land. Parthenium weed also spreads along irrigation canals, in a semi-natural way, as observed in the Punjab Province (Anwar et al., 2012; Shabbir et al., 2012). In the central districts of Punjab province, parthenium weed was found to be the most frequently encountered weed growing along the banks of rivers and canals, and the edges of water channels (Shabbir et al., 2012). These watercourses are generally not well managed, and as a result parthenium weed thrives and acts as a potential source of further spread. The Indus river system comprises five major rivers (i.e. the Jhelum, Chinab, Ravi, Sutluj and the Indus) flowing from north to south. On average, one major flooding event occurs every 5–6 years. This poses a serious threat of further spread of the weed into the southern areas of the country (Shabbir, 2012).

In Sri Lanka, parthenium weed seed has been observed to be dispersed in the Mahaweli River (Jayasuriya, 2005). It appears that water is the chief natural dispersal vector in Sri Lanka, with irrigation systems effectively dispersing parthenium weed seed into non-infested areas. Extreme events such as flooding potentially further exacerbate the spread of seed into new areas, especially in the wet-zone habitats of Sri Lanka, if not effectively controlled in the heavily infested areas (Jayasuriya, 2005).

In Ethiopia, Ayele (2007) conducted a survey on parthenium weed which revealed that 63% of respondents reported that flooding was the primary seed dispersal mechanism, followed by animals (dung and movement on fur; Fig. 3.3) and wind. Each dispersal mechanism certainly contributed to the effective dispersal of parthenium weed, especially during extreme climatic events such as those leading to storms and flooding.

3.4.4 Wind

The dispersal of parthenium weed seed by wind has been investigated in several studies. For example, in a study by Jayachandra (1971), the sterile florets (air sacs) of parthenium weed cypselas were found to reduce the overall density of the seed, which as a result enhanced its dispersal by wind. Haseler (1976) and Parsons and Cuthbertson (1992), however, expressed the view that in Australia wind dispersal of parthenium weed seed was limited, and that most seed would only be carried for a few metres, but these authors acknowledged that whirlwinds and cyclones could carry large numbers of the light seeds for considerable distances. In KwaZulu-Natal, South Africa, parthenium weed became very abundant after cyclone Domoina hit the region in 1984, causing widespread damage and creating ideal conditions for a pioneer like parthenium weed to take hold (Wise *et al.*, 2007; McConnachie *et al.*, 2011). Moreover, Auld *et al.* (1983) claimed that wind is important in transporting parthenium weed on a localized scale, but was in the order of a few metres only. However, strong winds and storms have been reported to carry the seed efficiently over longer distances (Maheshwari and Pandey, 1973; Haseler, 1976).

3.4.5 Animals

Parthenium weed can be spread by domestic and wild animals. Its seeds could potentially be dispersed both internally and externally by animals (Fig. 3.3), but this has not been extensively addressed in the literature. Auld *et al.* (1983), McFadyen (1984), Parsons and Cuthberson (1992) and Adkins *et al.* (2005) all hypothesized that parthenium weed seeds can be dispersed by the movement of livestock, especially when contained within the mud that adheres to the animal's coat or hooves. Spread of parthenium weed seed by animals such as cattle between infested and uninfested land has also been observed in southern Queensland (D. Chandler, Injune, 2010, personal communication). Seed dispersal in mud adhering to the hooves of cattle and human feet is also thought to occur (Jayasuriya, 2005). It has even been postulated that parthenium weed seed may survive gut passage, and therefore be spread in animal dung, but this has not been well studied. Abraham and Girija (2005) reported that parthenium weed seed could remain viable and be transported in cow dung brought from Tamil Nadu to Kerala, southern India. Similarly, in Sri Lanka, one idea for the original arrival of the weed into this country was within goats brought from India – viable seeds were passed in their dung and subsequently germinated in the fields of Sri Lanka (Jayasuriya, 2005).

More recently, Khan *et al.* (2012) collected dung samples from three different locations in Queensland (i.e. Toogoolawah, Injune and Clermont) and found that dry and fresh dung samples contained an average of 287 and 262 germinable weed seeds/kg, respectively, which came from 55 weed species and represented 20 families. There were significant differences in the number of seeds in samples from different locations, and over different seasons. In addition, noxious weed seeds (e.g. parthenium weed) were identified in both dry and fresh samples of dung collected from Clermont in May 2010, when the parthenium weed plants were mature and shedding seed in the field. Khan *et al.* (2012) also demonstrated that parthenium weed seeds could be found in fresh (54 seeds/kg) and aged dung (28 seeds/kg) of cattle grazing in infested pastures in Clermont during autumn.

In a study conducted in the Khyber Pakhtunkhwa province of Pakistan during winter (Khan, 2012), large numbers of weed seeds were found in the dung samples of different animals (e.g. cow, water buffalo, goat, sheep, horse and donkey). All of these animals were found to be responsible for spreading a large number of weed seeds, which were all introduced to that particular locality. However, there were specific associations between certain weed species and certain animal species. For example, a large number of winter grass (*Poa annua* L.) seeds were found in horse dung samples, but not in the dung of goats and sheep. Thus, different types of animals spread the seeds of different weed species.

In Australia, parthenium weed is thought to be dispersed by feral pigs, wallabies and some birds (Grice, undated). At a property near Kilcoy, south-east Queensland, parthenium weed was consistently found growing alongside the tracks of feral pigs (R. Wilkinson, Kilcoy, 2011, personal communication). This led to the hypothesis that the feral pigs may be responsible for spreading the weed, either by eating it or through the seed catching on their hairs while grazing or brushing past parthenium weed growing next to their tracks.

3.4.6 Fodder

Experiments on weed seed spread in lucerne (*Medicago sativa* L.) and Rhodes grass (*Chloris gayana* Kunth) fodder were also conducted in Queensland (Khan, 2012). Lucerne and Rhodes grass fodder samples were collected from the districts of Laidley North, Monto and Biloela and were analysed for the presence of weed seeds. The seeds of numerous species of weeds were found in the fodder samples, and the majority of these seeds were alien to Australia. In the lucerne hay collected from Monto, Biloela and Laidley North, all weed seeds identified were from introduced species; however, no noxious weed seeds were present. In a separate experiment conducted in Khyber Pakhtunkhwa, Pakistan, large numbers of introduced species were found to be spread in wheat grain but parthenium weed was not present (Khan, 2012).

3.5 Future Spread

When the effects of climate change (scenario +3°C) were considered in the CLIMEX model of Shabbir (2012; Fig. 3.2), the potential ecoclimatic suitability for parthenium weed was slightly contracted in both South Asia and Australia. In South Asia, the climate change scenario shifted the ecoclimatic suitability for parthenium weed towards the north, while the southern regions became less suitable (Fig. 3.2). The model, when run to simulate climate change, indicated that the weed could potentially expand into the northern regions of India, Pakistan, Nepal and Bhutan, while southern regions of these countries would become less suitable (Fig. 3.2). Further areas within Afghanistan would also become more ecoclimatically suitable for parthenium weed growth.

Various parts of Australia (NSW, Victoria and WA) will become more ecoclimatically suitable for parthenium weed growth under the climate change scenario. However, the ecoclimatic suitability of certain parts of northern Queensland and the Northern Territory were expected to contract from inland areas towards the coastal regions (Fig. 3.2). For Pakistan, it seemed that under a changing climate, parthenium weed may be able to expand the northerly limits of its invaded range, while contracting its potential distribution in the southern extremities of the Punjab and Baluchistan provinces. Azad Jammu and Kashmir, Khyber Pakhtunkhwa and Gilgit-Baltistan will become more suitable for the weed as suggested by an increase in the ecoclimatic index value in these provinces.

It is hypothesized that in Australia, the predicted climate change pattern for the next 50 years will promote the competitiveness of C3 plant species (which includes parthenium weed) but not that of C4 plant species (e.g. tropical pasture plants and grasses with which parthenium weed is often found competing). Earlier studies have indicated that an elevated CO_2 concentration might alter the present and future distribution, and the abundance of parthenium weed (Navie *et al.*, 2005). Large sections of Africa and southern Asia are currently at risk of parthenium weed invasion, as spread into these areas is likely to increase under future climate change. Moreover, parthenium weed being either a C3 (Navie *et al.*, 2005) or a C3/C4 intermediate plant (Moore *et al.*, 1987) should be able to gain a significant growth advantage from a CO_2-rich environment, compared with that for C4 plants (Navie *et al.*, 2005). It was also found that the reproductive potential of parthenium weed could be significantly enhanced when it was grown under an atmosphere of elevated CO_2 (Navie *et al.*, 2005). Khan (2011) tested a number of different suppressive pasture plants under elevated CO_2 concentrations in a glasshouse trial and found that the suppression indices of most of the C3 pasture plants increased in relation to parthenium weed, while remaining unchanged for C4 pasture plants. This response was further confirmed with increased water use efficiency (>500%) for parthenium weed plants when grown under an elevated CO_2 concentration (Shabbir *et al.*, 2014). This improved water use efficiency for parthenium weed under an elevated CO_2 scenario has important implications for the future spread and impact of the weed and the management practices that will need to be applied in the future under a changing climate.

Parthenium weed is likely to expand its geographical range within Australia and southern Asia, particularly into southern Pakistan where additional moisture is available in the form of irrigation. Under climate change, the northern parts of the African continent and most of eastern and northern Europe and the Mediterranean are under the threat of invasion.

Studies suggest that parthenium weed has not yet reached its full potential in terms of its invaded range and is likely to undergo a further range expansion and become more competitive in the future under climate change. A global effort is required to stop this weed's further spread and impact, and reduce its presence in already invaded areas using an integrated management approach.

3.6 Conclusion

Parthenium weed is a highly invasive plant species that has been accidentally introduced into 48 countries outside of its native range and is likely to expand its geographical distribution even more, particularly in the Mediterranean, and eastern European and South and South-east Asian regions. Under a climate change scenario of an increase of 3°C, the northern parts of the African continent, and more of the eastern and northern European and Mediterranean regions, come under the threat of invasion. Since the weed can easily be spread through various natural, human and animal vectored pathways, it is

likely that the weed will continue to extend its distribution within areas it has already reached. The ability of parthenium weed to benefit from future CO_2 enrichment, predicted to occur under a changing climate, will encourage greater seed production, which in turn will increase the rate of spread. If not contained effectively, weed will spread more aggressively in the future, with greater chances of new outbreaks occurring within and beyond its present introduced range.

Acknowledgements

The authors thank the members of the International Parthenium Weed Network for providing information on the spread of parthenium weed in different parts of the world. The authors are also very grateful to Drs Ian A.W. Macdonald, S.K. Guru and A. Witt, who provided the photographs of parthenium weed spread.

References

Abraham, C.T. and Girija, T. (2005) Status of parthenium in Kerala. In: Ramachandra Prasad, T.V., Nanjappa, H.V., Devendra, R., Manjunath, A., Subramanya, S.C., *et al.* (eds) *Proceedings of the Second International Conference on Parthenium Management.* University of Agricultural Sciences, Bangalore, India, pp. 48–51.

Adkins, S.W. and Shabbir, A. (2014) Biology, ecology and management of the invasive parthenium weed (*Parthenium hysterophorus* L.). *Pest Management Science* 70, 1023–1029.

Adkins, S.W., Navie, S.C. and Dhileepan, K. (2005) Parthenium weed in Australia: Research progress and prospects. In: Ramachandra Prasad, T.V., Nanjappa, H.V., Devendra, R., Manjunath, A., Subramanya, S.C., *et al.* (eds) *Proceedings of the Second International Conference on Parthenium Management.* University of Agricultural Sciences, Bangalore, India, pp. 11–27.

Akter, A. and Zuberi, M.I. (2009) Invasive alien species in Northern Bangladesh: Identification, inventory and impacts. *International Journal of Biodiversity Conservation* 15, 129–134.

Aneja, K.R., Dhawan, S.R. and Sharma, A.B. (1991) Deadly weed, *Parthenium hysterophorus* L. and its distribution. *Indian Journal of Weed Science* 23, 14–18.

Anonymous (1994) Landholders face fines over parthenium. *Queensland Grain Grower*, December, p. 6.

Anonymous (2003) Incursion of Parthenium Weed (*Parthenium hysterophorus*) in Papua New Guinea. *Pest Alert, Plant Protection Service, Secretariat of the Pacific Community* No. 30. ISSN 1727–8473. Available at: http://www.hear.org/pier/pdf/pest_alert_30_parthenium_weed.pdf (accessed 1 May 2017).

Anonymous (2009) Parthenium Weed Poses Danger to Crops. *Dhakka Mirror*, 30 December. Available at: http://www.dhakamirror.com/environment/parthenium-weed-poses-danger-to-crops/ (accessed 1 June 2017).

Anonymous (2011a) NSW Parthenium Strategy. New South Wales Department of Primary Industries. Available at: http://www.dpi.nsw.gov.au/data/assets/pdf_file/0003/355917/parthenium-weed.pdf (accessed 12 March 2013).

Anonymous (2011b) Keep your eyes peeled for parthenium weed. *The Farmweekly* 21 November. http://www.farmweekly.com.au/news/agriculture/cattle/general-news/keep-your-eyes-peeled-for-parthenium-weed/2365257.aspx (accessed 3 May 2018).

Anonymous (2018) Parthenium. https://nt.gov.au/environment/weeds/A-Z-list-of-weeds-in-the-NT/parthenium (accessed 3 May 2018).

Anwar, W., Khan, S.N., Tahira, J.J. and Suliman, R. (2012) *Parthenium hysterophorus*: An emerging threat for *Curcuma longa* fields of Kasur District, Punjab, Pakistan. *Pakistan Journal of Weed Science Research* 18, 91–97.

Auld, B.A., Hosking, J. and McFadyen, R.E. (1983) Analysis of the spread of tiger pear and parthenium weed in Australia. *Australian Weeds* 2, 56–60.

Ayele, S. (2007) Impact of parthenium (*Parthenium hysterophorus* L.) on the range ecosystem dynamics of the Jijiga Rangeland, Ethiopia. MSc thesis, Haramaya University, Ethiopia.

Beyene, H. and Tessema, T. (2015) Distribution, abundance and socio-economic impacts of parthenium (*Parthenium hysterophorus*) in Southern Zone of Tigray, Ethiopia. *Journal of Poverty, Investment and Development* 19, 22–29.

Bhowmik, P.C. and Sarkar, D. (2005) *Parthenium hysterophorus*: Its world status and potential management. In: Ramachandra Prasad, T.V., Nanjappa, H.V., Devendra, R., Manjunath, A., Subramanya, S.C., *et al.* (eds) *Proceedings of the Second International Conference on Parthenium Management*. University of Agricultural Sciences, Bangalore, India, pp. 1–5.

Blackmore, P.J. (2011) Parthenium weed in NSW – a model for continuing success In: van Klinken, R.D., Osten, V.A., Panetta, F.D. and Scanlan, J.C. (eds) *Proceedings of 16th NSW Weeds Conference* 18–21 July 2011, Pacific Bay Conference Centre. Weeds Society of New South Wales, Coffs Harbour, Australia, pp. 144–152.

Blackmore, P.J. and Johnson, S.B. (2010) Continuing successful eradication of parthenium weed (*Parthenium hysterophorus*) from New South Wales, Australia. In: Zydenbos, S.M. (ed.) *Proceedings of 17th Australasian Weeds Conference*. New Zealand Plant Protection Society, Christchurch, New Zealand, pp. 382–385.

Boulos, L. and El-Hadidi, M.N. (1984) *The Weed Flora of Egypt*. American University of Cairo Press, Cairo.

Callander, J.T. and Dhileepan, K. (2016) Biological control of parthenium weed: Field collection and redistribution of established biological control agents. In: Randall, R., Lloyd, S. and Borger, C. (eds) *Proceedings of the Twentieth Australasian Weeds Conference*. Weeds Society of Western Australia, Perth, Australia, pp. 242–245.

Chippendale, J.F. and Panetta, F.D. (1994) The cost of parthenium weed to the Queensland cattle industry. *Plant Protection Quarterly* 9, 73–76.

Dafni, A. and Heller, D. (1982) Adventive flora of Israel – phytogeographical, ecological and agricultural aspects. *Plant Systematics and Evolution* 140, 1–18.

Dale, I.J. (1981) Parthenium weed in the Americas. *Australian Weeds* 1, 8–14.

Day, M.D. and Bule, S. (2016) The status of weed biological control in Vanuatu. In: Daehler, C.C., van Kleunen, M., Pyšek, P. and Richardson, D.M. (eds) *Proceedings of 13th International EMAPi Conference*, Waikoloa, Hawaii. *NeoBiota* 30, 151–166. doi:10.3897/neobiota.30.7049.

Day, M.D. and Winston, R.L. (2016) Biological control of weeds in the 22 Pacific island countries and territories: Current status and future prospects. In: Daehler, C.C., van Kleunen, M., Pyšek, P. and Richardson, D.M. (eds) *Proceedings of 13th International EMAPi Conference*, Waikoloa, Hawaii. *NeoBiota* 30, 167–192. doi:10.3897/neobiota.30.7113.

Dhileepan, K. (2009) Managing parthenium weed across diverse landscapes: prospects and limitations. In: Inderjit (ed.) *Management of Invasive Weeds*. Springer, Dordrecht, the Netherlands, pp. 227–259.

Dhileepan, K. and Senaratne, K.A.D.W. (2009) How widespread is *Parthenium hysterophorus* and its biological control agent *Zygogramma bicolorata* in South Asia? *Weed Research* 49, 557–562.

Etana, A., Ensermu, K. and Teshome, S. (2015) Impact of *Parthenium hysterophorus* L. (Asteraceae) on soil chemical properties and its distribution in a reserve area: a case study in Awash National Park (ANP), Ethiopia. *Journal of Soil Science and Environmental Management* 6, 116–124.

European and Mediterranean Plant Protection Organization (EPPO) (2014) *Parthenium hysterophorus* L. Asteraceae – parthenium weed. *EPPO Bulletin* 44, 474–478.

Evans, H.C. (1997) *Parthenium hysterophorus*: A review of its weed status and the possibilities for biological control. *Biocontrol News and Information* 18, 89N–98N.

Everist, S.L. (1976) Parthenium weed. *Queensland Agricultural Journal* 10, 2.

Friend, E. (1983) *Queensland Weed Seeds*. Queensland Department of Primary Industries, Brisbane, Australia.

Grice, T. (undated) *Weeds of the Burdekin Rangelands: Managing Parthenium*. CSIRO Sustainable Systems, Townsville, Queensland, Australia.

Hammel, B. (1997) Parthenium weed in Costa Rica. *The Cutting Edge, News and Notes* IV. http://www.mobot.org/MOBOT/research/Edge/jan97/jan97new.shtml (accessed 5 June 2018).

Haseler, W.H. (1976) *Parthenium hysterophorus* L. in Australia. *PANS* 22, 515–517.

Henderson, L. (2001) Alien weeds and invasive plants. *Plant Protection Research Institute Handbook* No. 12. ARC, Plant Protection Research Institute, Pretoria, South Africa.

Jayachandra (1971) Parthenium weed in Mysore State and its control. *Current Science* 40, 568–569.

Jayasuriya, A.H.M. (2005) Parthenium weed – status and management in Sri Lanka. In: Ramachandra Prasad, T.V., Nanjappa, H.V., Devendra, R., Manjunath, A., Subramanya, S.C., *et al.* (eds) *Proceedings of the Second International Conference on Parthenium Management*. University of Agricultural Sciences, Bangalore, India.

Joel, D.M. and Liston, A. (1986) New adventive weeds in Israel. *Israel Journal of Botany* 35, 215–223.

Karim, S.M.R. (2014) Malaysia invaded: Weed it out before it is too late. *International Parthenium News* 9, 1–2.

Kawi, A. and Orapa, W. (2010) Status of parthenium weed in Papua New Guinea. *International Parthenium News* 2, 2–3.

Kelly, I. (1991) Parthenium weed in the Castlereagh Macquarie County Council district. In: Trounce, B. and Popovic, P. (eds) *Proceedings of the 6th Biennial Noxious Plants Conference*. New South Wales Agriculture and Fisheries, Sydney, Australia, pp. 69–73.

Khan, I. (2012) Spread of weed seeds and its prevention. PhD thesis, School of Agriculture and Food Sciences, the University of Queensland, Brisbane, Australia.

Khan, I., O'Donnell, O., Navie, S., George, D. and Adkins, S. (2012) Weed seed spread by vehicles: a case study from southeast Queensland, Australia. *Pakistan Journal of Weed Science Research, Special Issue* 18, 281–288.

Khan, N. (2011) Long term, sustainable management of parthenium weed (*Parthenium hysterophorus* L.) using suppressive plants. PhD thesis, the University of Queensland, Brisbane, Australia.

Khosla, S.N. and Sobti, S.N. (1979) Parthenium – a national health hazard, its control and utility – a review. *Pesticides* 13, 121–127.

Kilian, N., Peter, H. and Ali, H.M. (2002) New and noteworthy records for the flora of Yemen, chiefly of Hadhramout and Al-Mahra. *Willdenowia* 32, 239–269.

Kim, J.W. (2013) Parthenium weed in Korea. *International Parthenium Weed Newsletter* 7, 1.

Krishnamurthy, K., Ramachandra Prasad, T.V., Muniyappa, T.V. and Venkata Rao, B.V. (1977) Parthenium – a new pernicious weed in India. *UAS Technical Series*, No. 17. University of Agricultural Sciences (UAS), Bangalore, India.

Kriticos, D.J., Brunel, S., Ota, N., Fried, G., Oude Lansink, A.G.J.M., Panetta, F.D., Prasad, T.V.R., Shabbir, A. and Yaacoby, T. (2015) Downscaling pest risk analyses: identifying current and future potentially suitable habitats for *Parthenium hysterophorus* with particular reference to Europe and North Africa. *PLoS One* 10(9), e0132807.

Lewis, W., Dixit, A.B. and Wedner, H.J. (1988) Reproductive biology of *Parthenium hysterophorous* (Asteraceae). *Journal of Palynology* 23–24, 73–82.

Li, M. and Gao, X. (2012) Occurrence and management of parthenium weed in Shandong Province, China. *International Parthenium News* 6, 5–6.

MacKee, H.S. (1994) *Catalogue des Plantes Introduites et Cultivées en Nouvelle-Calédonie*. Muséum National d'Histoire Naturelle, Paris.

Maheshwari, J.K. and Pandey, R.S. (1973) Parthenium weed in Bihar State. *Current Science* 42, 733.

Mahmoud, T., Gairola, S. and El-Keblawy, A. (2015) *Parthenium hysterophorus* and *Bidens pilosa*, two new records to the invasive weed flora of the United Arab Emirates. *Journal on New Biological Reports* 41, 26–32.

Mainali, K.P., Warren, D.L., Dileepan, K., McConnachie, A.J., Strathie, L., Hassan, G., Karki, D., Shrestha, B.B. and Parmesan, C.

(2015) Projecting future expansion of invasive species: Comparing and improving methodologies for species distribution modelling. *Global Change Biology* 21, 4464–4480.

McConnachie, A.J., Strathie, L.W., Mersie, W., Gebrehiwot, L., Zewdie, K., *et al.* (2011) Current and potential geographical distribution of the invasive plant *Parthenium hysterophorus* (Asteraceae) in eastern and southern Africa. *Weed Research* 51, 71–84.

McFadyen, R.E. (1984) Annual ragweed in Queensland. In: *Proceedings of the 7th Australian Weeds Conference*, Perth, September. Weed Society of Western Australia, Perth, Australia, pp. 205–209.

Medhin, B.G. (1992) *Parthenium hysterophorus*, new weed problem in Ethiopia. *FAO Plant Protection Bulletin* 40, 49.

Mishra, K.K. (1991) *Parthenium hysterophorus* L. – a new record for Nepal. *Journal of the Bombay Natural History Society* 88, 466–467.

Mito, T. and Uesugi, T. (2004) Invasive alien species in Japan: the status quo and the new regulation for prevention of their adverse effects. *Global Environmental Research* 2, 171–791.

Motooka, P., Castro, L., Nelson, D., Nagai, N. and Ching, L. (2003) *Weeds of Hawaii's Pastures and Natural Areas: An Identification and Management Guide*. College of Tropical Agriculture and Human Resources, University of Hawaii Manoa, Honolulu.

Nath, R. (1981) Note on the effect of parthenium extract on seed germination and seedling growth in crops. *Indian Journal of Agricultural Sciences* 51, 601–603.

Nath, R. (1988) *Parthenium hysterophorus* L. – a review. *Agricultural Reviews* 9, 171–179.

Navie, S.C., McFadyen, R.E., Panetta, F.D. and Adkins, S.W. (1996) The biology of Australian weeds. 27. *Parthenium hysterophorus* L. *Plant Protection Quarterly* 11, 76–88.

Navie, S.C., McFadyen, R.E., Panetta, F.D. and Adkins, S.W. (2005) The effect of CO_2 enrichment on the growth of a C3 weed (*Parthenium hysterophorus* L.) and its competitive interaction with a C4 grass (*Cenchrus ciliaris* L.). *Plant Protection Quarterly* 20, 61–66.

Nguyen, T.L.T. (2011) Biology of parthenium weed (*Parthenium hysterophorus* L.) in Australia. PhD thesis, the University of Queensland, Brisbane, Australia.

Njoroge, J.M. (1991) Tolerance of *Bidens pilosa* L and *Parthenium hysterophorus* L to paraquat (Gramoxone) in Kenya coffee. *Kenya Coffee* 56, 999–1001.

Parker, C. (1992) *Weeds of Bhutan*. Royal Government of Bhutan, Thimpu.

Parsons, W.T. and Cuthbertson, E.G. (1992) *Noxious Weeds of Australia*. Inkata Press, Melbourne, Australia.

Peng, C.I., Hu, L.A. and Kao, M.T. (1988) Unwelcome naturalisation of *Parthenium hysterophorus* (Asteraceae) in Taiwan. *Journal of Taiwan Museum* 41, 95–101.

Queensland Government (2018) *Fact Sheet Index Parthenium hysterophorus* L. https://keyserver. lucidcentral.org/weeds/data/media/Html/ parthenium_hysterophorus.htm (accessed 1 February 2018).

Ramachandra Prasad, T.V., Denesh, G.R., Ananda, N., Sushilkumar and Varsheny Jay, G. (2010) *Parthenium hysterophorus* L. – a national weed, its menace and integrated management strategies in India. In: Gautam, R.D., Mahapatro, G.K., Shahsi Bhalla, Shankarganehs, K., Sudhida Gautam, Kaushal Verma and Gaur, H.S. (eds) *Proceedings of the 3rd International Conference on Parthenium*, 8–10 December 2010. Indian Agricultural Research Institute, New Delhi, India. pp. 13–20.

Ramaswami, P.P. (1997) Potential uses of parthenium weed. In: Patil, M.M.A.V.C. (ed.) *Proceedings of the First International Conference on Parthenium Management*. University of Agricultural Sciences, Dharwad, India, pp. 77–80.

Rao, R.S. (1956) Parthenium – a new record for India. *Journal of Bombay Natural History Society* 5, 218–220.

Razaq, Z.A., Vahidy, A.A. and Ali, S.I. (1994) Chromosome numbers in Compositae from Pakistan. *Annals of the Missouri Botanical Garden* 81, 800–808.

Riaz, T., Khan, S.N., Javaid, A. and Farhan, A. (2007) Weed flora of *Gladiolus* in Lahore, Pakistan. *Pakistan Journal of Weed Science Research* 13, 113–120.

Rollins, R.C. (1950) *The Guayule Rubber Plant and its Relatives*. Gray Herbarium of Harvard University No. 172. Harvard University, Cambridge, Massachusetts.

Shabbir, A. (2012) Towards the improved management of parthenium weed: Complementing biological control with plant suppression. PhD thesis, the University of Queensland, Brisbane, Australia.

Shabbir, A., Dhileepan, K. and Adkins, S.W. (2012) Spread of parthenium weed and its biological control agent in the Punjab, Pakistan. *Pakistan Journal of Weed Science Research* 18, 581–588.

Shabbir, A., Dhileepan, K., Khan, N. and Adkins, S.W. (2014) Weed–pathogen interactions and elevated CO_2: Growth changes in favour of the biological control agent. *Weed Research* 54, 217–222.

Shrestha, B.B., Shabbir, A. and Adkins, S.W. (2015) *Parthenium hysterophorus* in Nepal: A review of its weed status and possibilities for management. *Weed Research* 55, 132–144.

Strathie, L., Wood, A.R., van Rooi, C. and McConnachie, A.J. (2005) *Parthenium hysterophorus* (Asteraceae) in southern Africa, and initiation of biological control against it in South Africa. In: Ramachandra Prasad, T.V., Nanjappa, H.V., Devendra, R., Manjunath, A., Subramanya, S.C., *et al.* (eds) *Proceedings of the Second International Conference on Parthenium Management.* University of Agricultural Sciences, Bangalore, India, pp. 127–133.

Sushilkumar and Varshney, J.G. (2010) Parthenium infestation and its estimated cost management in India. *Indian Journal of Weed Science* 42, 73–77.

Tadesse, M. (2004) *Asteraceae (Compositae). Flora of Ethiopia and Eritrea*, Vol. 4(2). The National Herbarium, Addis Ababa University, Ethiopia and Uppsala University, Uppsala, Sweden.

Tamado, T. and Milberg, P. (2000) Weed flora in arable fields of eastern Ethiopia with emphasis on the occurrence of *Parthenium hysterophorus*. *Weed Research* 40, 507–521.

Tamado, T., Schutz, W. and Milberg, P. (2002) Germination ecology of the weed *Parthenium hysterophorus* in eastern Ethiopia. *Annals of Applied Biology* 140, 263–270.

Tang, S., Wei, C., Pan, Y. and Pu, G. (2010) Reproductive adaptability of the invasive weed *Parthenium hysterophorus* L. under different nitrogen and phosphorus levels. *Journal of Wuhan Botanical Research* 28, 213–217.

Thomas, J., Basahi, R., Al-Ansari, A.E., Sivadasan, M., El-Sheikh, M.A., Alfarhan, A.H. and Al-Atar,

A.A. (2015) Additions to the Flora of Saudi Arabia: two new generic records from the Southern Tihama of Saudi Arabia. *National Academy Science Letters* 38, 513–516.

USDA-ARS (2012) Germplasm Resources Information Network (GRIN). Online Database. National Germplasm Resources Laboratory, United States Department of Agriculture Agricultural Research Center (USDA-ARS), Beltsville, Maryland. Available at: https://npgsweb.ars-grin.gov/gringlobal/taxon/taxonomysearch.aspx (accessed 1 March 2017).

Wabuyele, E., Lusweti, A., Bisikwa, J., Kyenune, G., Clark, K., Lotter, W.D., McConnachie, A.J. and Mersie, W.A. (2014) Roadside survey of the invasive weed *Parthenium hysterophorus* (Asteraceae) in East Africa. *Journal of East African Natural History* 103, 49–57.

Wise, R.M., van Wilgen, B.W., Hill, M.P., Schulthess, F., Tweddle, D., Chabi-Olay, A. and Zimmermann, H.G. (2007) *The economic impact and appropriate management of selected invasive alien species on the African continent. Final report.* Global Invasive Species Programme. http://www.issg.org/pdf/publications/GISP/Resources/CSIRAISmanagement.pdf (accessed 5 June 2018).

Yaduraju, N.T., Sushilkumar, Prasad Babu, M.B.B. and Gogoi, A.K. (2005) *Parthenium hysterophorus* – distribution, problem and management strategies in India. In: Ramachandra Prasad, T.V., Nanjappa, H.V., Devendra, R., Manjunath, A., Subramanya, S.C., *et al.* (eds) *Proceedings of the Second International Conference on Parthenium Management.* University of Agricultural Sciences, Bangalore, India, pp. 6–10.

Zahran, M.A. and Willis, A.J. (1992) *The Vegetation of Egypt.* Chapman & Hall, London.

4 Interference and Impact of Parthenium Weed on Agriculture

Ali Ahsan Bajwa,[1]* Asad Shabbir[2] and Steve W. Adkins[1]

[1]The University of Queensland, Gatton, Queensland, Australia; [2]University of the Punjab, Lahore, Pakistan; current affiliation: Plant Breeding Institute, the University of Sydney, Narrabri, New South Wales, Australia

4.1 Introduction

Parthenium weed (*Parthenium hysterophorus* L.) has rapidly become one of the most damaging invasive weed species present in natural systems and agroecosystems (Bajwa *et al.*, 2016). It affects environmental stability by disturbing natural floral diversity and species distribution within rangelands, forests and fallow lands. In addition, it also infests a large number of crops across a diverse range of cropping systems around the world. Its environmental and health impacts, including those on the displacement of native species, on the promotion of land degradation, and on human and animal health are well known and are of great concern for communities in many countries (Bajwa *et al.*, 2016). However, its impacts upon crop production are not as well documented in a global perspective, and this may be due to the fact that parthenium weed historically has been considered predominantly to be a weed of rangelands.

The negative impacts of parthenium weed on agricultural production are multifaceted and include both direct and indirect impacts. Parthenium weed interferes with the production and management of a multitude of agricultural, horticultural and silvicultural crops (Fig. 4.1; Safdar *et al.*, 2015) and pastures (Fig. 4.2). It can suppress the

growth and development of these crops directly due to severe resource competition and also through allelopathy. The growth reduction in these crops ultimately leads to acute yield losses and quality deterioration of the harvested product (Masum *et al.*, 2012). Crop yield losses as high as 50% have been reported in different studies (Netsere, 2015). Such heavy losses in some crops due to parthenium weed have been reported in several countries. For instance, sorghum (*Sorghum bicolor* (L.) Moench), the most important grain crop in the dry lowland areas of Ethiopia, now suffers greatly from parthenium weed infestation (Tamado *et al.*, 2002). Similarly, farmers in the northern areas of Pakistan have had to face substantial yield losses in wheat (*Triticum aestivum* L.) and maize (*Zea mays* L.) crops which have resulted in regional food shortages (Khan *et al.*, 2013). Apart from grain crops, a large number of vegetable and fruit crops are also negatively affected by parthenium weed (Gupta and Narayan, 2010; Demissie *et al.*, 2013). Moreover, in many parts of the world the agroforestry industry is also affected significantly by this lethal invader (Swaminathan *et al.*, 1990), especially in young, establishing plantations.

Livestock production is the second major agricultural production system affected by parthenium weed, and is the

* a.bajwa@uq.edu.au

Fig. 4.1. Parthenium weed infestation in (a) maize in Ethiopia, (b) sorghum in Uganda, (c) sugarcane in Pakistan and (d) potato in Nepal. (a,b: A. Witt; c: A. Shabbir; d: R.B. Khadka.)

most important system affected in several countries. Parthenium weed has been long known to negatively affect this sector, both directly and indirectly. Its interference with forage production has contributed significantly towards losses in livestock production in many countries (Shabbir, 2012). Typical examples are the forage yield losses in India and pasture losses and community deterioration in Australia and Ethiopia (Kanchan and Jayachandra, 1979; McFadyen, 1992; Chippendale and Panetta, 1994; Navie, 2002). In addition to this, parthenium weed can affect the livestock sector by causing health challenges in animals (Tudor et al., 1982; Chippendale and Panetta, 1994).

Another indirect mode of impact on crop production is through the weed's ability to act as an alternative host to several important pests and pathogens (Shabbir, 2012, 2014). For instance, parthenium weed hosts

the cotton mealybug (*Phenacoccus solenopsis* Tinsley), which causes huge losses to several crops (Arif et al., 2009). All these impacts on agriculture contribute towards a bigger problem, one in the form of socio-economic instabilities. Parthenium weed has been shown to bring about a decline in the livelihoods of people, devaluation of produce and land, and disruption in supply chain and commodity marketing (Joshi, 1990; Kapoor, 2012; Beyene et al., 2013). Consequently, the agricultural sustainability and therefore the livelihoods of whole communities have become challenged in particular regions of the world by parthenium weed invasion.

The following provides a detailed account of the different impacts parthenium weed is having on food, fodder and fibre production systems around the world. A comprehensive discussion on the different aspects of the impacts will be presented in

Fig. 4.2. Parthenium weed infestation in (a) an agroforestry area in Australia, (b) open grazing land in Australia, (c) an agroforest in Kenya and (d) cattle grazing land in Ethiopia. (a,b: K. Pukallus; c,d: A. Witt.)

order to highlight the magnitude of the problem. This chapter not only provides a thorough literature review of the problems caused by parthenium weed invasion but also gives an insight into the broader subject of weed invasion biology and management. The quantitative and qualitative descriptions of different negative impacts of this particular weed on agricultural production will help to devise sustainable and improved management strategies for the future.

4.2 Impacts on Crop Production

Parthenium weed interferes with crop production directly through resource competition and by allelopathic growth inhibition (Bajwa *et al.*, 2016). A substantial literature exists that reports the growth and yield reduction caused in grain, fibre, horticultural, forage and cash crops due to parthenium weed in India and Ethiopia (e.g. Kohli *et al.*, 1996; Tamado *et al.*, 2002; Singh *et al.*, 2003; Maharjan *et al.*, 2007; Kumar *et al.*, 2010). Much of this research has been conducted under controlled conditions and has been directed to evaluate the effects of the weed on the early stages of crop establishment and growth (Bajwa *et al.*, 2004). As a matter of fact, these early impacts play a vital role in the overall crop development and eventually grain yield losses. So, an inferential approach is often adopted by many researchers to predict the ultimate effect on grain yield by observing the reduction in seed germination and early vigorous growth. However, sometimes the results of such laboratory bioassays may not be pronounced and so cannot be used to provide meaningful ecological predictions (Belgeri and Adkins, 2015). Here, we will present the negative effects of parthenium weed on several plant species; however, the kinds of mechanisms that may be involved include physiological disruption, biochemical restraint, physical suppression and pure resource competition.

Parthenium weed has been reported to have a strong allelopathic capacity as it contains a variety of allelochemicals, in different plant parts and in varying concentrations (Batish et al., 2002; Shabbir and Javaid, 2010). Moreover, it has the ability to release or express those chemicals in many different ways, including: (i) release through root exudation; (ii) release from above-ground plant parts; (iii) volatilization into the atmosphere; (iv) leaching from plant residues; and (v) decomposition of dead or decayed plant parts (Bajwa et al., 2016). The relative concentration of the allelochemicals in extracts or residues is also important, as higher concentrations than would have been expected have been reported to be more inhibitory (Kumar and Gautam, 2008; Kumar et al., 2010). So, the biochemical interference through the expression of harmful secondary metabolites by parthenium weed is a likely cause for crop growth suppression and ultimately yield reduction.

Parthenium weed has multiple morphological, physiological and biochemical features which makes it one of the most successful invaders of agroecosystems (Bajwa et al., 2016). The vigorous and tall growing habit, its deep tap root, the wide range of temperatures that stimulate its germination, its pubescent leaves, its compact inflorescence and large seed production, possibly its C3/C4 intermediate photosynthesis strategy, tolerances to many environmental stresses, efficient uptake of nutrients, and its multiple seed dispersal mechanisms are all distinct features of parthenium weed which make it a highly competitive weed species (Bajwa et al., 2016). The competition offered by parthenium weed will have significant negative impacts upon crop growth and yield due to the limitation of resources in the field as well as its ability to physically suppress growth. Although the interference of parthenium weed with crop species is very complex in nature and the physical or biochemical interaction cannot be differentiated clearly, certain parameters provide the comparative ability for resource acquisition. The interference of parthenium weed with crop production is a broader phenomenon and, for better understanding, further consideration must be given to its competitive ability, its biochemical interactions (allelopathy) and spatial configurations, simultaneously. In the following sections, specific information about the negative impacts of parthenium weed on different types of crop will be reviewed.

4.2.1 Grain crops

Multiple factors negatively affect the production of grain crops, including climatic catastrophes, agricultural pests and diseases, environmental stresses and agronomic mismanagement. However, among these constraints, weeds stand out as the most important reason for reduced yields and loss of quality of grain crops (Bajwa, 2014; Bajwa et al., 2015). Parthenium weed, being one of the most invasive weed species, has in the past 50 years become an important weed in the production of many grain crops. The negative impact of parthenium weed on grain crop production is through resource competition and allelopathy and has been reported in many countries of Asia, Africa and Australia. In most of these countries the weed has already infested susceptible crops, while in other countries it is still in the process of moving from the areas that it originally colonized into the cropping lands.

Parthenium weed has been reported to cause serious growth suppression and yield reductions in several grain crops, including cereals, pulses and oilseeds in different parts of the world (Table 4.1). Germination and early growth suppression are the most commonly affected stages due to the powerful dual action of resource competition and allelopathy at this early stage of crop growth. In dry-seeded rice crops in India, due to the production of allelochemicals, parthenium weed was detrimental to final yield

Table 4.1. Impact of parthenium weed on grain crop growth and production. (Bajwa, Shabbir and Adkins.)

Crop affected	Negative impact(s)	Associated mechanism	Reference
Cereals			
Rice (*Oryza sativa* L.)	Inhibition of germination	Leaf extracts at high concentration caused germination inhibition	Maharjan *et al.* (2007)
	Suppression of emergence and growth	Plant debris mixed in soil caused phytotoxic effects	Biswas *et al.* (2010)
	Growth suppression	Toxic effects of aqueous leaf extracts	Devi *et al.* (2013)
Wheat (*Triticum aestivum* L.)	Shoot and root growth suppression	Toxic and suppressive effects of shoot and whole plant extracts and residues	Mersie and Singh (1987)
	Germination and growth suppression	Extracts made from burnt and unburnt weed residues released toxic phenolic compounds that were active in amended soil	Singh *et al.* (2003)
	Germination and growth suppression	Leaf biomass applied to soil affected soil characteristics and reduced germination and subsequent growth	Gupta and Narayan (2010)
	Germination inhibition but little effect on growth	Toxic allelochemicals present in aqueous extracts	Dhole *et al.* (2011)
	Germination and growth suppression depending upon cultivar	Allelopathic activity of root, shoot and leaf extracts. Leaf extracts were more suppressive than root and shoot extracts	Khan *et al.* (2012)
Maize (*Zea mays* L.)	Shoot growth suppression	Suppressive effects of shoot and whole plant extracts and residues	Mersie and Singh (1987)
	Germination and growth suppression	Toxic allelochemicals present in extracts	Dhole *et al.* (2011)
	Germination and growth suppression	Toxic effects of aqueous leaf extracts	Maharjan *et al.* (2007)
	Germination and seedling vigour suppression	Toxic effects of aqueous leaf extracts	Devi and Dutta (2012)
	Germination and growth suppression	Extracted allelochemicals reduced germination and growth	Masum *et al.* (2012)
	Germination and growth suppression	Toxic allelochemicals present in aqueous extracts	Netsere (2015)
	Growth, nutrient uptake and yield suppression	Severe competition for resources especially macronutrients (N, P, K) which were inefficiently taken up by the crop	Safdar *et al.* (2015)
Sorghum (*Sorghum bicolor* (L.) Moench)	Yield reduction	Strong competition with crop at different weed densities	Tamado *et al.* (2002)
	Growth and yield reduction	Reductions due to competition and allelopathic interactions	Tamado and Milberg (2004)
	Moderate suppression of germination and growth	Toxic allelochemicals present in aqueous weed extracts	Dhole *et al.* (2011)
	Germination and growth suppression	Toxic allelochemicals present in aqueous weed extracts	Netsere (2015)

continued

Table 4.1. *continued*

Crop affected	Negative impact(s)	Associated mechanism	Reference
Barley (*Hordeum vulgare* L.)	Germination and growth suppression	Allelopathic effect	Srivastava *et al.* (2011)
Finger millet (*Eleusine coracana* Gaertn.)	Germination inhibition	Allelopathic effect of leachates	Bhatt *et al.* (1994)
Tef (*Eragrostis tef* (Zucc.) Trotter)	Germination and growth suppression	Toxic allelochemicals present in aqueous weed extracts	Tefera (2002)
Pulses and oilseed			
Chickpea (*Cicer arietinum* L.)	Germination and growth inhibition	Phytotoxic weed residues in soil leading to reduced growth, impaired nutrient acquisition and declined soil health	Batish *et al.* (2002)
	Strong germination and growth suppression	Soaking in fresh leaf leachates caused phytotoxic effects on germination	Kohli *et al.* (1996)
	Germination and growth suppression	The extracts from burnt and unburnt weed residues released toxic phenolic compounds in soil	Singh *et al.* (2003)
Lentil (*Lens culinaris* L.)	Germination inhibition	Soaking in fresh leaf leachates caused phytotoxic effects on germination	Kohli *et al.* (1996)
Mungbean (*Vigna radiata* L.)	Germination and seedling vigour suppression	Soaking in fresh leaf leachates caused phytotoxic effects on germination	Kohli *et al.* (1996)
	Germination and growth suppression	Phytotoxic activity of aqueous extracts	Shabbir and Javaid (2010)
Cowpea (*Vigna unguiculata* L.)	Germination inhibition	Soaking in fresh leaf leachates caused phytotoxic effects on germination	Kohli *et al.* (1996)
	Germination inhibition	Weed extracts reduced water absorption in germinating cowpea seeds	Srivastava *et al.* (2011)
Mustard (*Brassica rapa* L.)	Strong germination and growth suppression	Soaking in fresh leaf leachates caused phytotoxic effects on germination	Kohli *et al.* (1996)
	Germination and growth suppression	Presence of phytotoxic soluble phenolics in aqueous extracts and weed residues	Singh *et al.* (2003)
Pigeon pea (*Cajanus cajan* (L) Millsp.)	Strong germination and growth suppression	Soaking in fresh leaf leachates caused phytotoxic effects on germination	Kohli *et al.* (1996)

Crop	Effect	Mechanism	Reference
Common bean (*Phaseolus vulgaris* L.)	Growth suppression depending upon plant part used in extracts	Varying concentration of allelochemicals in different weed parts having phytotoxic effects on growth	Netsere and Mendesil (2012)
	Germination and shoot elongation reduced	Phytotoxic allelopathic extracts	Demissie *et al.* (2013)
Black gram (*Vigna mungo* L.)	Germination and growth suppression	Weed ash in the field reduced crop performance when present in higher amount	Kumar *et al.* (2010)
	Germination and growth suppression	Aqueous leaf extracts having phytotoxic effects on physiological processes	Parthasarathi *et al.* (2012)
	Strong germination and growth suppression	Soaking in fresh leaf leachates caused phytotoxic effects on seeds during imbibition	Kohli *et al.* (1996)
Soybean (*Glycine max* L.)	Shoot growth suppression	Toxic and suppressive effects of shoot and whole plant extracts and residues	Mersie and Singh (1987)
	Germination and root and shoot growth suppression	Toxic allelochemicals present in aqueous extracts	Dhole *et al.* (2011)
	Growth suppression depending upon plant part used in extracts	Varying concentration of allelochemicals in different plant parts having phytotoxic effects on growth	Netsere and Mendesil (2012)
	Germination and growth suppression	Allelopathic extracts reducing germination and growth	Masum *et al.* (2012)
Sunflower (*Helianthus annuus* L.)	Germination and early growth inhibition	Allelopathic toxicity from aqueous extracts	Bajwa *et al.* (2004)
	Cytomorphological behaviour affected negatively	Higher concentration of allelochemicals present in leaf extracts had a negative effect on growth and pollen mother tube formation in the crop	Kumar and Gautam (2008)
Groundnut (*Arachis hypogaea* L.)	Germination and growth suppression	Aqueous leaf extracts having phytotoxic effects on physiological processes	Parthasarathi *et al.* (2012)

(Maharjan et al., 2007; Devi et al., 2013). Biswas et al. (2010) showed that even when parthenium weed residues were present, the emergence of rice seedlings and subsequent growth was dramatically reduced.

In other studies, the germination and growth of wheat was also reduced when an extract made from the whole plant of parthenium weed or its residues was applied to the soil (Mersie and Singh, 1987). The burning of rice residues before the sowing of the wheat crop, a very common practice in the rice–wheat cropping system of Indo-Gangetic Plains, has shown that the germination and yield of the crop will significantly reduce if parthenium weed residues are also burnt. Interestingly, the magnitude of suppression was greater from burnt than from unburnt weed residues (Singh et al., 2003). Gupta and Narayan (2010) reported that soil-incorporated parthenium weed leaf residues caused phytotoxicity to the growth of wheat in India. The negative effects of leaf residues on wheat growth were found to be more than those from stem or root and this may be due to higher amount of allelochemicals present in the leaf residues (Khan et al., 2012).

Parthenium weed infestation in maize crops has been very problematic in different regions. The growth suppression has been reported to be through phytotoxic leachates, physical smothering by residues, and through biochemicals released from surface or incorporated residues (Dhole et al., 2011; Devi and Dutta, 2012; Masum et al., 2012; Netsere, 2015). Parthenium weed has been shown to strongly compete for soil nutrients and moisture, with maize sown at different densities. In a 2-year study in Pakistan, Safdar et al. (2015) found that increasing the density of parthenium weed in a maize field from 0 to 20 plants/m^2 increased the relative competition index of parthenium weed by 145%, concurrently with a remarkable increase in the weed's dry biomass of 433%. At the higher densities of parthenium weed plants, greater amounts of nitrogen (N), phosphorus (P) and potassium (K) (336%, 180% and 295%, respectively) were taken up. Due to such severe competition, maize plant height, number of cobs per plant, cob weight, grain weight, biological yield, grain

yield and harvest index were all decreased significantly. A massive 50% yield loss was recorded at a moderate infestation rate of 20 parthenium weed plants/m^2. A predictive model showed that the economic threshold of parthenium weed for maize was only 1.0–1.2 plants/m^2. Such a quantitative illustration of parthenium weed's impact on crop production are alarming and sufficient to predict that immediate management is necessary.

Sorghum is another important cereal crop which is used for human and animal consumption worldwide, but especially in Africa. Parthenium weed has the ability to interfere with this resilient cereal crop under diverse climatic and geographic planting conditions (Tamado and Milberg, 2004). Although sorghum itself is perhaps one of the most allelopathic crops (Farooq et al., 2013), parthenium weed allelochemicals have been shown to strongly inhibit the germination and growth of sorghum (Dhole et al., 2011; Netsere, 2015). In the lowlands of eastern Ethiopia, sorghum grain yield losses were up to 97% under a high density of 100 plants/m^2 (Tamado et al., 2002). Even at a density of three plants/m^2, parthenium weed caused 69% grain yield loss. The critical period for parthenium weed control to avoid the substantial sorghum yield loss was up to 60 days after crop emergence. These figures for yield losses and for critical control period are extreme and indicate why this weed is a concern to any small grain crop farmer in any part of the world. Another African cereal suffering from parthenium weed infestation is tef (Eragrostis tef (Zucc.) Trotter), which is grown as a staple in many parts of Ethiopia and some other African countries. Parthenium weed has invaded many tef fields and has been shown to suppress its growth and its grain production severely. For example, Tefera (2002) reported that, due to allelopathic interactions, parthenium weed significantly reduced tef germination and early growth.

In addition to cereal grains, many pulses and oilseed crops have also encountered problems from parthenium weed invasion (Table 4.1). Chickpea (Cicer arietinum L.) is one of the most important cool-season legumes with a highly nutritious profile. Parthenium weed has been shown to suppress the performance of chickpea in India through the presence of allelopathic residuals and through direct competition for soil nutrients (Batish *et al.*, 2002; Singh *et al.*, 2003). Similarly, growth reduction and yield losses were also observed in lentil (*Lens culinaris* L.), common bean (*Phaseolus vulgaris* L.) (Kohli *et al.*, 1996), pigeon pea (*Cajanus cajan* (L) Millsp.) (Singh *et al.*, 2003), mustard (*Brassica rapa* L.) and mungbean (*Vigna radiata* L.) (Shabbir and Javaid, 2010; Netsere and Mendesil, 2012). Parthenium weed leaf leachates completely inhibited the germination of cowpea (*Vigna unguiculata* L.) as they contained allolochemicals (Kohli *et al.*, 1996). Srivastava *et al.* (2011) reported the reduction of water uptake in germinating cowpea seeds as one mechanism of germination inhibition by parthenium weed. Leaf leachates, root exudates and ash from burnt parthenium weed plants have all been shown to inhibit the germination of black gram (*Vigna mungo* L.) through water uptake inhibition (Kumar *et al.*, 2010; Parthasarathi *et al.*, 2012). The subsequent crop growth was also severely affected because of poor germination and seedling vigour. In addition, liquid extracts and residues of various parthenium weed plant parts have been shown to have a negative impact upon germination and growth of soybean (*Glycine max* L.) (Mersie and Singh, 1987; Netsere and Mendesil, 2012; Masum *et al.*, 2012). Interestingly, parthenium weed has been shown to reduce pollination of sunflower (*Helianthus annuus* L.) and hence seed set and yield (Bajwa *et al.*, 2004; Kumar and Gautam, 2008).

In summary, the productivity of all grain crops is under serious threat from the ever-spreading parthenium weed. Another important consideration is that such crops are mostly grown on marginal, less fertile and non-irrigated lands. These are the kinds of land that parthenium weed commonly invades. So, the likelihood of getting heavier infestations in these crops suggests that immediate action should take place to develop better management strategies for these crops.

4.2.2 Vegetables and other horticultural crops

It has been observed that parthenium weed may cause serious losses to many susceptible vegetables and fruit crops due to competition and allelopathic interference (Table 4.2). Parthenium weed residues in soil have been shown to severely hinder the growth of radish (*Raphanus sativus* L.) in India (Batish *et al.*, 2002). Leaf extracts and root exudates have also been shown to have a negative impact upon the germination and growth of radish plants (Paudel *et al.*, 2009). The germination and growth of pea (*Pisum sativum* L.) and onion (*Allium cepa* L.) were also inhibited by parthenium weed litter, plant residues or through direct competition (Gupta and Narayan, 2010; Demissie *et al.*, 2013). In a recent field study in Sri Lanka, Nishanthan *et al.* (2013) reported parthenium weed infestations in tomato (*Solanum lycopersicum* L.) caused substantial yield losses due to resource competition. The average fruit weight was 8% higher in parthenium weed-free plots than in those having the weed. It was also reported in the same study that the weed seed rain in the infested fields causes the formation of a seed bank which then initiates a several-year-long reduction in tomato yield. Similar kinds of growth suppression due to parthenium weed have also been observed for spinach (*Spinacea oleracea* L.) and okra (*Abelmoschus esculentus* Moench) (Dhawan and Dhawan, 1995) in India. Parthenium weed also inhibited yield from a number of vegetables in India (namely tomato, okra and aubergine (*Solanum melongena* L.)) due to pollen allelopathy (Sukhada and Jayachandra, 1980). In addition to vegetables, some fruit orchards, such as mandarin orange (*Citrus reticulata* Blanco) in Ethiopia, have been negatively affected by parthenium weed invasion (Navie *et al.*, 1996).

Parthenium weed has been shown to directly interfere in the production of certain kinds of ornamental grasses, herbs and shrubs. The weed not only suppresses their growth but also deteriorates the aesthetic appeal of such landscapes. For example, parthenium weed infestation has been

Table 4.2. Impact of parthenium weed on vegetable crop growth and production. (Bajwa, Shabbir and Adkins.)

Crop	Negative impact(s)	Associated mechanism	Reference
Radish (*Raphanus sativus* L.)	Germination and growth inhibition	Residues in soil leading to direct toxicity, impaired nutrient acquisition and declined soil health	Batish *et al.* (2002)
	Germination and shoot length suppression	Phytotoxic effects of leaf extracts containing allelochemicals	Maharjan *et al.* (2007)
	Germination inhibition	Diffusates from seedlings and rhizosphere soil reducing amylase activity in seeds	Paudel *et al.* (2009)
Pea (*Pisum sativum* L.)	Germination and growth suppression	Leaf biomass applied to soil reducing germination and subsequent growth	Gupta and Narayan (2010)
Tomato (*Solanum lycopersicum* L.)	Growth and yield suppression	Resource competition and negative impact of weed infestation on soil but depending upon management strategy used	Nishanthan *et al.* (2013)
Onion (*Allium cepa* L.)	Germination and growth suppression	Phytotoxic allelopathic extracts	Demissie *et al.* (2013)
Lettuce (*Lactuca sativa* L.)	Germination and root growth suppression	Phytotoxic allelopathic extracts	Wakjira *et al.* (2005)
Onion	Germination and growth suppression	Phytotoxic chemicals in compost prepared from weed	Wakjira *et al.* (2009)
Broccoli (*Brassica oleracea* L.) Bell pepper (*Capsicum annuum* L.) Carrot (*Daucus carota* L.) Tomato Radish Aubergine (*Solanum melongena* L.)	Germination, seedling vigour and seedling growth suppressed	Soaking in fresh leaf leachates caused phytotoxic effects on seeds during imbibition	Kohli *et al.* (1996)

Spinach (*Spinacea oleracea* L.)	Dhawan and Dhawan (1995)
Okra (*Abelmoschus esculentus* Moench)	
Bottle gourd (*Lagenaria siceraria* L.)	Shi (2016)
Cucurbit (*Cucurbit* sp.)	
Garlic (*Allium sativum* L.)	
Potato (*Solanum tuberosum* L.)	
Sweet gourd (*Cucurbita maxima* L.)	
Taro (*Colocasia esculenta* L.)	

reported in ornamental nurseries in Florida, USA (Stamps, 2011). In such locations the weed was reported to cause growth suppression through competition for water, light, space and nutrients, and management was extremely expensive to implement. In some cases, the quality of the most expensive ornamental plants was reduced so much that it led to purchase rejection (Stamps, 2011).

4.2.3 Fibre and other field crops

Parthenium weed infestation has been reported in a diverse range of field crops, including, in India, cotton (*Gossypium hirsutum* L.), one of the world's most important fibre crops (Madhu *et al.*, 1995). Aqueous leaf extracts of parthenium weed completely inhibited the germination of cotton seed (Dhole *et al.*, 2011). In another study, significant reductions in cotton germination and seedling growth were attributed to allelopathic interference from the weed (Masum *et al.*, 2012). Negative impacts of parthenium weed on the growth and yield of an important sugar crop, sugarcane (*Saccharum officinarum* L.), have also been reported in Australia (Parsons and Cuthbertson, 1992) and the USA (Fernandez *et al.*, 2015). In the USA, it was shown that parthenium weed residues reduced germination, cane height and biomass in mineral soils; however, no significant reduction was observed in crops grown on organic soils. Some other crops, such as tea (*Camellia sinensis* (L.) Kuntze; Njoroge, 1986) and coffee (*Coffea arabica* L.; Njoroge, 1991) in Kenya and tobacco (*Nicotiana rustica* L.; Khan, 2011) in Pakistan have also been reported to be negatively affected by parthenium weed infestation. Further research is required to estimate the current status of parthenium weed infestation in these and other crops.

4.2.4 Agroforestry

Parthenium weed interference with agroforestry production systems has been a common occurrence. For instance, the growth and nodulation capacity of leucaena (*Leucaena leucocephala* (Lam.) de Wit.), a beneficial multipurpose tree species grown on agricultural land, has been reduced by parthenium weed in India (Dayama, 1986). The germination and growth suppression of some other important trees, including the coastal she-oak (*Casuarina equisetifolia* L.), white-bark acacia (*Acacia leucophloea* (Roxb.) Willd.) and forest red-gum (*Eucalyptus tereticornis* Sm.), was also observed to occur in response to parthenium weed's allelopathic capacity (Swaminathan *et al.*, 1990). It is important to note that, unlike crops, the establishment phase of tree growth may not be significantly affected by parthenium weed. Nevertheless, the interference of parthenium weed with juvenile trees in agroecosystems and plant nurseries may cause severe growth suppression of these tree species.

4.3 Impacts on Livestock Production

Livestock production is another important component of agriculture in which parthenium weed infestation has a great impact, either through direct or indirect pathways. Having immense ability to invade a wide range of geographic and climatic conditions, parthenium weed is commonly found in grasslands and pastures outside of its native range. Direct effects are often through its ability to reduce grass productivity, but it also affects animal health, which can lead to deterioration in milk and meat quality. A summation of all of these actions is a dramatic reduction in livestock production. In the following sections, a detailed account of the negative impacts of parthenium weed on fodder plant production and meat and milk quality is presented.

4.3.1 Fodder crops

Parthenium weed is known to suppress the germination and growth of many important fodder crops around the world (Shabbir, 2012). Parthenium weed leaf leachates significantly reduced the germination and growth of a number of fodder crops, including cluster bean (*Cyamopsis tetragonoloba* L.) in India, Egyptian clover (*Trifolium alexandrinum* L.), yellow alfalfa (*Medicago falcata* L.) and alfalfa (*Medicago sativa* L.) (Kohli *et al.*, 1996). The suppression of the growth of fodder oats (*Avena sativa* L.) has also been observed and thought to occur due to the phytotoxic effects of the weed's root exudates (Kanchan and Jayachandra, 1979). The growth suppression of fodder crops reduces the overall carrying capacity of the land, which in turn reduces livestock production. Another important issue is the contamination of fodder seed lots with parthenium weed, which aids its further dispersal to uninfested areas.

4.3.2 Grazing lands

The ability of parthenium weed to invade grasslands and pastures as well as forage production is remarkable (Khan *et al.*, 2013). It has the tendency to first establish in disturbed sites then to move into the grazing lands by displacing the native plant species. Parthenium weed outperforms the existing native flora through its strong resource-capturing ability and its allelopathic capacity (Shi, 2016). It has been shown that parthenium weed is capable of spreading very efficiently into new grazing lands once it has been introduced there (Navie, 2002). It causes rapid decline in plant biodiversity by its suppressive displacement and changes species composition. It grows more vigorously than many existing grazing land plant species and, thus, disturbs the overall ecosystem. In this way, the direct impact of parthenium weed on grazing lands affects the productivity of the livestock feeding on it. In some grazing lands of Central Queensland, Australia, parthenium weed has replaced almost all of the original species present, resulting in a lack of suitable vegetation for grazing (Navie, 2002). The overgrazed lands of Central Queensland were highly susceptible to parthenium weed invasion as they contained vast areas of disturbed land (McFadyen, 1992). As a result of this susceptibility to invasion, parthenium weed spread over an area of 170,000 km^2 (about 10% of the total area of state) in just a matter of 20 years (Chippendale and Panetta, 1994).

An increasing density of parthenium weed in most grazing lands will reduce the diversity and stability of that grassland community. Nigatu *et al.* (2010) studied the impact of different densities of parthenium weed on grazing lands in Ethiopia. They found that the density of different grass species, including Bermuda grass (*Cynodon dactylon* L.), liverseed grass (*Urochloa panicoides* P. Beauv.) and Rhodes grass (*Chloris gayana* Kunth) was significantly reduced as the density of parthenium weed was increased. The mean cover abundance of parthenium weed became 33%, while that of the grasses and other species was 41% and 26%, respectively (Nigatu *et al.*, 2010). This study indicates how parthenium weed can negatively affect the spatial diversity in a grazing land. In addition to the decrease in the above-ground diversity, the below-ground viable seed bank of all species was also reduced in the presence of parthenium weed (Nigatu *et al.*, 2010). The seed bank of parthenium weed dominated the grazing land (68%) in comparison to that of the grasses (26%) and other species (6%). In another study in Ethiopia, Ayele *et al.* (2013) studied the influence of parthenium weed infestation on the composition and diversity of species existing in grazing land. It was observed that the majority of plants present were *Poaceae* (20) followed by *Asteraceae* species (9). The above-ground cover of *Poaceae* species was reduced to 47% at the maximum parthenium weed density, as compared with plots that had no parthenium weed (Ayele *et al.*, 2013). The biomass of the above-ground *Poaceae* species' vegetation was also reduced drastically from 428 g/m^2, for a 'no parthenium weed' situation, to 30 g/m^2, for a situation with a high density of parthenium weed. The remarkable decrease in seed banks of grassland species was also observed during this study (Ayele *et al.*, 2013).

A similar kind of growth suppression of grazing land species under varying densities of parthenium weed and different management regimes was reported in Kilcoy, Queensland, Australia. A high parthenium weed infestation was correlated with a low native species diversity and abundance (Belgeri et al., 2014). The shift in species composition and diversity was mainly attributed to the direct decline in the above-ground species richness and the below-ground seed bank size and diversity. In a further study, Belgeri and Adkins (2015) highlighted the role parthenium weed can play in the suppression of grazing land species. It was found that parthenium weed seedlings had an allelopathic capacity which could inhibit the growth of grazing land species that were in their vicinity; however, the impact was species specific. The growth reduction of native grasses was much higher than that of introduced grass species. Growth reduction in native species such as curly windmill grass (Enteropogon acicularis L.) and cotton panic grass (Digitaria brownii L.) was 59% and 54%, respectively. However, only 0%, 8% and 9% reductions in growth were observed for the introduced Rhodes grass, buffel grass (Cenchrus ciliaris L.) and siratro (Macroptilium atropurpureum (DC.) Urb.), respectively (Belgeri and Adkins, 2015). The conclusion was that parthenium weed was found to be more harmful to native than introduced grazing land species. This is a further reason for the rapid spread of parthenium weed in grazing lands worldwide.

Finally, it should be noted that parthenium weed infestation not only causes a decline in species composition and diversity but also deterioration in the quantity of feed available to livestock. In a recent laboratory study, the seedling growth of nine introduced pasture species (African lovegrass, Eragrostis curvula L.; buffel grass; creeping blue grass, Bothriochloa insculpta L.; giant rats' tail grass, Sporobolus pyramidalis L.; green panic grass, Panicum maximum Jacq.; lambsquarters, Chenopodium album L.; liverseed grass; Rhodes grass; and tall finger grass, Digitaria milanjiana Stapf) and eight native pasture grasses (bull Mitchell grass, Astrebla squarrosa C.E. Hubb; cotton panic grass; curly windmill grass; forest blue grass, Bothriochloa bladhii (Retz.) S.T. Blake; kangaroo grass, Themeda triandra Forssk.; pitted blue grass, Bothriochola desipens (Hack.) C.E. Hubb; Queensland blue grass Dichanthium sericeum R.Br; and weeping grass, Microlaena stipoides L.) was shown to be inhibited by the presence of a single parthenium weed seedling (Shi, 2016). It was interesting to note that the inhibitory effects were more readily apparent in the native grass seedlings than in introduced pasture species. Thus, pastures comprised of mainly native grass species may be more susceptible to parthenium weed invasion than pastures comprised largely of introduced pasture species.

4.3.3 Meat and milk quality

Livestock production is directly related to the quality of the grazing land the animals are being raised on. Parthenium weed infestation of grazing lands not only disturbs the natural equilibrium of species in a particular region but also causes a deterioration in the quality of the forage produced there. In addition to that, livestock directly grazing parthenium weed-infested communities may become severely affected by the weed, developing a range of health and productivity concerns. Although parthenium weed is not palatable and is disliked by most animals, consumption does occur when the weed is mixed with other grass or fodder species. In a study from Central Queensland, Australia, Tudor et al. (1982) compared the meat quality of sheep from parthenium weed-infested and non-infested grazing lands, immediately after their slaughter. It was revealed that the meat from lambs that had been feeding on parthenium weed 2–4 weeks before their slaughter had a strong odour which not only spoiled the natural aroma of the meat but also affected its taste (Tudor et al., 1982). The meat and milk quality from cattle can also be reduced as a result of consumption of parthenium weed, which also affects the animals' blood circulation and impairs their immune system. The direct negative impact of parthenium weed on meat and milk quality reduces productivity within the livestock sector and causes serious losses to the rural economy. Moreover, the consumption of tainted meat and infected milk by people, particularly babies, may also cause serious illness.

In addition to non-life-threatening actions, parthenium weed can have life-threatening impacts on animals' health. Parthenium weed has been shown to cause severe dermatitis in buffalo (*Bubalus bubalis* L.) in India (Narasimhan *et al.*, 1977). Calves fed only on parthenium weed for 2 days developed diarrhoea within 1 day and within 2–4 weeks showed symptoms of acute muscular contraction and excitability (Narasimhan *et al.*, 1977). The main cause revealed after autopsy was that the dermatitis had led to ulcer formation in the alimentary canal with resulting damage to liver and kidney function. It was suggested that the sesquiterpene parthenin might have played a key role in this toxic effect. Chippendale and Panetta (1994) revealed that the decline in cattle health and productivity due to parthenium weed infestation in Central Queensland resulted in substantial productivity and income losses to the graziers of that region.

4.4 Role as Alternate Host to Plant Pests and Pathogens

Parthenium weed not only affects crop production through its direct interference but can also cause substantial yield and quality losses by acting as an alternate host to a wide range of serious crop pests and pathogens (Shabbir, 2012). Insects, bacteria, fungi and virus pests of several crops have been reported to use parthenium weed as the secondary plant host and then subsequently move onto the crop once it has established in the field (Table 4.3).

Initially, within the agricultural landscape parthenium weed will be found near water channels, along field banks and pathways as well as along roadsides. From these locations it then invades the cropping lands. Therefore, parthenium weed can carry any number of new pests and diseases into a cropping area, and this increases the chances of a new pest outbreak on the crops planted there. For instance, a serious pest of cotton, the cotton mealybug has been consistently reported to spend part of its life cycle on parthenium weed in different regions of Pakistan (Arif *et al.*, 2009; Shabbir, 2014). In addition to cotton, this insect can also destroy many grain and vegetable crops grown around the world. Parthenium weed has been reported to be an alternate host for the cotton mealybug in India (Shabbir, 2014). It is thought that the insect starts its life cycle on parthenium weed plants and, after substantial multiplication, moves onto the cotton crop. In this way, the crop is attacked rapidly by a large number of insects, causing immediate serious damage with catastrophic losses in terms of both yield and quality (Shabbir, 2014).

Parthenium weed has also been identified as a major secondary host for the tobacco streak virus (TSV), which is a causal agent of disease in several crops, including sunflower and mungbean (Sharman *et al.*, 2015). Sharman *et al.* (2009) reported that TSV was frequently present on parthenium weed in Central Queensland, Australia; however, it does not create any symptoms of disease in the parthenium weed plants. Seed transmission of TSV to subsequent generations of the weed was between 7% and 48% and this rate of TSV seed transmission remained unchanged even after 2 years of seed storage. It was suggested that due to the higher propensity for TSV, parthenium weed growing in the vicinity of crops is a major threat for disease outbreak (Sharman *et al.*, 2009).

In addition to insect pests and viral diseases, parthenium weed may assist some other serious crop-damaging agents. For instance, the root knot nematode group, which can cause serious damage to many crops, was reported to use parthenium weed as a secondary host in Pakistan (Shahina *et al.*, 2012). *Meloidogyne incognita* and *Meloidogyne javanica* strains of the root knot nematode were found using parthenium weed as an alternate host before moving onto their main crop hosts (Shahina *et al.*, 2012).

Table 4.3. Pests and pathogens that employ parthenium weed as an alternate host and subsequently are a cause of crop production loss. (Modified from Shabbir, 2012.)

Pest or pathogens	Main host crops	References
Insects		
Black burn (*Pseudoheteronyx* spp.)	Sunflower	Robertson and Kettle (1994)
American serpentine leafminer (*Liriomyza trifolii* Burgess)	Bell pepper	Chandler and Chandler (1988)
Black bean aphid (*Aphis fabae* Scopoli)	Black bean	Rajulu *et al.* (1976)
Jute hairy caterpillar (*Diacrisia obliqua* Walker.)	Multiple crops	Shabbir (2012)
Cotton mealybug (*Phenacoccus solenopsis* Tinsley)	Cotton and other crops	Arif *et al.* (2009)
Bacteria		
Pseudomonas solanacearum E.F. Smith	Multiple crops	Kishun and Chand (1987)
Xanthomonas campestris pv. *phaseoli*	Bean	Ovies and Larrinaga (1988)
Fungi		
Alternaria zinniae M.B. Ellis.	Several crops and tree species	Kumar *et al.* (1979), Sharma and Gupta (1998)
Colletotrichum gloeosporioides (Penz.) Sacc.		
Fusarium oxysporum Schlect.		
Fusarium moniliforme Shield.		
Myrothecium roridum Tode ex Fr.		
Rhizoctonia solani Kuhn		
Sclerotium rolfsii Sacc.		
Viruses		
Tomato leaf curl virus	Tomato	Govindappa *et al.* (2005)
Tobacco leaf curl virus	Tobacco	Reddy *et al.* (2002)
Tobacco streak virus	Cotton, sunflower, mungbean, groundnut	Sharman *et al.* (2009)
Potato virus X and Y	Potato	Cordero (1983)

In most cases the occurrence of these multiple crop pests and pathogens on parthenium weed does not affect the weed plant itself due to its distinct biochemical and physiological features, and its fast-growing habit. However, when such pests and pathogens start their early life stages on parthenium weed plants, they still have exceptional vigour when they move onto the main host crop. In this way of acting as a secondary host, parthenium weed acts as a pathway to crop growth and yield suppression, and in some cases this results in considerable agricultural productivity losses.

4.5 Socio-economic Impacts

Parthenium weed causes huge economic losses to the agriculture sector every year and in many countries where it has invaded. Economic losses also aggravate the stability of the social set-up of the farming community and the individual household. The socio-economic impacts of parthenium weed on agriculture can be categorized as: (i) direct negative impact on livelihoods such as income reduction; (ii) devaluation of commodities, enterprise and land due to invasion; and (iii) problems in marketing products produced on invaded lands. However, these impacts are deeply interconnected and therefore will be discussed jointly.

The agricultural industry has been directly hampered by parthenium weed infestations in many and varied parts of the world. For instance, Joshi (1990) estimated that parthenium weed occupied a huge agricultural area in India and was reducing fodder production in certain areas of northern India by 90%. Sushilkumar and Varshney (2010) reported that about 35 million ha of land had been infested with parthenium weed in India, where the estimated cost of manual and chemical control would have been about 3.0 billion Indian rupees (IR) if implemented since its arrival in 1956. Furthermore, costs for the treatment of parthenium weed-induced health problems was estimated to be IR88 million (Sushilkumar and Varshney, 2010). Similarly in Australia, by one estimation, parthenium weed has now spread over *c.*600,000 km^2 of pasture land (Anonymous, 2011), and parthenium weed on pasture land was estimated to cost AUS$16.8 million in lost beef production per annum in the 1970s (Chippendale and Panetta, 1994), with a predicted cost to Queensland beef producers set at AUS$69 million per annum in the 2000s (Adamson, 1996; Adamson and Bray, 1999). Adamson (1996) then estimated the cost to the beef industry by 2050 to be closer to AUS$110 million per annum, if the spread continues.

Due to the rapid spread, parthenium weed has caused serious allergy consequences for farm workers (Chippendale and Panetta, 1994; Goldsworthy, 2005). Goldsworthy (2005) reported that about 73% of farm workers in Clermont, Queensland had parthenium weed allergy symptoms. To combat the adversity of parthenium weed infestation in the 1990s, every year farmers had to spend money to: (i) purchase extra forage (AUS$285,000); (ii) undertake chemical weed control (AUS$1,441,280); and (iii) purchase extra machinery (AUS$171,000) (Chippendale and Panetta, 1994). The authors concluded that parthenium weed not only caused huge losses to the beef industry but also contributed towards land devaluation. Moreover, the costs added to control parthenium weed increased the overall cost of production and, therefore, reduced farm profits. At a social level, the farm workers suffered to a greater degree due to health problems caused by parthenium weed (Chippendale and Panetta, 1994). Lost work days and medical expenses for these farm workers increased substantially. In areas of heavy parthenium weed

infestation, the relocation of families into uninfested areas was also undertaken, which then had an impact on the financial status of the people involved. Similarly, the quality of fodder seed lots was reduced and farmers had many problems in trying to market potentially infested seed lots of lucerne and other forage crops (Chippendale and Panetta, 1994).

Parthenium weed has had a substantial impact on the socio-economic set-up of households in its introduced range. In a survey carried out in India, it was revealed that farmers were often aware of the weed's appearance, its growth cycle, and particularly the crop yield losses it caused; however, they were less aware of its impacts on environmental biodiversity (Kapoor, 2012). Furthermore, parthenium weed was often left uncontrolled, as there was a lack of knowledge in how to implement control measures. Acute socio-economic problems have also been attributed to parthenium weed in Ethiopia. Thousands of hectares have been infested by parthenium weed across the country and further spread is predicted. About 22% of the Oromiya Regional State, covering 22 districts in Ethiopia, is already affected by severe parthenium weed infestation (Beyene *et al.*, 2013). Due to unchecked spread, parthenium weed has now seriously impacted upon the lifestyle of the small landholders of those regions. More than 70% of farmers in Ethiopia held a view that parthenium weed caused crop yield losses as high as 50%. Similarly, between 24% and 71% of farmers recognized it as major threat to human and animal health (Beyene *et al.*, 2013). Further, about 73% farmers thought that parthenium weed affected soil fertility negatively and therefore reduced crop yield (Beyene *et al.*, 2013). So, the most severe impacts of parthenium weed in such regions were considered to be on food security, which was already a big problem.

The distribution and socio-economic impacts of parthenium weed were also studied in Khyber Pakhtunkhwa province of Pakistan (Khan *et al.*, 2013). It was revealed that parthenium weed had adverse, multifaceted impacts upon society. Farmers reported different negative impacts caused

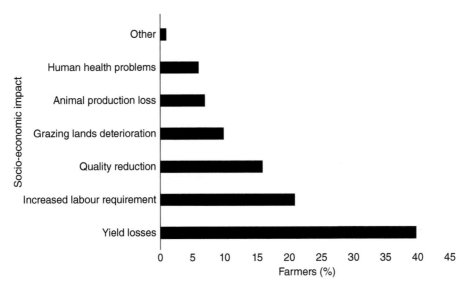

Fig. 4.3. The relative percentage of farmers reporting negative socio-economic impacts of parthenium weed in Khyber Pakhtunkhwa province, Pakistan. (Adapted from Khan *et al.*, 2013.)

by parthenium weed (Fig. 4.3), with significant declines in yields of their major field crops stated as the most important problem. Consequently, the farmers' incomes, health, education and overall lifestyles were negatively affected by parthenium weed infestations in the region (Khan *et al.*, 2013).

4.6 Conclusions

Parthenium weed is known to have drastic impacts on agricultural productivity and sustainability. However, little information is available on its direct impacts on agriculture, especially with reference to crop yield losses. With the increasing invasion of this noxious weed, more cropping systems are suffering production losses. The agricultural impacts of parthenium weed are multilateral and therefore require dynamic broad-spectrum solutions. Preventive measures are important to check for new infestations in cropping areas. The integrated management strategies involving chemical, biological and mechanical control methods are likely to be the most effective. However, proper legislation and policy making, and coordination of management (see Shabbir

et al., Chapter 9, this volume) to help control parthenium weed in many agricultural areas are necessary if the severe impacts of this weed across the world are to be avoided.

Acknowledgements

The authors are thankful to their colleagues: Arne Witt (Kenya), Ram Bahadur Khadka (Nepal) and Kelli Pukallus (Australia) who contributed Figures 4.1 and 4.2. Mr Ali Bajwa is thankful to the University of Queensland, Australia, for the provision of Research Training Program and UQ Centennial Scholarships for his PhD studies.

References

Adamson, D.C. (1996) Introducing dynamic considerations when economically evaluating weeds. MSc thesis, the University of Queensland, Brisbane, Australia.

Adamson, D.C. and Bray, S. (1999) The economic benefit from investing in insect biological control of parthenium weed (*Parthenium hysterophorus*). RDE Connections, Natural and Rural Systems Management (NRSM), the University of Queensland, Brisbane, Australia.

Anonymous (2011) New South Wales Parthenium Strategy. New South Wales (NSW) Department of Primary Industries. Available at: http://www.dpi.nsw.gov.au/data/assets/pdf_file/0003/355917/parthenium-weed.pdf (accessed 12 March 2013).

Arif, M.I., Rafiq, M. and Ghaffar, A. (2009) Host plants of cotton mealy bug (*Phenacoccus solenopsis*): a new menace to cotton agroecosystem of Punjab, Pakistan. *International Journal of Agriculture and Biology* 11, 163–167.

Ayele, S., Nigatu, L., Tana, T. and Adkins, S.W. (2013) Impact of parthenium weed (*Parthenium hysterophorus* L.) on the above-ground and soil seed bank communities of rangelands in Southeast Ethiopia. *International Research Journal of Agricultural Science and Soil Science* 3, 262–274.

Bajwa, A.A. (2014) Sustainable weed management in conservation agriculture. *Crop Protection* 65, 105–113.

Bajwa, R., Shafique, S., Shafique, S. and Javaid, A. (2004) Effect of foliar spray of aqueous extract of *Parthenium hysterophorus* on growth of sunflower. *International Journal of Agriculture and Biology* 6, 474–478.

Bajwa, A.A., Mahajan, G. and Chauhan, B.S. (2015) Nonconventional weed management strategies for modern agriculture. *Weed Science* 63, 723–747.

Bajwa, A.A., Chauhan, B.S., Farooq, M., Shabbir, A. and Adkins, S.W. (2016) What do we really know about alien plant invasion? A review of the invasion mechanism of one of the world's worst weeds. *Planta* 244, 39–57.

Batish, D.R., Singh, H.P., Pandher, J.K., Arora, V. and Kohli, R.K. (2002) Phytotoxic effect of parthenium residues on the selected soil properties and growth of chickpea and radish. *Weed Biology and Management* 2, 73–78.

Belgeri, A. and Adkins, S.W. (2015) Allelopathic potential of invasive parthenium weed (*Parthenium hysterophorus* L.) seedlings on grassland species in Australia. *Allelopathy Journal* 36, 1–14.

Belgeri, A., Navie, S.C., Vivian-Smith, G. and Adkins, S.W. (2014) Early recovery signs of an Australian grassland following the management of *Parthenium hysterophorus* L. *Flora* 209, 587–596.

Beyene, H., Gebrehiwot, L., Nigatu, L., Zewdie, K. and Regassa, S. (2013) Impacts of parthenium (*Parthenium hysterophorus*) on the life of Ethiopian farmers. *Ethiopian Journal of Weed Management* 6, 1–14.

Bhatt, B.P., Chauhan, D.S. and Todaria, N.P. (1994) Effect of weed leachates on germination and radicle extension of some food crops. *Indian Journal of Plant Physiology* 37, 177–179.

Biswas, O., Paul, K.P., Ghosh, S. and Karim, S.R. (2010) Allelopathic effects of parthenium weed debris in soil on the emergence and development of rice. *Journal of Agroforestry and Environment* 4, 193–196.

Chandler, L.D. and Chandler, J.M. (1988) Comparative host suitability of bell pepper and selected weed species for *Liriomyza trifolii* (Burgess). *Southwestern Entomologist* 13, 137–146.

Chippendale, J.F. and Panetta, F.D. (1994) The cost of parthenium weed to the Queensland cattle industry. *Plant Protection Quarterly* 9, 73–76.

Cordero, M. (1983) *Parthenium hysterophorus* (bitter broom), undesirable plant, natural reservoir of potato viruses X and Y. *Ciencia y Tecnica en la Agricultura* 2, 23–32.

Dayama, O.P. (1986) Allelopathic potential of *Parthenium hysterophorus* L. on the growth, nodulation, and nitrogen content of *Leucaena leucocephala*. *Leucaena Research Reports* 7, 36–37.

Demissie, A.G., Ashenafi, A., Arega, A., Etenash, U., Kebede, A. and Tigist, A. (2013) Effect of *Parthenium hysterophorus* L. on germination and elongation of onion (*Allium cepa*) and bean (*Phaseolus vulgaris*). *Research Journal of Chemical and Environmental Sciences* 1, 17–21.

Devi, O.I. and Dutta, B.K. (2012) Allelopathic effect of the aqueous extract of *Parthenium hysterophorus* and *Chromolaena odorata* on the seed germination and seedling vigour of *Zea mays* L. in vitro. *Academic Journal of Plant Sciences* 5, 110–113.

Devi, O.I., Dutta, B.K. and Choudhury, P. (2013) Effect of allelopathic plant extracts (*Parthenium hysterophorus* and *Chromolaena odorata*) on the seed germination and seedling vigour of rice (*Oryza sativa* L.) in vitro. *International Journal of AgriScience* 3, 766–774.

Dhawan, P. and Dhawan, S.R. (1995) Allelochemic effect of *Parthenium hysterophorus* L. on the germination behaviour of some agricultural crops. *Flora and Fauna* 1, 59–60.

Dhole, J.A., Bodke, S.S. and Dhole, N.A. (2011) Allelopathic effect of aqueous leaf extract of *Parthenium hysterophorus* L. on seed germination and seedling emergence of some cultivated crops. *Journal of Research in Biology* 1, 15–18.

Farooq, M., Bajwa, A.A., Cheema, S.A. and Cheema, Z.A. (2013) Application of allelopathy in crop production. *International Journal of Agriculture and Biology* 15, 1367–1378.

Fernandez, J.V., Odero, D.C. and Wright, A.L. (2015) Effects of *Parthenium hysterophorus* L.

residue on early sugarcane growth in organic and mineral soils. *Crop Protection* 72, 31–35.

Goldsworthy, D. (2005) Parthenium weed and health. Technical report. University of Central Queensland, Rockhampton, Australia.

Govindappa, M.R., Shankarappa, K.S., Rangaswamy, K.T. and Muniyappa, V. (2005) *In vitro* evaluation of plant protection chemicals for compatibility with potential bioagent *Paecilomyces farinosus* of whitefly (*Bemisia tabaci*), an insect vector of tomato leaf curl virus disease. *Environment and Ecology* 23, 517–522.

Gupta, S. and Narayan, R. (2010) Effects of applied leaf biomass of *Parthenium hysterophorus*, *Cassia obtusifolia* and *Achyranthes aspera* on seed germination and seedling growth of wheat and pea. *Allelopathy Journal* 26, 59–70.

Joshi, S. (1990) An economic evaluation of control methods for *Parthenium hysterophorus* L. *Biological Agriculture and Horticulture* 6, 285–291.

Kanchan, S.D. and Jayachandra (1979) Allelopathic effects of *Parthenium hysterophorus* L. exudation of inhibitors through roots. *Plant and Soil* 34, 27–35.

Kapoor, R.T. (2012) Awareness related survey of an invasive alien weed, *Parthenium hysterophorus* L. in Gautam Budh Nagar district, Uttar Pradesh, India. *Journal of Agricultural Technology* 8, 1129–1140.

Khan, H., Marwat, K.B., Hassan, G. and Khan, M.A. (2013) Socio-economic impacts of parthenium (*Parthenium hysterophorus* L.) in Peshawar Valley, Pakistan. *Pakistani Journal of Weed Science Research* 19, 275–293.

Khan, N. (2011) Long term, sustainable management of parthenium weed (*Parthenium hysterophorus* L.) using suppressive plants. PhD thesis, the University of Queensland, Brisbane, Australia.

Khan, N., Hashmatullah, K., Hussain, Z. and Khan, S.A. (2012) Assessment of allelopathic effects of parthenium (*Parthenium hysterophorus* L.) plant parts on seed germination and seedling growth of wheat (*Triticum aestivum* L.) cultivars. *Pakistani Journal of Weed Science Research* 18, 39–50.

Kishun, R. and Chand, R. (1987) New collateral hosts for *Pseudomonas solanacearum*. *Indian Journal of Mycology and Plant Pathology* 17, 237.

Kohli, R.K., Rani, D., Singh, H.P. and Kumar, S. (1996) Response of crop seeds towards the leaf leachates of *Parthenium hysterophorus* L. *Indian Journal of Weed Science* 282, 104–106.

Kumar, G. and Gautam, N. (2008) Allelotoxicity of parthenium leaf extracts on cytomorphological

behaviour of sunflower (*Helianthus annuus*). *Journal of Environmental Biology* 29, 243–247.

Kumar, M., Kumar, S. and Sheikh, A.M. (2010) Influence of *Parthenium hysterophorus* ash on the growth of *Phaseolus mungo*. *American-Eurasian Journal of Agricultural and Environmental Science* 8, 692–694.

Kumar, S., Jayaraj, S. and Muthukrishnan, T.S. (1979) Natural enemies of *Parthenium hysterophorus* Linn. *Journal of Entomological Research* 3, 32–35.

Madhu, M., Nanjappa, H.V. and Ramachandrappa, B.K. (1995) Allelopathic effect of weeds on crops. *Mysore Journal of Agricultural Sciences* 29, 106–112.

Maharjan, S., Shrestha, B.B. and Jha, P.K. (2007) Allelopathic effects of aqueous extract of leaves of *Parthenium hysterophorus* L. on seed germination and seedling growth of some cultivated and wild herbaceous species. *Scientific World Journal* 5, 33–39.

Masum, S.M., Ali, M.H., Mandal, M.S., Haque, M.N. and Mahto, A.K. (2012) Influence of *Parthenium hysterophorus*, *Chromolaena odorata* and PRH on seed germination and seedling growth of maize, soybean and cotton. *Bangladesh Journal of Weed Science* 3, 83–90.

McFadyen, R.E. (1992) Biological control against parthenium weed in Australia. *Crop Protection* 11, 400–407.

Mersie, W. and Singh, M. (1987) Allelopathic effect of parthenium (*Parthenium hysterophorus* L.) extract and residue on some agronomic crops and weeds. *Journal of Chemical Ecology* 13, 1739–1747.

Narasimhan, T.R., Ananth, M., Swamy, M.N., Babu, M.R., Mangala, A. and Rao, P.S. (1977) Toxicity of *Parthenium hysterophorus* L. to cattle and buffaloes. *Experientia* 33, 1358–1359.

Navie, S.C. (2002) The biology of *Parthenium hysterophorus* L. in Australia. PhD thesis, the University of Queensland, Brisbane, Australia.

Navie, S.C., McFadyen, R.E., Panetta, F.D. and Adkins, S.W. (1996) The biology of Australian weeds. 27. *Parthenium hysterophorus* L. *Plant Protection Quarterly* 11, 76–88.

Netsere, A. (2015) Allelopathic effects of aqueous extracts of an invasive alien weed *Parthenium hysterophorus* L. on maize and sorghum seed germination and seedling growth. *Journal of Biology, Agriculture and Healthcare* 5, 120–124.

Netsere, A. and Mendesil, E. (2012) Allelopathic effects of *Parthenium hysterophorus* L. aqueous extracts on soybean (*Glycine max* L.) and haricot bean (*Phaseolus vulgaris* L.) seed germination, shoot and root growth and dry matter production. *Journal of Applied Botany and Food Quality* 84, 219–222.

Nigatu, L., Hassen, A., Sharma, J. and Adkins, S.W. (2010) Impact of *Parthenium hysterophorus* on grazing land communities in north-eastern Ethiopia. *Weed Biology and Management* 10, 143–152.

Nishanthan, K., Sivachandiran, S. and Marambe, B. (2013) Control of *Parthenium hysterophorus* L. and its impact on yield performance of tomato (*Solanum lycopersicum* L.) in the northern province of Sri Lanka. *Tropical Agricultural Research* 25, 56–68.

Njoroge, J.M. (1986) New weeds in Kenya coffee, a short communication. *Kenya Coffee* 51, 333–335.

Njoroge, J.M. (1991) Tolerance of *Bidens pilosa* L. and *Parthenium hysterophorus* L. to paraquat (Gramoxone) in Kenya coffee. *Kenya Coffee* 56, 999–1001.

Ovies, J. and Larrinaga, L. (1988) Transmission of *Xanthomonas campestris* pv. *phaseoli* by a wild host. *Ciencia y Tecnica en la Agricultura, Proteccion de Plantas* 11, 23–30.

Parsons, W.T. and Cuthbertson, E.G. (1992) *Noxious Weeds of Australia*. Inkata, Melbourne, Australia.

Parthasarathi, T., Suganya, V. and Sivakumar, R. (2012) Allelopathic effect of aqueous leaf extract of *Parthenium hysterophorus* L. on seed germination and seedling growth in green gram, black gram and groundnut. *Madras Agricultural Journal* 99, 514–517.

Paudel, V.R., Gupta, V.N.P. and Agarwal, V.P. (2009) Effect of diffusates of *Parthenium hysterophorous* on seed germination of *Raphanus sativus* L. *Scientific World Journal* 7, 29–32.

Rajulu, G.S., Gowri, N. and Perumal, S.S. (1976) Biological control for the pernicious weed *Parthenium hysterophorus* L. *Current Science* 45, 624–625.

Reddy, A.S., Rao, R.D., Thirumala-Devi, K., Reddy, S.V., Mayo, M.A., *et al.* (2002) Occurrence of Tobacco streak virus on peanut (*Arachis hypogaea*) in India. *Plant Disease* 86, 173–178.

Robertson, L.N. and Kettle, B.A. (1994) Biology of *Pseudoheteronyx* sp. (Coleoptera: Scarabaeidae) on the central highlands of Queensland. *Journal of the Australian Entomological Society* 33, 181–184.

Safdar, M.E., Tanveer, A., Khaliq, A. and Riaz, M.A. (2015) Yield losses in maize (*Zea mays*) infested with parthenium weed (*Parthenium hysterophorus* L.). *Crop Protection* 70, 77–82.

Shabbir, A. (2012) Towards the improved management of parthenium weed: Complementing biological control with plant suppression. PhD thesis, the University of Queensland, Brisbane, Australia.

Shabbir, A. (2014) Ragweed emerging a major host for the cotton mealy bug in Pakistan. *Indian Journal of Weed Science* 46, 291–293.

Shabbir, A. and Javaid, A. (2010) Effect of aqueous extracts of alien weed *Parthenium hysterophorus* and two native Asteraceae species on germination and growth of mungbean, *Vigna radiata* L. Wilczek. *Journal of Agricultural Research* 48, 483–488.

Shahina, F., Firoza, K., Mehreen, G., Salma, J. and Bhatti, M.I. (2012) Molecular characterization of root-knot nematodes with five new host records from Pakistan. *Pakistani Journal of Nematology* 30, 129–141.

Sharma, D.D. and Gupta, A.K. (1998) *Alternaria zinniae* on *Parthenium hysterophorus*. *EPPO Bulletin* 28, 217.

Sharman, M., Persley, D.M. and Thomas, J.E. (2009) Distribution in Australia and seed transmission of tobacco streak virus in *Parthenium hysterophorus*. *Plant Disease* 93, 708–712.

Sharman, M., Thomas, J.E. and Persley, D.M. (2015) Natural host range, thrips and seed transmission of distinct tobacco streak virus strains in Queensland, Australia. *Annals of Applied Biology* 167, 197–207.

Shi, B. (2016) Invasive potential of the weed *Parthenium hysterophorus*: the role of allelopathy. PhD thesis, the University of Queensland, Brisbane, Australia.

Singh, H.P., Batish, D.R., Pandher, J.K. and Kohli, R.K. (2003) Assessment of allelopathic properties of *Parthenium hysterophorus* residues. *Ecosystem and Environment* 95, 537–541.

Srivastava, J., Raghava, N. and Raghava, R.P. (2011) Allelopathic potential of parthenium to reduce water absorption in germinating cowpea seeds. *Indian Journal of Scientific Research* 2, 59–65.

Stamps, R.H. (2011) Identification, impacts, and control of ragweed parthenium (*Parthenium hysterophorus* L.). University of Florida Institute of Food and Agricultural Sciences (IFAS) Extension Report. University of Florida, Gainsville, Florida.

Sukhada, K. and Jayachandra (1980) Pollen allelopathy – a new phenomenon. *New Phytologist* 84, 739–746.

Sushilkumar and Varshney, J.G. (2010) Parthenium infestation and its estimated cost management in India. *Indian Journal of Weed Science* 42, 73–77.

Swaminathan, C., Rai, R.V. and Suresh, K.K. (1990) Allelopathic effects of *Parthenium hysterophorus* L. on germination and seedling growth of a few multipurpose trees and arable crops. *International Tree Crops Journal* 6, 143–150.

Tamado, T. and Milberg, P. (2004) Control of parthenium (*Parthenium hysterophorus*) in grain sorghum (*Sorghum bicolor*) in the smallholder farming system in eastern Ethiopia. *Weed Technology* 18, 100–105.

Tamado, T., Ohlander, L. and Milberg, P. (2002) Interference by the weed *Parthenium hysterophorus* L. with grain sorghum: influence of weed density and duration of competition. *International Journal of Pest Management* 48, 183–188.

Tefera, T. (2002) Allelopathic effects of *Parthenium hysterophorus* extracts on seed germination and seedling growth of *Eragrostis tef.*

Journal of Agronomy and Crop Science 188, 306–310.

Tudor, G.D., Ford, A.L., Armstrong, T.R. and Bromage, E.K. (1982) Taints in meat from sheep grazing *Parthenium hysterophorus*. *Animal Production Science* 22, 43–46.

Wakjira, M., Berecha, G. and Bulti, B. (2005) Allelopathic effects of *Parthenium hysterophorus* extracts on seed germination and seedling growth of lettuce. *Tropical Science* 45, 159–162.

Wakjira, M., Berecha, G. and Tulu, S. (2009) Allelopathic effects of an invasive alien weed *Parthenium hysterophorus* L. compost on lettuce germination and growth. *African Journal of Agricultural Research* 4, 1325–1330.

5 Impacts on the Environment

Arne Witt[1]* and Amalia Belgeri[2]

[1]CABI, Nairobi, Kenya; [2]AGROTERRA S.A., Montevideo, Uruguay

5.1 Introduction

Invasive species can alter the production, maintenance and quality of ecosystem services (Levine *et al.*, 2003; Brooks *et al.*, 2004; Kimbro *et al.*, 2009), which are broadly defined as the benefits provided to people by natural ecosystems and include provisioning, regulating, cultural and supporting services (MEA, 2005). Invasive plant species change community structure and function through exploitation (resource use) and interference (competition for resources and possibly allelopathy). According to Vilà *et al.* (2010), invasive plants decrease species diversity and abundance by *c*.51% and *c*.44%, respectively, also reducing fitness and growth of native plant species by *c*.42% and 22%, respectively. By reducing native species diversity, abundance, fitness and distribution and by altering community structure, alien invasive plant species have a significant impact on ecosystem services (Hulme, 2006; Gabbard and Fowler, 2007; Hickman *et al.*, 2010). The loss of plant species or changes in vegetation structure, as a result of alien plant invasions, may have cascading trophic effects (Sakai *et al.*, 2001), especially on other species at higher trophic levels. For example, alien plant species have been reported to decrease animal fitness and abundance both by *c*.17% (Vilà *et al.*, 2010). By altering community structure, invasive plant species may also act as ecosystem engineers or transformers (Jones *et al.*, 1997), affecting a multitude of other organisms (Maerz *et al.*, 2005; Brown *et al.*, 2006; Canhoto and Laranjeira, 2007).

Invasive plant species can also alter natural cycles such as those of energy, nutrients and water (Bock *et al.*, 1986; Christian and Wilson, 1999). By changing evapotranspiration rates, invasive plants can impact negatively on water resources (Zavaleta, 2000). For example, in South Africa's Mpumalanga province, the replacement of grassland catchment areas with pines and *Eucalyptus* spp. led, within 6–12 years, to the drying up of streams (van Lill *et al.*, 1980). Other studies in South Africa have found that pine plantations on the Drakensberg in KwaZulu-Natal, South Africa, have reduced stream flows by 82% (Bosch, 1979), while in the Western Cape stream flows from invaded fynbos water catchments have declined by 55%. Invasive plants such as giant sensitive plant (*Mimosa pigra* L.; *Fabaceae*) have also been reported to affect water flow and sedimentation. Many of these and other invasive species have a negative impact on water quality and purification (Carlsson *et al.*, 2004). Invasive species can also reduce soil nutrient levels (Christian and Wilson, 1999; Evans *et al.*, 2001). Changes in decomposition rates, brought about by plant invasions, can also affect nutrient cycling, as can the introduction of nitrogen-fixing species or species that leach chemicals that inhibit nitrogen (N) fixation by other species (Vitousek and Walker, 1989). Alien plant species can enhance microbial activity by *c*.32%, and available nitrogen (N), phosphorous (P) and carbon (C) pools by *c*.22%, 20% and 12%, respectively (Vilà *et al.*, 2010). A number of invasive plant species may also change certain community disturbance

* a.witt@cabi.org

regimes, including those of fire, flooding and erosion, which may contribute to changes in weather patterns and even local climate (Butler and Fairfax, 2003; Rossiter *et al.*, 2003). For example, invasive grasses may alter fire frequencies and intensities, which may contribute to increases in atmospheric C and contribute to global warming. Invasive plants can also impact on climate and atmospheric composition. Kudzu vine (*Pueraria montana* (Lour.) Merr.; *Fabaceae*), for example, can increase nitric oxide emissions and thereby increase ozone pollution. In parts of the USA, kudzu vine could, under future scenarios, increase nitrogen oxide emissions by as much as 28% when compared with present-day emissions (Hickman *et al.*, 2010). Invasive plant species can also make habitats unsuitable for other species by releasing salt, by acidifying the soil or by releasing allelopathic chemical compounds.

Parthenium weed (*Parthenium hysterophorus* L.) is no different to a multitude of other invasive plant species in terms of its environmental impacts. It is allelopathic, unpalatable to most grazing animals (see Allan *et al.*, Chapter 6, this volume), especially in its introduced range, and it can survive and reproduce in a wide range of soil types and climatic conditions. The species can complete its life cycle in as little as 4–6 weeks, and produce thousands of seeds, attributes that have made parthenium weed one of the most successful invasive plants in the world (see Adkins *et al.*, Chapter 2, this volume). Once introduced, parthenium weed is quick to establish, provided there is sufficient soil moisture, and then spreads rapidly. Based on observations made in Africa, parthenium weed is probably, up to now, one of the fastest-spreading invasive plants on the continent. It has now been accidentally introduced into multiple countries in eastern Africa, southern Africa, North Africa, the Middle East, South and South-east Asia, North-west Asia, the Pacific and Australia, where it is affecting ecosystem services in croplands, grasslands, savannahs, riparian vegetation, woodlands (provided there is sufficient light penetration), and even in urban and peri-urban environments (see Shabbir *et al.*, Chapter 3,

this volume), where it also impacts negatively on human and animal health.

Despite the fact that parthenium weed is so widespread and abundant, there have been few qualitative studies on its impact on biodiversity. Much of the evidence regarding its negative impacts on biodiversity is largely anecdotal with very few quantitative studies available on its impacts on wildlife, native plant diversity and abundance, and soil nutrients (e.g. Osunkoya *et al.*, 2017). Although some research has been undertaken on its impacts on livestock health, there have been no studies on its impacts on wildlife, and that includes insects, birds and other organisms. Therefore in this chapter we have largely inferred impacts from studies that have been undertaken on other invasive plants, especially on impacts on native species at higher trophic levels. This chapter is therefore largely a review of the impacts of invasive plants on biodiversity in general, but we have linked it back to parthenium weed whenever possible.

5.2 Impacts on Biodiversity: Below Ground

5.2.1 Soil properties

Invasive plant species may change soil properties, including those of: (i) temperature; (ii) moisture content (Zavaleta, 2000; Belnap and Phillips, 2001); (iii) soil organic matter; (iv) C, N and P content (Ehrenfeld, 2003; Koutika *et al.*, 2007; Timsina *et al.*, 2011); (v) pH (D'Antonio, 1993; Kourtev *et al.*, 1998); and (vi) soil microbial activity (Kourtev *et al.*, 2003; Chacon *et al.*, 2008; Qin *et al.*, 2014). Changes in the species composition of a plant community, as a result of an invasion, result also in changes to nutrient cycling processes (Ehrenfeld, 2003). The soil microbial community may be altered based on changes in the quantity and quality of inputs to the soil from introduced plant species (Garland, 1996; Grayston and Campbell, 1998; Osunkoya *et al.*, 2017), while the physical properties of the soil may also be altered, which in turn will influence nutrient dynamics (Ehrenfeld and Scott, 2001). Plant traits which are thought to affect soil nutrient cycling processes include those to do with plant structure, life cycle, photosynthetic pathway, physiology, symbiotic relationships, vegetative spread, roots, and tissue type and chemistry (Ehrenfeld, 2003). Changes in plant functional types such as herbaceous versus woody plants, grasses versus woody plants, N-fixing versus non-fixing, C3 versus C4, and others will impact on soil nutrient dynamics (Gill and Burke, 1999). Impacts on nutrient cycling will also depend on how different the traits or characteristics of the invasive species are as compared with those of species present within the existing plant community (Ehrenfeld, 2003).

Soil pools of C, N and water change as a result of plant invasions. For instance, sites invaded by ragweed (*Ambrosia artemisiifolia* L.; *Asteraceae*) in southern China had significantly higher total P, available N and P than non-invaded sites (Qin *et al.*, 2014). Microbial biomass, C, N and P were also significantly higher in soils from invaded sites. Mikania (*Mikania micrantha* Kunth; *Asteraceae*), an invasive plant species from tropical America, also affected soil nutrients and N transformation in southern China (Chen *et al.*, 2008). The invasive shrub Crofton weed (*Ageratina adenophora* (Spreng.) King & Rob.; *Asteraceae*) alters soil chemistry and soil biota in such a way as to benefit itself while inhibiting indigenous plants through positive plant–soil feedback loops (Niu *et al.*, 2007). In this manner, introduced plants may change soil biota to facilitate further invasions, contributing to increased dominance and significant changes in vegetation composition and structure (Vitousek, 1990).

Parthenium weed has also been reported to change the physical and chemical properties of the soil such as its texture, pH, organic matter and N, K and P content (Karki, 2009; Terfa, 2009; Timsina *et al.*, 2011; Osunkoya *et al.*, 2017). Impacts of parthenium weed on soil pH have been found to be variable, but in general they have remained unchanged or increased as compared with uninvaded sites (Table 5.1). Karki (2009) and Upadhyay *et al.* (2013) found that invaded plots had

significantly higher soil pH than non-invaded areas, while Batish *et al.* (2002a) observed that parthenium weed decreased soil pH. Joshi (2005) found parthenium weed in both acidic and neutral soils in the Kathmandu Valley, Nepal. In contrast, observations made in Australia suggest that the rhizosphere soil around parthenium weed roots becomes more acidic than soil in nearby uninfested sites (S.W. Adkins, Australia, 2016, personal communication). According to Karki (2009), organic matter was slightly lower at one of two invaded sites with organic C levels decreasing with an associated increase in parthenium weed densities at both sites. In contrast, Batish *et al.* (2002a) and Timsina *et al.* (2011) found higher organic matter in invaded sites. Significant decreases in soil C content in invaded sites at Hetauda, Nepal, might have been due to biomass (fodder) removal by the local community (Karki, 2009). There was no real difference in soil N levels between uninvaded and invaded areas in Nepal, although Karki (2009) found that soil N decreased under increased parthenium weed density at the Hetauda site, while Batish *et al.* (2002a) found that N levels decreased (Table 5.1). Upadhyay *et al.* (2013) found that parthenium weed did not appear to affect organic matter and N levels with any real discernible pattern. In a study by Timsina *et al.* (2011) it was concluded that for both N and organic matter content, the effect of locality was stronger than the effect of the plot. This is largely supported by Terfa (2009), who found that impacts of parthenium weed invasion varied across altitudes and soil types in Ethiopia. For example, soil organic content was higher in uninvaded soils at lower altitudes than at higher altitudes, while soil N was lower in invaded soils at 973 m above sea level (asl) but higher than that in uninvaded soils at 1014 m asl. However, in general most soil chemical parameters measured were higher in infested areas (Terfa, 2009).

It is widely recognized that impacts of plant invasions on soil properties are extremely variable. In a review of various studies Vilà *et al.* (2010) found that invasion generally decreased litter decomposition by

Table 5.1. Differences between areas invaded and uninvaded by parthenium weed with regard to soil pH, organic matter, nitrogen (N), potassium (K) and phosphorous (P) as determined by various studies.[a,b]

Region/area and/or country	Soil pH	Organic matter	N	K	P	Reference
Hetauda Municipality, Nepal	Higher	Slightly lower	Similar	–	–	Karki (2009)
Bharatpur Municipality, Nepal	Higher	Similar	Similar	–	–	
Gorkha, Nuwakot and Kathmandu districts, Nepal	Higher	Higher	Higher	Higher	Higher	Timsina et al. (2011)
Kathmandu Valley, Nepal	Present in acidic and neutral soils	–	–	–	–	Joshi (2005)
India	Present in alkaline to neutral or slightly acidic soils	–	–	–	–	Bhowmik et al. (2007)
	Present in alkaline to acidic soils	Higher	Lower	–	–	Batish et al. (2002a)
	Higher	Similar	Similar	–	–	Upadhyay et al. (2013)
Ethiopia – 973 m asl	Lower	Slightly lower	Lower	Higher	Similar	Terfa (2009)
Ethiopia – 1014 m asl	Similar	Similar	Higher	Slightly higher	Higher	
Ethiopia – 1035 m asl		Higher	Slightly lower	Higher	Similar	
Ethiopia – 1044 m asl			Higher	Higher	Slightly higher	

[a] Only differences in invaded areas as compared with uninvaded areas are shown.
[b] asl, Above sea level; –, indicates no data collected.

c.16%, although results were mixed, with many showing increases in litter decomposition. In many areas parthenium weed is a recent invader and according to Collins (2005) it takes time before changes in soil nutrient pools in invaded areas manifest themselves. Variation in results may be due to differences in the length of time that study areas have been colonized and invaded by parthenium weed.

5.2.2 Soil invertebrates

Certain abiotic and biotic soil properties are known to associate with vegetation type (Wardle, 2002). Feedback loops or interactions between plants and soil microbes influence both plant and soil community composition and ecosystem processes (Bever, 2003; van der Putten, 2007). Plants interact with mutualists, pathogens and herbivores in the areas surrounding their roots, whereas interactions with decomposers are more indirect, through root exudates, litter and mineralized nutrients (van der Putten, 2007). As such, plant species composition and abundance impacts on soil properties, as discussed above, but also on a range of ground-dwelling organisms.

There are six general trophic categories of below-ground arthropods namely predators, fungivores, bacteriovores, detritivores, herbivores and omnivores (Moore et al., 1988). Predators of arthropods include pursuit predators such as rhagidiid mites and ambush predators. Soils also provide habitats for a host of detritivores which include millipedes, isopods, termites and some oribatid mites, all of which scavenge on dead organic matter. Herbivores feed on roots, bulbs and other underground plant parts, while some species of springtails, mites and insects also prey on nematodes and other ground-dwelling invertebrates (Moore *et al.*, 1988). These organisms influence N flows within the soil, for instance soil fauna accounted for 37% of N mineralization in native short grass steppe in north-eastern Colorado (Hunt *et al.*, 1987). Soil-inhabiting arthropods play a key role in nutrient cycling and decomposition of litter (Moore *et al.*, 1988). These processes may be altered through the introduction and proliferation of invasive plant species.

Kourtev *et al.* (1999) found higher earthworm densities under the introduced Japanese barberry (*Berberis thunbergii* DC.; *Berberidaceae*) and Japanese stilt grass (*Microstegium vimineum* (Trin.) A. Camus; *Poaceae*) than under the native *Vaccinium* spp. (*Ericaceae*) species. This could be related to the fact that soils below Japanese barberry had higher nitrate concentrations and a smaller litter layer. There were also higher earthworm densities under the introduced fire tree (*Myrica faya* Ait.) in Hawaii (Aplet, 1990). Belnap and Phillips (2001) found lower species richness and lower numbers of fungi and invertebrates, and higher abundances of active bacteria in sites invaded by cheat grass (*Bromus tectorum* L.; *Poaceae*). In areas invaded by Himalayan balsam (*Impatiens glandulifera* Royle; *Balsaminaceae*) there was an increased abundance of springtails (Collembola) and detritivores in the summer months of 2007 (Tanner *et al.*, 2013). By 2008, the abundance of invertebrate larvae was significantly higher in those areas (Tanner *et al.*, 2013).

Jeyalakshmi *et al.* (2011) isolated 13 different microorganism species from soils which were dominated by parthenium weed in Tamil Nadu, India, but no comparative data was collected from uninvaded areas. In another study in India it was found that parthenium weed had a positive impact on the rhizopheric microorganisms (Cibichakravarthy *et al.*, 2011). Few other studies that looked at the impacts of parthenium weed on below-ground organisms could be sourced. However, based on other research, and the fact that parthenium weed alters nutrient cycling, we are of the opinion

that this species has an impact on soil-living organisms.

5.3 Impacts on Biodiversity: Above Ground

5.3.1 Plants

There have been a number of studies on the impacts of invasive plants on the components of plant communities such as native species diversity and abundance. These studies have all shown that invasive species have significant effects on diversity, structure and biomass production of plant communities (Chippendale and Panneta, 1994; Levine *et al.*, 2003; Hejda *et al.*, 2009; Olsson *et al.*, 2012). Hejda *et al.* (2009) looked at the impact of 13 invasive plant species, including giant knotweed (*Fallopia sachalinensis* F. Schmidt; *Polygonaceae*), Japanese knotweed (*Fallopia japonica* (Thunb.) Decne. & Planch.), monk's rhubarb (*Rumex alpinus* L.; *Polygonaceae*), Jerusalem artichoke (*Helianthus tuberosus* L.; *Asteraceae*) and Himalayan balsam in the Czech Republic. They found that the impact of invasions differed markedly among the species, with significant differences between invaded and non-invaded plots for five species only.

The fact that parthenium weed possesses allelopathic properties contributes, among other invasive traits, to its ability to displace plant species (Evans, 1997; Bajwa *et al.* 2016), and as such impacts on the goods and services provided by ecosystems. The allelopathic nature of parthenium weed has been proposed through the observation of water-soluble phenolic and sesquiterpene lactones, mainly parthenin, in roots, stems, leaves, inflorescences, pollen and seeds. These chemicals can inhibit the growth of plants (Picman and Picman, 1984; Batish *et al.*, 2005; Belz *et al.*, 2007; Rashid *et al.*, 2008). Parthenium weed seedlings in pots showed strong allelopathic effects on seed germination in chir pine (*Pinus roxburghii* Sarg.; *Pinaceae*) with one seedling in a pot reducing germination rates by *c.*18%, two seedlings by 36%, and three seedlings by

43% (Huy and Seghal, 2004). The allelo-chemicals released by parthenium weed inhibit the germination and growth of a wide range of plant species (Oudhia, 2000; Batish *et al.*, 2002a, 2002b; Singh *et al.*, 2003) and often displace most of the associated herbaceous flora (Bhowmik *et al.*, 2007: Belgeri *et al.*, 2014). However, it is important to recognize that an effective and consistent allelopathic inhibition of one species to another is more likely to occur in communities with low species richness than in species-rich communities (Kruse *et al.*, 2000). In communities with high plant species diversity, it is less likely that one species would reach sufficient dominance for its allelochemicals to dominate the soil biochemistry (Wardle *et al.*, 2006). However, parthenium weed does appear to dominate in many habitats.

Parthenium weed is an aggressive colonizer that can establish in natural and disturbed ecosystems, grasslands, open woodlands, riparian zones, flood plains, protected areas, human settlements, fallow lands, croplands and even gardens (Chippendale and Panetta, 1994; Tamado and Milberg, 2000; Pandey *et al.*, 2003). Although there is considerable variability in the invasibility of these plant communities or habitats, species diversity is a key community attribute that is related to their susceptibility to invasion. Diverse communities (i.e. those with high species richness) are more resistant to invasion (Meiners and Cadenasso, 2005) than communities with low species diversity. Although parthenium weed may be less likely to invade species-rich habitats, evidence demonstrates that it can be a significant factor in determining the botanical composition within managed grasslands (Belgeri *et al.*, 2014) and other habitats. According to Khan *et al.* (2013), 'little to no other vegetation' can be seen in areas dominated by parthenium weed (Fig. 5.1). This species has the ability to invade and establish in new habitats, where it reduces the number of native plants (Huy and Seghal, 2004; Shabbir and Bajwa, 2006; Akter and Zuberi, 2009). In Australia, the weed has been reported to change habitats (Evans, 1997), especially in native grasslands (Belgeri *et al.*, 2014), open woodlands, flood plains and along river banks. In India, the weed has been reported to replace the native vegetation in a number of ecosystems (Yaduraju *et al.*, 2005). Other studies have shown that its prolonged presence might have greatly reduced the diversity of seed banks, thereby reducing the ability of native species to regenerate (Navie *et al.*, 2004; Belgeri *et al.*, 2014; Shabbir, 2015).

In India, parthenium weed has contributed to the loss of plant diversity in many ecosystems (Rajwar *et al.*, 1998; Kumar and Rohatgi, 1999; Sushilkumar, 2005). It threatens the forest biodiversity in sal (*Shorea robusta* Gaertn; *Dipterocarpaceae*; Pandey and Saini, 2002) and teak (*Tectona grandis* L.; *Verbenaceae*) forests of Pench National Park in Madhya Pradesh. Parthenium weed has also replaced important and abundant native grasses such as Guria grass (*Chrysopogon fulvus* (Spreng.) Chiov.; *Poaceae*) and bluestem (*Dicanthium annulatum* (Forssk.) Stapf.; *Poaceae*) in chir (*Pinus roxburghii* Sarg.; *Pinaceae*) forests and significantly changed the natural distribution patterns of herbs (Huy and Seghal, 2004). In the northern Himalayas (India), invasions of parthenium weed, goatweed (*Ageratum conyzoides* L.; *Asteraceae*) and lantana (*Lantana camara* L.; *Verbenaceae*) significantly decreased species richness (Kohli *et al.*, 2004). Parthenium weed is also considered to be the greatest threat to biodiversity in the Einasleigh Uplands bioregion in Australia (Sattler and Williams, 1999). Despite the observation that parthenium weed has a dramatic impact on biodiversity, there have been few detailed studies that have actually demonstrated this. A few studies measuring impacts have been undertaken in Nepal (Karki, 2009; Timsina *et al.*, 2011; Table 5.2), India (Huy and Seghal, 2004; Upadhyay *et al.*, 2013; Table 5.2), Australia (Nguyen *et al.*, 2010; Belgeri *et al.*, 2014; Table 5.2) and Ethiopia (Terfa, 2009; Nigatu *et al.*, 2010; Ayele *et al.*, 2013; Table 5.2).

Parthenium weed has been reported to reduce plant species diversity in the districts of Bharatpur and Hetauda in Nepal, but only marginally so (Karki, 2009). The most abundant plant species on both sites were found

Fig. 5.1. Following parthenium weed invasion there is a reduction in plant species diversity and community evenness as parthenium weed densities increase. (A. Witt.)

to be other weeds, many of them introduced, an indication that the study sites were probably highly disturbed, even prior to the invasion by parthenium weed. An introduced weed, common cocklebur (*Xanthium strumarium* L.; *Asteraceae*) was absent in non-invaded areas in Bharatpur, whereas madanaghanti (*Borreria articularis* (L.f.) Williams; *Rubiaceae*), Indian pennywort (*Centella asiatica* (L.) Urban; *Apiaceae*) and gripeweed (*Phyllanthus urinaria* L.; *Phyllanthaceae*) were absent in areas invaded by parthenium weed (Karki, 2009). Plant species richness and abundance of most species, at both sites, declined with increasing parthenium weed density (Karki, 2009).

In Nepal, parthenium weed did not seem to have a negative impact on plant diversity, although it did influence species composition at a landscape level (Timsina *et al.*, 2011). In fact, there were fewer plant species per square metre in non-invaded

versus invaded plots, albeit only marginally so (Table 5.2). Parthenium weed did displace native grasses such as *Acrachne racemosa* Ohwi (unresolved name) (*Poaceae*) and lesser spear grass (*Chrysopogon aciculatus* (Retz.) Trin.; *Poaceae*), a non-native grass, rats tail grass (*Sporobolus* sp.; *Poaceae*), and an exotic forb white clover (*Trifolium repens* L.; *Fabaceae*). However, native grass species such as Bermuda grass (*Cynodon dactylon* (L.) Pers.; *Poaceae*), Egyptian crowfoot grass (*Dactyloctenium aegyptium* (L.) Willd.; *Poaceae*), native woody plants such as sickle senna (*Senna tora* (L.) Roxb.; *Fabaceae*) and the invasive plants cocklebur and arrow leaf sida (*Sida rhombifolia* L.; *Malvaceae*) could coexist with parthenium weed (Timsina *et al.*, 2011). This supports other studies which indicate that species in India such as senna (*Senna sericea* (Symon) Albr. & Symon; *Fabaceae*), sickle senna, matura tea tree (*Senna auriculata* (L.) Roxb.; *Fabaceae*), sicklepod

Table 5.2. Impacts of parthenium weed (PW) on plant species (native and exotic combined) as determined by various studies across the world.

Region/district and/or country	Number of plant species under different PW densities/ impacts of PW	Reference
Hetauda, Nepal	No PW = 27 spp.; PW present = 23 spp.	Karki (2009)
Bharatpur, Nepal	No PW = 20 spp.; PW present = 19 spp.	
Gorkha District, Nepal	Low or no PW = 6 spp.; medium PW = 6 spp.; high PW = 6 spp.	Timsina et al. (2011)
Nuwakot District, Nepal	Low or no PW = 5 spp.; medium PW = 6 spp.; high PW = 5 spp.	
Kathmandu District, Nepal	Low or no PW = 5 spp.; medium PW = 5 spp.; high PW = 5 spp.	
India	Reduced abundance of herbs and replaced some grass species	Huy and Seghal (2004)
Chaukiya, India	No PW = 27 spp.; PW present = 12 spp.	Upadhyay et al. (2013)
Khetsarai, India	No PW = 16 spp.; PW present = 9 spp.	
Kuthan, India	No PW = 22 spp.; PW present = 11 spp.	
Sahgang, India	No PW = 13 spp.; PW present = 8 spp.	
Sipah, India	No PW = 12 spp.; PW present = 8 spp.	
Lower Himalayas, India	No PW = 25 spp.; PW present = 12 spp.	Kohli et al. (2004)
North Wello Zone, Ethiopia	Low PW = 48 spp.; medium PW = 46 spp.; high PW = 37 spp.; grass species density decreased significantly as invasion levels increased	Nigatu et al. (2010)
Jijiga Zone, Ethiopia	As PW levels increased, the percentage cover of both grasses and plant taxa decreased	Ayele et al. (2013)
Ethiopia	Similar species diversity between invaded and uninvaded sites but species were encountered more frequently and were also more abundant in uninvaded compared with invaded quadrats	Terfa (2009)
Fentale District, Ethiopia	94%, 91% and 78% variation in density of broadleaved plants, grass and sedges, respectively, were accounted for by the density of PW	Urga et al. (2008)
Gamo Gofa, Ethiopia	Reduction in plant diversity as dominance of PW increased	Gebrehiwot and Berhanu (2015)
Queensland, Australia	Plant community diversity was highest in uninvaded areas	Nguyen et al. (2010)
Queensland, Australia	PW reduced overall species diversity	Belgeri et al. (2014)

(*Senna obtusifolia* (L.) Irwin & Barneby; *Fabaceae*), and others can coexist and in some situations out-compete parthenium weed (Wahab, 2005; Yaduraju et al., 2005; Wegari, 2008). In contrast to the findings of Timsina et al. (2011), plant species diversity was consistently lower in areas invaded by parthenium weed in Juanpur District, India, with some uninvaded areas having twice the number of plant species (Upadhyay et al., 2013). This is supported by Kohli et al. (2004), who found 12 plant species in invaded compared with 25 in uninvaded

areas in the lower Himalayas, India (Table 5.2).

In Ethiopia, data on species diversity was collected from five areas, each having sites with low, medium and high levels of parthenium weed cover (Nigatu et al., 2010). A total of 72 species of plants, many exotic, consisting of 23 grass and 49 other species, including parthenium weed, were found across all five areas (Nigatu et al., 2010). There was a reduction in plant species diversity and community evenness as parthenium weed densities increased (Fig. 5.1).

Grass species density decreased significantly as parthenium weed densities increased, but there was no such trend for other taxa (Nigatu *et al.*, 2010). Thirteen species of grass were found across all three invasion levels with andropogon (*Andropogon abyssinicus* Fresen.; *Poaceae*), crab grass (*Digitaria ternata* (A. Rich.) Stapf.; *Poaceae*) and common thatching grass (*Hyparrhenia hirta* (L.) Stapf; *Poaceae*) only present under the low and medium invasion levels. The percentage above-ground dry biomass produced by the grass species was *c.*71%, 41% and 10% in the low, medium and high parthenium weed invasion levels, respectively, while the percentage above-ground biomass that was produced by the other species, mainly dicots, was *c.*22%, 17% and 12% in the respective invasion levels (Nigatu *et al.*, 2010).

Another study, conducted by Ayele *et al.* (2013) in south-eastern Ethiopia, found similar impacts of parthenium weed (Table 5.2). Fifty-six plant species were recorded across all sites, dominated by species in the families *Poaceae* and *Asteraceae*. Grasses dominated in areas where parthenium weed was not present, accounting for 63% of total cover. As the parthenium weed invasion levels increased, the percentage cover of both grasses and other species decreased (Ayele *et al.*, 2013). The dry biomass of the grass species was significantly higher in the uninfested areas compared with areas where there was no parthenium weed. Species richness and evenness indices of the above-ground vegetation were significantly lower at those sites with high levels of parthenium weed.

In the Fentale District in the Central Rift Valley, Ethiopia, 111 herbaceous plant species were recorded in a $1\,km^2$ area that had been invaded by parthenium weed (Urga *et al.*, 2008; Table 5.2). Of the 100 quadrats sampled in this area, 75 had severe invasions of parthenium weed (>20 plants/m²). The results of a polynomial regression indicated that *c.*94%, 91% and 78% variation in density of broadleaved plants, grass and sedges, respectively, accounted for the density of parthenium weed. In other words, the density of parthenium weeds significantly influenced the density of broadleaved plants,

grasses and sedges (Urga *et al.*, 2008). In a study undertaken in Awash National Park, Ethiopia, 86 species of plants were found in parthenium weed-infested and uninfested quadrats (Terfa, 2009). Of the species recorded, 14 and 13 were limited to non-invaded and invaded quadrats, respectively, while 59 species were common to both (Terfa, 2009), an indication that parthenium weed was not only having an impact on species diversity but also on species composition. Species were encountered more frequently and were also more abundant in uninvaded compared with invaded quadrats.

The impact of parthenium weed on above-ground vegetation and the soil seed bank has been assessed in Kilcoy, Queensland, Australia (Nguyen *et al.*, 2010). Pastoral areas infested with high (16 plants/m²), low (two plants/m²) and no parthenium weed were surveyed using $1\,m^2$ quadrats. A total of 65 species of plants were found in the above-ground and soil seed bank communities (19 grasses, nine flowering plants, and 37 species from 22 other families). The most dominant species in the above-ground plant community across all invasion levels were three grasses: Queensland blue couch (*Digitaria didactyla* Willd.; *Poaceae*), Bermuda grass (*Cynodon dactylon* L. (Pers.); *Poaceae*) and pitted bluegrass (*Bothriochloa decipiens* L.; *Poaceae*). The uninvaded transects had the highest above-ground plant diversity, those areas where parthenium weed was present at low densities had slightly lower diversity and the heavily invaded transects had the lowest diversity. A similar trend was seen in the soil seed bank community, with the species diversity being lowest under the highest density of parthenium weed.

Belgeri *et al.* (2014) assessed shifts in plant community composition following management of parthenium weed in Australian grasslands. A baseline plant community survey, prior to management, showed that the above-ground community was dominated by parthenium weed, stoloniferous grasses and a high frequency of species from the *Malvaceae*, *Chenopodiaceae* and *Amaranthaceae* families. In heavily invaded areas, parthenium weed abundance and biomass was found to negatively correlate with species diversity and native species abundance. Queensland blue couch was present in high abundance when parthenium weed was not, with these two species contributing most to the dissimilarity seen between areas. The application of selective broadleaved weed herbicides significantly reduced parthenium weed biomass under ungrazed conditions, but this management did not result in an immediate increase in species diversity during the 2-year study period. In the above-ground community, parthenium weed was partly replaced by the introduced Bermuda grass 1 year after management began, increasing the above-ground forage biomass production, while Queensland blue couch replaced parthenium weed in the below-ground community. This improvement in forage availability continued to strengthen

over the time of the study, resulting in a total increase of 80% after 2 years in the ungrazed treatment, demonstrating the stress that grazing was imposing upon this grassland-based agroecosystem and showing that it is necessary to remove grazing to obtain the best results from the chemical management approach (Belgeri et al., 2014).

There is considerable evidence to suggest that parthenium weed displaces a number of native and exotic plant species, and impacts on the abundance of others (Fig. 5.2). This is largely as a result of direct competition for resources or indirectly through the release of allelochemicals into the soil which prevent the germination and growth of other plant species. However, allelopathic pollen may also impact on fertilization in other plant species. The transfer of hetero-specific pollen from one plant species to another, resulting in reduced fertilization,

Fig. 5.2. Parthenium weed displaces native and exotic plant species, and impacts on the abundance of others. This is largely as a result of direct competition for resources and through the release of allelo-chemicals into the soil which prevent the germination and growth of other plant species. Photograph taken in Kruger National Park South Africa. (L. Strathie.)

has been considered as a form of allelopathy between plant species (Levin and Anderson, 1970; Wissel, 1977; Waser, 1978). For example, there is a significant decrease in northern bush honeysuckle (*Diervilla lonicera* Mill; *Caprifoliaceae*) ovule development in the presence of even a small amount of hawkweed (*Hieracium floribundum* Wimmer and Grab; *Asteraceae*) pollen (Thomson *et al.*, 1982). Allelopathic pollen from Timothygrass (*Phleum pratense* L.; *Poaceae*) reduced seed set in common couch (*Elymus repens* (L.) Gould; *Poaceae*) in the field (Murphy and Aarssen, 1995), while mixed pollen from the invasive purple loosestrife (*Lythrum salicaria* L.; *Lythraceae*) reduced seed set in the native swamp-loosestrife (*Decodon verticillatus* (L.) Elliott; *Lythraceae*) by 33% (Da Silva and Sargent, 2011). Pollen from parthenium weed has shown to inhibit seed set in a number of plant species when deposited on their stigmas (Sukada and Jayachandra, 1980), and reduced or prevented spore germination in a number of pathogens (Ganeshan and Jayachandra, 1990). The pollen of parthenium weed also inhibited the pollen germination and tube growth of other species in artificial cultures of mixed pollen, and could reduce the chlorophyll content of foliage on which it is deposited (Sukada and Jayachandra, 1980). As such, parthenium weed may be reducing seed set in a number of native species with potentially significant long-term impacts. However, this has not been studied in any detail under field conditions, but based on what we do know, it could be assumed that the presence of parthenium weed will alter native species recruitment and vegetation structure and composition wherever it is present.

5.3.2 Fauna

Insects

Many introduced organisms arrive in new localities without their native associates (Elton, 1958; Strong *et al.*, 1984), and persist and proliferate, not because they necessarily possess a suite of extraordinary traits but because they don't have natural enemies

and competitors in their new environment (Mack *et al.*, 2000). The 'enemy-release' hypothesis argues that many introduced species become invasive as a result of the absence or reduction of natural enemies in the introduced environment (Keane and Crawley, 2002; Joshi and Vrieling, 2005).

In an analysis of data on 15 plant species whose herbivore diversity was compared in the native and introduced range, it was found that plants have significantly higher numbers of phytophagous insect species in their native than in their introduced ranges (Liu and Stiling, 2006). For example, 156 species of phytophagous insects were recorded on giant sensitive plant in its native range in Mexico, while only 114 were recorded in its introduced range in northern Australia (Wilson *et al.*, 1990). Similar figures were recorded for other species (Liu and Stiling, 2006), with 56 herbivorous insects recorded on common thistle (*Cirsium vulgare* (Savi) Ten.; *Asteraceae*) in Europe and only 19 being recorded in its introduced range in the USA and Australia (Zwolfer, 1965; Briese, 1989). More than 260 phytophagous species were collected on parthenium weed in its native range in Mexico, with 144 of these feeding on the plant at some or other stage in their life cycle (McClay *et al.*, 1995). Although no studies have been undertaken on the natural enemies associated with parthenium weed in its introduced range, we can assume, based on evidence from other studies, and given the high invasion levels achieved by the weed, that few natural enemies are also associated with parthenium weed in areas where it has been introduced.

There is further evidence that native plant species host more phytophagous insects than introduced species. Darke and Tallamy (2014) found significantly more butterfly and moth larvae in suburban gardens dominated by native plants as compared with gardens of predominantly introduced plants. Perre *et al.* (2011) also found that exotic species supported a small subset of the herbivore assemblage found on native plants. In a study in forests in the Azores, species richness of plants and insects declined in invaded areas with a dramatic decrease in insect biomass (Heleno et al., 2008). A meta-analysis of 56 studies found that invaded habitats had 29% fewer arthropods and 17% lower diversity as compared with uninvaded habitats (Hengstum et al., 2014). Riparian sites in the UK, invaded by Japanese knotweed, supported fewer plant species and had lower abundance, morphospecific richness and biomass of invertebrates, as compared with native grassland or shrub-dominated plots (Gerber et al., 2008). Also in the UK, the foliage community of plots invaded by Himalayan balsam had 64% and 58% fewer beetle (Coleoptera) and true bug (Heteroptera) species than uninvaded plots, respectively (Tanner et al., 2013). Many of these impacts on invertebrates are not necessarily as a result of invasive plants being unsuitable as a food source for a host of native organisms, but rather because they bring about changes in the composition and structure of the vegetation.

Parthenin, a sesquiterpene lactone found in parthenium weed, is also known to be toxic to termites, cockroaches (Tilak, 1977) and migratory grasshoppers (Picman et al., 1981; Fagoonee, 1983). Plant extracts of parthenium weed also affected insect growth regulatory activity in the cotton stainer (Dysdercus angulatus F.; Pyrrhocoridae) (Kareem, 1984) and Oriental leafworm moth (Spodoptera litura F.; Noctuidae) larvae (Ranjandran and Gopalan, 1979; Balasubramanian, 1982), and had toxic effects on the cabbage leaf webber (Crocidolomia binolalis Zell; Crambidae), pulse beetle (Callosobruchus maculatus F.: Chrysomelidae) (Bhaduri et al., 1985), mites (Gupta, 1968) and mustard aphid (Lipaphis erysimi Kaltenbach; Aphididae) (Sohal et al., 2002). The fact that it is unpalatable, and even toxic, to these phytophagous insects indicates that parthenium weed is unlikely to support many herbivorous insects in its introduced range. Heavily invaded areas with reduced plant diversity will therefore, in all likelihood, have a depauperate insect fauna.

It is also widely acknowledged that many of the changes in abundance and richness of native species are probably also a result of changes in the structure of the invaded vegetation in terms of height, cover and density (Samways and Moore, 1991; Jones et al., 1997). The displacement of native plant species, especially grasses, by parthenium weed has significant impacts on the structure of plant communities, which could have knock-on impacts on a myriad of other organisms. Invasive plants generally establish thick monocultures that not only displace native plant species and alter their distribution but also decrease associated animal diversity as a result (Cohen et al., 2012).

Amphibians and reptiles

A reduction in insect quality and quantity can have significant consequences for those taxa at higher trophic levels such as predatory invertebrates, amphibians, reptiles, birds and mammals (Scheiman et al., 2003; Flanders, 2006; Capinera, 2010). Maerz et al. (2005) found that green frogs (Lithobates clamitans Latreille; Ranidae) gained less mass along transects from areas of native vegetation into areas invaded by Japanese knotweed given a decrease in insect abundance as a result of the plant invasion. The black legless lizard (Anniella pulchra nigra Gray; Anniellidae) is now threatened, partly due to a decrease in its natural prey base, mainly insects, as a result of the invasion by the invasive succulent pig face (Carpobrotus edulis (L.) N.E. Br; Aizoaceae) on sand dunes in the USA (Rutherford and Rorabaugh, 1995). Amphibian species richness and evenness were lower in forest plots with high densities of the invasive shrub Amur honeysuckle (Lonicera maackii (Rupr.) Maxim; Caprifoliaceae). This was probably because mean daily maximum temperatures were lower in invaded plots, demonstrating how changes in vegetation structure can influence microclimate, resulting in knock-on impacts on other species. Field observations of reptiles in a habitat invaded by rubber vine (Cryptostegia grandiflora Roxb. Ex R.Br; Asclepiadaceae) in Australia recorded only a single lizard as compared with 131 lizards in nearby native vegetation (Valentine et al., 2006). Fewer reptiles were also recorded in areas invaded by giant sensitive plant (Braithwaite et al., 1989), while fewer insects, reptiles and birds were found in areas dominated by Athel pine (*Tamarix aphylla* (L.) Karst.; *Tamaricaceae*) in Australia (Griffin *et al.*, 1989).

Although we have not yet found similar studies for parthenium weed, based on the demonstrated shifts in vegetation structure and composition that these other invasive species cause, we can assume that its presence will also have a significant negative impact on native amphibian and reptile populations.

Birds

Other than amphibians and reptiles, which are largely dependent on insects, approximately 34% and 36% of mammal and bird families, respectively, also feed on insects (Fleming, 1991 cited in Capinera, 2010). About 80% of the birds breeding in Central Europe feed on insects, at one or other stage in life, while 61% and 28% of North American birds are primarily insectivorous and partially insectivorous, respectively (Capinera, 2010). Even the chicks of granivorous, nectarivorous, frugivorous, fungivorous and folivorous birds are often fed mostly insects, with food being a major limitation to breeding success as these rapidly growing birds need a diet that is rich in protein and fat (Nagy and Holmes, 2005; Granbom and Smith, 2006; Capinera, 2010). A number of studies show a positive correlation between average clutch size and/or fledgling success with insect abundance (Strehl and White, 1986; Duguay *et al.*, 2000).

In North America, the density of a ground-foraging bird was lower on sites dominated by leafy spurge (*Euphorbia esula* L.; *Euphorbiaceae*) as a result of a reduction in the amount of seeds and insect resources in invaded areas (Scheiman *et al.*, 2003). In western North America, spotted knapweed (*Centaurea maculosa* Lam.; *Asteraceae*) invasions had a negative effect on the chipping sparrow (*Spizella passerina* Bechstein; *Emberizidae*), a migratory songbird which feeds on seeds and insects. The initiation of nests was delayed at knapweed versus native sites, probably as a result of lower food availability (Ortega *et al.*, 2006). There were also

significantly fewer breeding adults and reduced reproductive success in invaded versus uninvaded areas. Grasshoppers, an important food source for the sparrows, were much reduced in invaded areas (Ortega *et al.*, 2006). Lloyd and Martin (2005) compared the reproductive success of the chestnut-collared longspurs (*Calcarius ornatus* Townsend; *Calcariidae*) in patches of native prairie and in monocultures of the invasive crested wheatgrass (*Agropyron cristatum* (L.) Gaertn.; *Poaceae*), in eastern Montana, USA. In the exotic habitat, the odds of a nest surviving in a given day were 17% lower; nestlings gained mass at a slower rate, fledged at a smaller mass and took 1 day longer to fledge (Lloyd and Martin, 2005).

In Africa, various bird guilds were also negatively impacted by the invasive shrub/tree honey mesquite (*Prosopis glandulosa* Torr.; *Fabaceae*). Dense *Prosopis* species woodlands had less herbaceous understorey cover than the uninvaded *Acacia* woodlands, while the latter was also botanically more diverse (Dean *et al.*, 2002). Bird communities in native woodlands were also consistently more species rich and more diverse with fewer frugivores and insectivores recorded in areas dominated by *Prosopis* species (Dean *et al.*, 2002). Shanungu (2009) only recorded 24 bird species in areas dominated by the giant sensitive plant in Lochinvar National Park, Zambia, compared with 46 species in uninvaded areas. Negative impacts on birds may be as a result of changes in vegetation structure or decreases in the amount of insects associated with the introduced species.

Archer's lark (*Heteromirafra archeri* Clarke; *Alaudidae*), an Ethiopian endemic confined to the grassy plains of Wajaale in north-east Ethiopia, is thought to have been driven to extinction as a result of parthenium weed invasions together with the expansion of large-scale agricultural schemes (Spottiswoode *et al.*, 2013). Insects are an important food source for larks, although they also feed on seeds, grasses, leaves, fruits and flowers. Studies in rangelands in Colorado, USA, have shown that various lark and closely related species consume 64–78 grasshoppers/day plus hundreds of other insects. Grasshoppers are known to be particularly vulnerable to changes in grassland management (van Wingerden *et al.*, 1992), and decreases in several bird species in Europe that rely on large insects have been linked to decreasing grasshopper populations (van Nieuwenhuyse *et al.*, 1999). Both a reduction in available food in the form of insects and native plant species, especially grasses, and changes in vegetation structure have probably contributed to the loss of Archer's lark. Loss of grasslands to commercial tree plantations, invasive species and land-use intensification have also been implicated in the decline of Rudd's lark (*Heteromirafra ruddi* Grant; *Alaudidae*) in South Africa (Hockey *et al.*, 1988; Barnes, 2000). The loss and modification of grasslands has also been implicated in the decline of Botha's lark (*Spizocorys fringillaris* Sundevall; *Alaudidae*) in South Africa (Barnes, 2000). At a global level, the decline of grassland passerine populations, such as Archer's lark, can largely be attributed to factors reducing the quality of grassland habitats such as presence of invasive plants (Samson and Knopf, 1994; Vickery *et al.*, 2000; Scheiman *et al.*, 2003). Although we cannot be sure if the invasion of grasslands by parthenium weed in Ethiopia has been responsible for the demise of Archer's lark, evidence from other studies indicates that a reduction in food availability and changes in the structure and species composition of habitats as a result of plant invasions can affect bird populations.

Given studies showing decreases in grass richness (Nguyen *et al.*, 2010; Upadhyay *et al.*, 2013; Belgeri *et al.*, 2014) when parthenium weed is present at high densities, the species is likely to have a dramatic impact on many birds that are dependent on these grasslands, especially larks. In Queensland, Australia, grasses established in areas cleared of parthenium weed (Belgeri *et al.*, 2014), may lead to the re-establishment of bird populations.

Mammals

Other taxa will also be affected by a reduction in the insect fauna and changes in plant diversity and vegetation structure. A reduction in insect populations may also affect many small mammals such as bats and rodents that feed predominantly on insects (Capinera, 2010). A reduction in available forage will also reduce the number of large herbivores present in grasslands and savannah ecosystems. Studies (see above) have indicated a significant reduction in the biomass of grasses in areas invaded by parthenium weed. A reduction in available forage, as a result of parthenium weed invasions, is also likely to reduce the abundance of large herbivores and this has been demonstrated for other invasive species (Fig. 5.3). In Theodore Roosevelt National Park, North Dakota, USA, researchers found that bison (*Bison bison* L.; *Bovidae*) grazing of leafy spurge-infested grassland habitats averaged 83% less than that for uninvaded sites. Other than the displacement of native forage species, leafy spurge is also toxic and as such wildlife generally do not feed on it (Halaweish *et al.*, 2002). This also seems to be the case for parthenium weed. Sesquiterpene lactones found in parthenium weed exhibit a wide spectrum of biological activities, which include cytotoxic, antitumour, allergenic, antimicrobial, antifeedant, phytotoxic and insecticidal properties (Rodriguez *et al.*, 1976). Buffalo calves fed with parthenium weed during experimental trials exhibited signs of acute toxicity with some dying (Narasimhan *et al.*, 1977; Ahmed *et al.*, 1987). As such, wildlife are likely to avoid areas dominated by parthenium weed until such time as natural forage is significantly reduced.

5.3.3 Water systems

Invasive plants can also reduce the quality of habitats by changing the chemical make-up of the environment (Vilà *et al.*, 2010). Invasive plants growing alongside water bodies can change the chemical nature of the aquatic environment by altering organic matter inputs and C:N ratios (Bottollier-Curtet *et al.*, 2011; McNeish *et al.*, 2012). These chemical changes can have a negative impact on the foraging, growth and survival

Fig. 5.3. A reduction in available forage, as a result of parthenium weed invasions, is reducing the abundance of large herbivores. (L. Strathie.)

of aquatic organisms, including invertebrates (Canhoto and Laranjeira, 2007; Going and Dudley, 2008; Leonard, 2008), larval amphibians (Brown *et al.*, 2006; Leonard 2008; Watling *et al.*, 2011) and fishes (Schultz and Dibble, 2012). For example, aquatic amphibian larvae exposed to the leachates of purple loosestrife and Amur honeysuckle have shown reduced survival compared with exposure to native plants, probably as a result of exposure to phytochemicals (Maerz *et al.*, 2005; Brown *et al.*, 2006; Watling *et al.*, 2011). Tadpole survival is compromised when exposed to Chinese tallow leaf litter (*Triadica sebifera* (L.) Small; *Euphorbiaceae*) (Leonard, 2008; Cotten *et al.*, 2012), an invasive tree which replaces native vegetation, especially in wetland habitats (Cameron and Spencer, 1989; Bruce *et al.*, 1997). There is some evidence that parthenium weed may have similar impacts (Patel *et al.*, 2008; Ashraf *et al.*, 2010).

Aqueous extracts from parthenium weed flowers and leaves have been reported to have lethal impacts on frog tadpoles (Patel *et al.*, 2008). Most of the pollen allergens are water soluble, which enhances their degree of diffusion in mucoid tissues (Stanley and Linskens, 1974). Parthenium weed-seed extracts have shown potential to alter fish metabolism (Ashraf *et al.*, 2010). Sublethal seed extracts stimulate the respiratory muscles in fish, resulting in an increased rate of oxygen uptake, whereas higher concentrations exhibited neurotoxic effects causing paralysis of the respiratory muscles 'manifested by a decreased rate of oxygen uptake, cessation of the respiratory pumps and casualty by asphyxiation' (Ashraf *et al.*, 2010). Indian and Chinese carp (*Cyprinus* sp.) displayed abnormal behaviour such as 'disturbed swimming, rapid opercular movements, loss of balance, incessant gulping of air, darkening of the whole body and fish settling at the bottom motionless' (Ashraf et al., 2010). Leaching of parthenin and other chemicals into the soil and water bodies may therefore have negative impacts on a host of organisms, although this has never been demonstrated under field conditions.

5.3.4 Pollinators

The toxicity of parthenium weed may also have an impact on pollinators. Nectar from a number of plant species contain alkaloids which are toxic to bees (Huang, 2010). For example, nectar from species such as azalea (*Rhododendron molle* (Blume) G. Don; *Ericaeae*), black hellebore (*Veratrum nigrum* L.; *Melanthiaceae*), California buckeye (*Aesculus californica* (Spach.) Nutt.; *Sapindaceae*), jimson weed (*Datura stramonium* L.; *Asteraceae*) and tea (*Camellia sinensis* (L.) Kuntze; *Theaceae*) is usually toxic to both adult bees and brood (Huang, 2010). In some cases pollen merely retards brood development. Larvae of orchard mason bee (*Osmia lignaria* Say; Megachilidae) and alfalfa leafcutter bee (*Megachile rotundata* Fabricius; *Megachilidae*), two pollen generalists, failed to develop on pure diets of *Asteroideae* pollen (Levin and Haydak, 1957). Pollen of sunflower (*Helianthus* spp.) was considered to be a poor diet for the honeybee (*Apis* sp.; *Apidae*) and the bumblebee (*Bombus terrestris* L.; Apidae), probably due to its low protein content (Rasmont et al., 2005).

Despite the fact that parthenium weed pollen contains a host of toxic chemicals, these do not appear to impact negatively on pollinators, especially bees. In Maharashtra, India, parthenium weed was found to be the most common source of nectar and pollen for the Indian honey bee (*Apis florea* F.; *Apidae*) (Bhusari et al., 2005). Although parthenium weed pollen may not have a negative impact on bees, the abundance of parthenium weed in the landscape may result in native pollinators visiting them more frequently, thus lowering pollination of native plants. For example, Himalayan balsam produces more nectar than the native marsh woundwort (*Stachys palustris* (L.) Walter;

Lamiaceae) and as a result receives more visitations by European bumblebees (Chittka and Schurkens, 2001). It is also acknowledged that diet diversity increases immunocompetence levels. Studies on bees have revealed that polyfloral diets induced higher glucose oxidase activity compared with monofloral diets, including protein-rich diets. In other words, if honeybees have access to pollen from a large number of plant species, they are less susceptible to pathogens/diseases. By reducing plant diversity, parthenium weed is reducing the diversity of pollen sources available to honeybees and as such compromising their immunity and long-term survival.

5.4 Conclusions

Although we have detected some key gaps with regard to determining the impacts of parthenium weed on biodiversity, there is sufficient evidence to conclude that parthenium weed is having a dramatic impact on biodiversity and ecosystem function throughout its introduced range in Africa, Asia and Australia. It has been shown to alter nutrient cycling, reduce plant species diversity and abundance, change vegetation structure, and as such it can be inferred that parthenium weed also impacts on other organisms such as invertebrates, amphibians, reptiles, birds and mammals. It has already invaded more than 50 protected areas in southern and East Africa, China, India, Nepal, Pakistan, Vietnam and Australia, including the Serengeti-Mara ecosystem (Fig. 5.4) in East Africa, where invasions may threaten the biggest annual wildlife migration on earth. A large number of other protected areas in Africa, Europe, Asia and Australia are at risk of invasion based on ecoclimatic models (Figs 5.5–5.7). Invasions will proliferate over time as a result of increased disturbance and climate change. However, many studies have also shown that parthenium weed can be effectively controlled through the implementation of an integrated management strategy, which should always include biological control as a

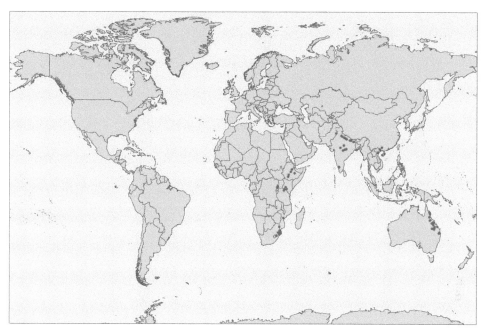

Fig. 5.4. Protected areas in Africa, Asia and Australia that have been invaded by parthenium weed. (Map by T. Beale. Data from WDPA. Available at: http://www.protectedplanet.net (accessed 20 October 2016).)

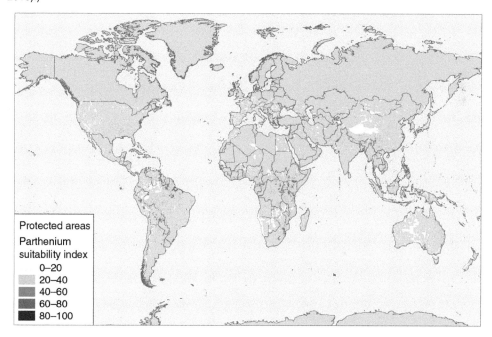

Fig. 5.5. Protected areas around the world that are ecoclimatically suitable for parthenium weed, and as such are at risk of invasion should the species be introduced and establish. Many of these protected areas have already been invaded as indicated in Fig. 5.4. Note that parthenium weed is native to large parts of the Americas. Data for map development is based on a CLIMEX ecoclimatic map developed by McConnachie *et al.* (2011).

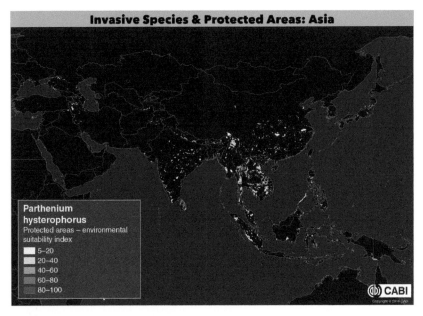

Fig. 5.6. A large number of other protected areas in Asia are at risk of invasion based on ecoclimatic models. Invasions will proliferate over time as a result of increased disturbance and climate change. Data for map development is based on a CLIMEX ecoclimactic map developed by McConnachie *et al*. (2011).

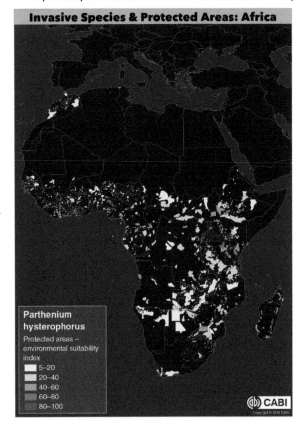

Fig. 5.7. A large number of other protected areas in Africa are at risk of invasion based on ecoclimatic models. Invasions will proliferate over time as a result of increased disturbance and climate change.

critical component. Thus, we firmly believe that its management requires immediate and coordinated efforts both from communities affected and from governments.

Acknowledgements

Thanks to Tim Beale for developing the distribution maps and Andrew McConnachie and Asad Shabbir for reviewing this book chapter. Funding for the writing of this manuscript was provided by the CABI Development Fund, which is supported by contributions from the Australian Centre for International Agricultural Research, Australia, the UK's Department for International Development, and others.

References

Ahmed, N.M., Rao, R.P. and Mahendar, M. (1987) Experimental introduction of acute toxicity in buffalo calves by feeding *Parthenium hysterophorus* L. *Indian Journal of Animal Sciences* 58, 731–734.

Akter, A. and Zuberi, M.I. (2009) Invasive alien species in Northern Bangladesh: Identification, inventory and impacts. *International Journal of Biodiversity and Conservation* 15, 129–134.

Aplet, G.H. (1990) Alteration of earthworm community biomass by the alien *Myrica faya* in Hawai'i. *Oecologia* 82, 414–416.

Ashraf, M., Ayub, M., Sajjad, T., Elahi, N., Ali, I. and Ahmed, Z. (2010) Replacement of rotenone by locally grown herbal extracts. *International Journal of Agriculture and Biology* 12, 77–80.

Ayele, S., Nigatu, L., Tana, T. and Adkins, S.W. (2013) Impact of parthenium weed (*Parthenium hysterophorus* L.) on the above-ground and soil seed bank communities of rangelands in Southeast Ethiopia. *International Research Journal of Agricultural Science and Soil Science* 3, 262–274.

Bajwa, A.A., Chauhan, B.S., Farooq, M., Shabbir, A. and Adkins, S.W. (2016) What do we really know about alien plant invasion? A review of the invasion mechanism of one of the world's worst weeds. *Planta* 244, 39–57.

Balasubramanian, M. (1982) *Plant Species Reportedly Possessing Pest Control Properties: an EWC/UH Database.* University of Hawaii, Hawaii.

Barnes, K.N. (2000) *The Eskom Red Data Book of Birds of South Africa, Lesotho and Swaziland.* Bird Life South Africa, Johannesburg, South Africa.

Batish, D.R., Singh, H.P., Pandher, J.K., Arora, V. and Kohli, R.K. (2002a) Phytotoxic effect of *Parthenium* residues on the selected soil properties and growth of chickpea and radish. *Weed Biology and Management* 2, 73–78.

Batish, D.R., Singh, H.P., Saxena, D.B. and Kohli, R.K. (2002b) Weed suppressing ability of parthenin – a sesquiterpene lactone from *Parthenium hysterophorus. New Zealand Plant Protection* 55, 218–221.

Batish, D.R., Singh, H.P., Pandher, J.K. and Kholi, R.K. (2005) Allelopathic interference of *Parthenium hysterophorus* residues in soil. *Allelopathy Journal* 15, 267–274.

Belgeri, A., Navie, S.C., Vivian-Smith, G. and Adkins, S.W. (2014) Early recovery signs of an Australian grassland following the management of *Parthenium hysterophorus* L. *Flora* 209, 587–596.

Belnap, J. and Phillips, S. (2001) Soil biota in an ungrazed grassland: response to annual grass (*Bromus tectorum*) invasion. *Ecological Applications* 11, 1261–1275.

Belz, R.G., Reinhardt, C.F., Foxcroft, L.C. and Hurle, K. (2007) Residue allelopathy in *Parthenium hysterophorus* L. – does parthenin play a leading role? *Crop Protection* 26, 237–245.

Bever, J.D. (2003) Soil community feedback and the coexistence of competitors: Conceptual frameworks and empirical tests. *New Phytology* 157, 465–473.

Bhaduri, N., Ram, S. and Patil, B.D. (1985) Evaluation of some plant extracts as protectants against the pulse beetle, *Callosobruchus maculates* Fabr, infesting cowpea seeds. *Journal of Entomological Research* 9, 183–187.

Bhowmik, P.C., Sakar, D. and Yaduraju, N.T. (2007) The status of *Parthenium hysterophorus* L. residues on three *Brassica* species. *Weed Biology and Management* 14, 1–17.

Bhusari, V.N., Mate, M.D. and Makde, H.K. (2005) Pollen of *Apis* honey from Maharashtra. *Grana* 44, 216–224.

Bock, C.E., Bock, J.H., Jepson, K.L. and Ortega, J.C. (1986) Ecological effects of planting African lovegrasses in Arizona. *National Geographic Research* 2, 456–463.

Bosch, J.M. (1979) Treatment effects on annual and dry period streamflow at Cathedral peak. *South African Forestry Journal* 108, 29–38.

Bottollier-Curtet, M., Charcosset, J.Y., Planty-Tabacchi, A.M. and Tabacchi, E. (2011) Degradation of native and exotic riparian plant leaf litter in a floodplain pond. *Freshwater Biology* 56, 1798–1810.

Braithwaite, R.W., Lonsdale, W.M. and Estbergs, J.A. (1989) Alien vegetation and native biota in tropical Australia: The impact of *Mimosa pigra*. *Biological Conservation* 48, 189–210.

Briese, D.T. (1989) Natural enemies of carduine thistles in New South Wales. *Journal of the Australian Entomological Society* 28, 125–134.

Brooks, M.L., D'Antonio, C.M., Richardson, D.M., Grace, J.B., Keeley, J.E., *et al.* (2004) Effects of invasive alien plants on fire regimes. *BioScience* 54, 677–688.

Brown, C.J., Blossey, B., Maerz, J.C. and Joule, S.J. (2006) Invasive plant and experimental venue affect tadpole performance. *Biological Invasions* 8, 327–338.

Bruce, K.A., Cameron, G.N., Harcombe, P.A. and Jubinsky, G. (1997) Introduction, impact on native habitats, and management of a woody invader, the Chinese tallow tree *Sapium sebiferum* (L.) Roxb. *Natural Areas Journal* 17, 255–260.

Butler, D.W. and Fairfax, R.J. (2003) Buffel grass and fire in a gidgee and Brigalow woodland: A case study from Central Queensland. *Ecological Management and Restoration* 4, 120–125.

Cameron, G.N. and Spencer, S.R. (1989) Rapid leaf decay and nutrient release in Chinese tallow forest. *Oecologia* 80, 222–228.

Canhoto, C. and Laranjeira, C. (2007) Leachates of *Eucalyptus globulus* in intermittent streams affect water parameters and invertebrates. *International Review of Hydrobiology* 92, 173–182.

Capinera, L.J. (2010) *Arthropods and their Relationships with Wild Vertebrate Animals*. Wiley Blackwell, Hoboken, New Jersey.

Carlsson, N.O.L., Brönmark, C. and Hansson, L.-A. (2004) Invading herbivory: The golden apple snail alters ecosystem functioning in Asian wetlands. *Ecology* 85, 1575–1580.

Chacon, N., Herrera, H., Flores, S., Gonzalez, J.A. and Nassar, J.M. (2008) Chemical, physical and biochemical soil properties and plant roots as affected by native and exotic plants in Neotropical arid zones. *Biology and Fertility of Soils* 45, 321–328.

Chen, C.Y., Hathaway, K.M., Thompson, D.G. and Folt, C.L. (2008) Multiple stressor effects of herbicide, pH, and food on wetland zooplankton and a larval amphibian. *Ecotoxicology and Environmental Safety* 71, 209–218.

Chippendale, J.F. and Panetta, F.D. (1994) The cost of parthenium weed to the Queensland cattle industry. *Plant Protection Quartely* 9, 73–76.

Chittka, L. and Schurkens, S. (2001) Successful invasion of a floral market – an exotic Asian plant has moved in on Europe's river-banks by bribing pollinators. *Nature* 411, 653.

Christian, J.M. and Wilson, S.D. (1999) Long-term ecosystem impacts of an introduced grass in the northern Great Plains. *Ecology* 80, 2397–2407.

Cibichakravarthy, B., Preetha, R., Sundaram, S.P., Kumar, K. and Balachandar, D. (2011) Diazotrophic diversity in the rhizosphere of two exotic weed plants, *Prosopis juliflora* and *Parthenium hysterophorus*. *World Journal of Microbiology and Biotechnology* 28, 605–613.

Cohen, J.S., Maerz, J.C. and Blossey, B. (2012) Traits, not origin explain impacts of plants on larval amphibians. *Ecological Applications* 22, 218–228.

Collins, A.R. (2005) Implications of plant diversity and soil chemical properties for Congrass (*Imperata cylindrica*) invasion in northwest Florida. MSc thesis, University of Florida, Gainsville, Florida.

Cotten, T.B., Kwiatkowski, M.A., Saenz, D. and Collyer, M. (2012) Effects of an invasive plant, Chinese tallow (*Triadica sebifera*), on development and survival of Anuran larvae. *Journal of Herpetology* 46, 186–193.

D'Antonio, C.M. (1993) Mechanisms controlling invasion of coastal plant communities by the alien succulent *Carpobrotus edulis*. *Ecology* 74, 83–95.

Darke, R. and Tallamy, D. (2014) *The Living Landscape. Designing for Beauty and Biodiversity in the Home Garden*. Timber Press, Portland, Oregon.

Da Silva, E.M. and Sargent, R.D. (2011) The effect of invasive *Lythrum salicaria* pollen deposition on seed set in the native species *Decodon verticillatus*. *Botany* 89, 141–146.

Dean, J.R.W., Anderson, D.M., Milton, S.J. and Anderson, T.A. (2002) Avian assemblages in *Acacia* and *Prosopis* drainage line woodland in the Kalahari, South Africa. *Journal of Arid Environments* 51, 1–19.

Duguay, J.P., Wood, P.B. and Miller, G.W. (2000) Effects of timber harvests on invertebrate biomass and avian nest success. *Wildlife Society Bulletin* 28, 1123–1131.

Ehrenfeld, J.G. (2003) Plant–soil interactions. In: Levin S. (ed.) *Encyclopedia of Biodiversity*. Academic Press, San Diego, California, pp. 689–709.

Ehrenfeld, J.G. and Scott, N. (2001) Invasive species and the soil: Effects on organisms and ecosystem processes. *Ecological Applications* 11, 1259–1260.

Elton, C.S. (1958) *The Ecology of Invasions by Animals and Plants*. Chapman & Hall, London.

Evans, H.C. (1997) *Parthenium hysterophorus*: A review of its weed status and the possibilities for biological control. *Biocontrol* 18, 89–98.

Evans, R.D., Rimer, R., Sperry, L. and Belnap, J. (2001) Exotic plant invasion alters nitrogen dynamics in an arid grassland. *Ecological Applications* 11, 1301–1310.

Fagoonee, I. (1983) Natural pesticides from neem tree (*Azadirachta indica* A. Juss) and other tropical plants. In: Schmutterer, H. and Ascher, K.R.S. (eds) *Proceedings of the II International Neem Conference*. German Agency for Technical Cooperation (GTZ), Eschborn, Germany, pp. 211–223.

Flanders, A.A., Kuvlesky Jr, W.P., Ruthven III, D.C., Zaiglin, R.E., Bingham, R.L., *et al.* (2006) Effects of invasive exotic grasses on south Texas rangeland breeding birds. *The Auk* 123, 171–182.

Gabbard, B.L. and Fowler, N.L. (2007) Wide ecological amplitude of a diversity reducing invasive grass. *Biological Invasions* 9, 149–160.

Ganeshan, G. and Jayachandra, J. (1990) Antifungal activity of parthenin. *Indian Phytopathology* 46, 193–194.

Garland, J. (1996) Patterns of potential C source utilization by rhizosphere communities. *Soil Biology and Biochemistry* 28, 23–30.

Gebrehiwot, N. and Berhanu, L. (2015) Impact of *Parthenium* on species diversity in Gamo Gafo, Ethiopia. *Journal of Agricultural Science* 5, 226–231.

Gerber, E., Krebs, C., Murrell, C., Moretti, M., Rocklin, R. and Schaffner, U. (2008) Exotic invasive knotweeds (*Fallopia* spp.) negatively affect native plant and invertebrate assemblages in European riparian habitats. *Biological Conservation* 141, 646–654.

Gill, R.A. and Burke, I.C. (1999) Ecosystem consequences of plant life form changes at three sites in the semi-arid United States. *Oecologia* 121, 551–563.

Going, B.M. and Dudley, T.L. (2008) Invasive riparian plant litter alters aquatic insect growth. *Biological Invasions* 10, 1041–1051.

Granbom, M. and Smith, H.G. (2006) Food limitation during breeding in a heterogeneous landscape. *The Auk* 123, 97–107.

Grayston, S.J. and Campbell, C.D. (1998) Functional biodiversity of microbial communities in the rhizosphere of hybrid larch (*Larix eurolepis*) and Sitka spruce (*Piceas itchensis*). *Tree Physiology* 16, 1031–1038.

Griffin, G.F., Stafford-Smith, D.M., Morton, S.R., Allan, G.E. and Masters, K.A. (1989) Status and implications of tamarisk (*Tamarisk aphylla*) on the Finke River, Northern Territory, Australia.

Journal of Environmental Management 29, 297–315.

Gupta, R.K. (1968) Studies of the curative effects of *Cedrus deodara* oil against sarcoptic mange in buffalo calves. *Indian Journal of Veterinary Science* 38, 203–209.

Halaweish, F.T., Kronberg, S., Hubert, M.B. and Rice, A.J. (2002) Toxic and aversive diterpenes of *Euphorbia esula*. *Journal of Chemical Ecology* 28, 1599–1611.

Hejda, M., Pysek, P. and Jarosik, V. (2009) Impact of invasive plants on the species richness, diversity and composition of invaded communities. *Journal of Ecology* 97, 393–403.

Heleno, R.H., Ceia, R.S., Ramos, J.A. and Memmott, J. (2008) Effects of alien plants on insect abundance and biomass: A food-web approach. *Conservation Biology* 23, 410–419.

Hengstum, T., Hooftman, D.A., Oostermeijer, J.G.B. and Tienderen, P.H. (2014) Impact of plant invasions on local arthropod communities: A meta-analysis. *Journal of Ecology* 102, 4–11.

Hickman, J.E., Wub, S., Mickley, L.J. and Lerdauc, M.T. (2010) Kudzu (*Pueraria montana*) invasion doubles emissions of nitric oxide and increases ozone pollution. *Proceedings of the National Academy of Sciences of the United States of America* 107, 10115–10119.

Hockey, P.A.R., Allan, D.G., Rabelo, A.G. and Dean, W.R.J. (1988) The distribution, habitat requirements and conservation status of Rudd's lark *Heteromirafra ruddi* in South Africa. *Biological Conservation* 45, 255–266.

Huang, Z. (2010) Honey Bee Nutrition. Available at: http://www.beeccdcap.uga.edu/documents/caparticle10.html (accessed 20 April 2016).

Hulme, P.E. (2006) Beyond control: Wider implications for the management of biological invasions. *Journal of Applied Ecology* 43, 835–847.

Hunt, H.W., Coleman, D.C., Ingham, E.R., Ingham, R.E., Elliott, E.T., Moore, J.C., Rose, S.L., Reid, C.P.P. and Morley, C.R. (1987) The detrital food web in a shortgrass prairie. *Biology and Fertility of Soils* 3, 57–68.

Huy, L.Q. and Seghal, R.N. (2004) Invasion of *Parthenium hysterophorus* L. in Chir-Pine forests and its allelopathic effects. In: *Abstracts of an International Workshop on Protocols and Methodologies in Allelopathy*. CSK HP Agricultural University, Palampur, India. International Allelopathy Society.

Jeyalakshmi, C., Doraisamy, S. and Valluvaparidasan, V. (2011) Occurrence of soil microbes under parthenium weed in Tamil Nadu. *Indian Journal of Weed Science* 43, 222–223.

Jones, C.G., Lawton, J.H. and Shachak, M. (1997) Positive and negative effects of organisms as physical ecosystem engineers. *Ecology* 78, 1946–1957.

Joshi, J. and Vrieling, K. (2005) The enemy release and EICA hypothesis revisited: Incorporating the fundamental difference between specialist and generalist herbivores. *Ecology Letters* 8, 704–714.

Joshi, S. (2005) Reproductive efficiency and biomass allocation of invasive weed, *Parthenium hysterophorus* L. MSc thesis, Tribhuvan University, Kathmandu, Nepal.

Kareem, A.A. (1984) *Progress in the Use of Neem and Other Plant Species in Pest Control in India.* Research Planning Works on Botanical Pest Control Project, International Rice Research Institute (IRRI), Los Banos, Philippines.

Karki, D. (2009) Ecological and socio-economic impact of *Parthenium hysterophorus* L. invasion in two urban cities of South-Central Nepal. MSc thesis, Tribhuvan University, Kathmandu, Nepal.

Keane, M.R. and Crawley, J.M. (2002) Exotic plant invasions and the enemy release hypothesis. *Trends in Ecology and Evolution* 17, 164–169.

Khan, H., Marwat, K.B., Hassan, G. and Muhammad, A.K. (2013) Socio-economic impacts of parthenium (*Parthenium hysterophorus* L.) in Peshawar Valley, Pakistan. *Pakistani Journal of Weed Science Research* 19, 275–293.

Kimbro, D.L., Grosholz, E.D., Baukus, A., Nesbitt, N., Travis, N., Attoe, S. and Coleman-Hulbert, C. (2009) Invasive species cause large-scale loss of native California oysters by disrupting trophic cascades. *Oecologia* 160, 563–575.

Kohli, K.R., Dogra, K.S., Daizy, R.B. and Singh, R.B. (2004) Impact of invasive plants on the structure and composition of natural vegetation of North-western India, Himalayas. *Weed Technology* 18, 1296–1300.

Kourtev, P., Ehrenfeld, J.G. and Huang, W. (1998) Effects of exotic plant species on soil properties in hardwood forests of New Jersey. *Water, Air and Soil Pollutants* 105, 493–501.

Kourtev, P., Ehrenfeld, J.G. and Huang, W. (1999) Differences in earth-worm densities and nitrogen dynamics under exotic and native plant species. *Biological Invasions* 1, 237–245.

Kourtev, P.S., Ehrenfeld, J.G. and Haggblom, M. (2003) Experimental analysis of the effect of exotic and native plant species on the structure and function of soil microbial communities. *Soil Biology and Biochemistry* 35, 895–905.

Koutika, L.S., Vanderhoeven, S., Chapuis-Lardy, L., Dassonville, N. and Meerts, P. (2007) Assessment of changes in soil organic matter after invasion by exotic plant species. *Biology and Fertility of Soils* 44, 331–341.

Kruse, M., Strandberg, M. and Strandberg, B. (2000) *Ecological Effects of Allelopathic Plants. A Review.* Rep. No. 315. Department of Terrestrial Ecology, Silkeborg, Denmark.

Kumar, S. and Rohatgi, N. (1999) The role of invasive weeds in changing floristic diversity. *Annals of Forestry* 7, 147–150.

Leonard, N.E. (2008) The effects of the invasive exotic Chinese tallow tree (*Triadica sebifera*) on amphibians and aquatic invertebrates. PhD thesis, University of New Orleans, New Orleans, Louisiana.

Levin, D.A. and Anderson, W.W. (1970) Competition for pollinators between simultaneously flowering species. *American Naturalist* 104, 455–467.

Levin, M.D. and Haydak, H.M. (1957) Comparative value of different pollens in the nutrition of *Osmia lignaria. Bee World* 38, 221–226.

Levine, J.M., Vilá, M., D'Antonio, C.M., Dukes, J.S., Grigulis, K. and Lavorel, S. (2003) Mechanisms underlying the impacts of exotic plant invasions. *Proceedings of the Royal Society of London B* 270, 775–781.

Liu, H. and Stiling, P. (2006) Testing the enemy release hypothesis: A review and meta-analysis. *Biological Invasions* 8, 1535–1545.

Lloyd, J.D. and Martin, T.E. (2005) Reproductive success of chestnut-collared longspurs in native and exotic grassland. *The Condor* 107, 363–374.

Mack, R.N., Simberloff, D., Lonsdale, W.M., Evans, H., Clout, M. and Bazzaz, F.A. (2000) Biotic invasions: Causes, epidemiology, global consequences and control. *Ecological Applications* 10, 689–710.

Maerz, C.J., Blossey, B. and Nuzzo, V. (2005) Green frogs show reduced foraging success in habitats invaded by Japanese knotweed. *Biodiversity and Conservation* 14, 2901–2911.

McClay, A.S., Palmer, W.A., Bennett, F.D. and Pullen, R.K. (1995) Phytophagous arthropods associated with *Parthenium hysterophorus* (Asteraceae) in North America. *Environmental Entomology* 24, 796–809.

McNeish, R.E., Benbow, M.E. and McEwan, R.W. (2012) Riparian forest invasion by a terrestrial shrub (*Lonicera maackii*) impacts aquatic biota and organic matter processing in headwater streams. *Biological Invasions* 14, 1881–1893.

Meiners, J.S. and Cadenasso, L.M. (2005) The relationship between community diversity and exotic plants: Cause or consequence of invasion? In: Inderjit (ed.) *Ecological and Agricultural Aspects of Invasive Plants.* Birkhäuser Verlag, Switzerland, pp. 97–114.

Millennium Ecosystem Assessment (MEA) (2005) Available at: http://www.unep.org/maweb/en/index.aspx (accessed 5 March 2016).

Moore, J.C., Walter, D.E. and Hunt, H.W. (1988) Arthropod regulation of micro- and mesobiota in below-ground detrital food webs. *Annual Review of Entomology* 33, 419–439.

Murphy, S.D. and Aarssen, L.D. (1995) Allelopathic pollen extract from *Phleum pratense* L. (Poaceae) reduces germination, *in vitro*, of pollen of sympatric species. *International Journal of Plant Science* 156, 425–434.

Nagy, L.R. and Holmes, R.T. (2005) Food limits annual fecundity of a migratory songbird: An experimental study. *Ecology* 86, 675–681.

Narasimhan, T.R., Ananth, M., Swamy, M.N., Babu, M.R., Mangala, A. and Rao, P.V.S. (1977) Toxicity of *Parthenium hysterophorus* L. to cattle and buffalo. *Experientia* 33, 1358–1359.

Navie, S.C., Panetta, F.D., McFadyen, R.E. and Adkins, S.W. (2004) Germinable soil seedbanks of central Queensland rangelands invaded by the exotic weed *Parthenium hysterophorus* L. *Weed Biology and Management* 4, 154–167.

Nguyen, T., Navie, S.C. and Adkins, S.W. (2010) The effect of parthenium weed (*Parthenium hysterophorus* L.) on plant diversity in pastures in Queensland, Australia. In: Zydenbos, P. (ed.) *Proceedings of the 17th Australasian Weeds Conference.* New Zealand Plant Protection, Christchurch, New Zealand, p. 138.

Nigatu, L., Hassena, A., Sharma, J. and Adkins, S.W. (2010) Impact of *Parthenium hysterophorus* on grazing land communities in north-eastern Ethiopia. *Weed Biology and Management* 10, 143–152.

Niu, H.B., Liu, W.X., Wan, F.H. and Liu, B. (2007) An invasive aster (*Ageratina adenophora*) invades and dominates forest understories in China: Altered soil microbial communities facilitate the invader and inhibit natives. *Plant Soil* 294, 1–2.

Olsson, A.D., Betancourt, J., McClaran, M.P. and Marsh, S.E. (2012) Sonoran desert ecosystem transformation by a C4 grass without the grass/fire cycle. *Diversity and Distributions* 18, 10–21.

Ortega, K.V., McKelvey, S.K. and Six, D.L. (2006) Invasion of an exotic forb impacts reproductive success and site fidelity of a migratory songbird. *Oecologia* 149, 340–351.

Osunkoya, O.O., Akinsanmi, O.A., Lim, S.A.S., Perret, C., Callander, J. and Dhileepan, K. (2017) *Parthenium hysterophorus* L. (Asteraceae) invasion had limited impact on major soil nutrients and enzyme activity: Is the null effect real or reflects data insensitivity? *Plant and Soil* 420, 177–194.

Oudhia, P. (2000) Allelopathic effects of *Parthenium hysterophorus* L. and *Ageratum conyzoides* on wheat var. Sujata. *Crop Research* 20, 563–566.

Pandey, P.K. and Saini, S.K. (2002) Parthenium: threat to biological diversity of natural forest ecosystem in Madhya Pradesh. *Vaniki Sandesh* 26, 21–29.

Pandey, D.K., Palni, L.M.S. and Joshi, S.C. (2003) Growth, reproduction and photosynthesis of ragweed parthenium (*Parthenium hysterophorus* L.). *Weed Science* 51, 191–201.

Patel, V., Chitra, V., Prasanna, P. and Krishnaraju, V. (2008) Hypoglycemic effects of aqueous extract of *Parthenium hysterophorus* L. in normal and alloxan induced diabetic rats. *Indian Journal of Pharmacology* 40, 183–185.

Perre, P., Loyola, R.D. and Lewinsohn, T.M. (2011) Insects in urban plants: Contrasting the flower head feeding assemblages on native and exotic hosts. *Urban Ecosystems* 14, 711–722.

Picman, A.K., Elliott, R.H. and Towers, G.H.N. (1981) Cardiac inhibiting properties of sesquiterpene lactone, parthenin, in the migratory grasshopper, *Melanoplus sanguinipes*. *Canadian Journal of Zoology* 59, 285–292.

Picman, J. and Picman, A.K. (1984) Autotoxicity in *Parthenium hysterophorus* L. and its possible role in control of germination. *Biochemistry, Systematics and Ecology* 12, 287–297.

Qin, Z., Xie, J., Quan, G., Zhang, J., Mao, D. and DiTommaso, A. (2014) Impacts of the invasive annual herb *Ambrosia artemisiifolia* L. on soil microbial carbon source utilization and enzymatic activities. *European Journal of Soil Biology* 60, 58–66.

Rajandran, B. and Gopalan, M. (1979) Note on juvenomimetic activity of some plants. *Indian Journal of Agricultural Science* 49, 295–297.

Rajwar, G.S., Haigh, M.J., Krecek, J. and Kilmartin, M.P. (1998) Changes in plant diversity and related problems for environmental management in the headwaters of the Garhwal Himalaya. In: Haigh, M.J., Krecek, J., Rajwar, G.S. and Kilmartin, M.P. (eds) *Proceedings of the Fourth International Conference on Headwater Control.* A.A. Balkema, Rotterdam, Netherlands, pp. 335–343.

Rashid, H., Khan, M.A., Amin, A., Nawab, K., Hussain, N. and Bhowmik, P.K.P. (2008) Effect of *Parthenium hysterophorus* L. root extracts on seed germination and growth of maize and barley. *American Journal of Plant Science and Biotechnology* 2, 51–55.

Rasmont, P., Alain, P., Patiny, S., Michez, D., Iserbyt, S., Barbier, Y. and Haubruge, E. (2005) *The Survey of Wild Bees (Hymenoptera, Apoidea) in Belgium and France.* Food and Agriculture Organization of the United Nations (FAO), Rome, pp. 1–18.

Rodriguez, E., Dillon, M.O., Mabry, T.J., Mitchell, J.C. and Towers, G.H.N. (1976) Dermatologically active sesquiterpene lactones in trichomes of *Parthenium hysterophorus* L. (Compositae). *Experientia* 15, 236–238.

Rossiter, N.A., Setterfield, S.A., Douglas, M.M. and Hutley, L.B. (2003) Testing the grass–fire cycle: alien grass invasion in the tropical savannas of northern Australia. *Diversity and Distributions* 9, 169–176.

Rutherford, C. and Rorabaugh, J. (1995) Endangered and threatened wildlife and plants: Proposed rule to determine five plants and a lizard from Monterey County, California, as endangered or threatened. *Federal Register* 60, 39326–39337.

Sakai, A.K., Allendorf, F.W., Holt, J.S., Lodge, D.M., Molofsky, J., *et al.* (2001) The population biology of invasive species. *Annual Review of Ecology and Systematics* 32, 305–332.

Samson, E.B. and Knopf, L.F. (1994) Prairie conservation in North America. *BioScience* 44, 418–421.

Samways, M.J. and Moore, S.D. (1991) Influence of exotic conifer patches on grasshopper (Orthoptera) assemblages in a grassland matrix at a recreational resort, Natal, South Africa. *Biological Conservation* 57, 117–137.

Sattler, P.S. and Williams, R.D. (eds) (1999) *The Conservation Status of Queensland's Bioregional Ecosystems.* Environmental Protection Agancy, Brisbane, Australia.

Scheiman, D.M., Bollinger, E.K. and Johnson, D.H. (2003) Effects of leafy spurge infestation on grassland bird. *Journal of Wildlife Management* 67, 115–121.

Shabbir, A. and Bajwa, R. (2006) Distribution of parthenium weed (*Parthenium hysterophorus* L.), an alien invasive weed species threatening the biodiversity of Islamabad. *Weed Biology and Management* 6, 89–95.

Shabbir, A. (2015) Soil seed bank studies on a riparian habitat invaded by *Parthenium hysterophorus*. *Indian Journal Weed Science* 47, 38–40.

Shanungu, G.K. (2009) Management of the invasive *Mimosa pigra* L. in Lochinvar National Park, Zambia. *Biodiversity* 10(2&3), 56–60.

Shultz, R. and Dibble, E (2012) Effects of invasive macrophytes on freshwater fish and macroinvertebrate communities: the role of invasive plant traits. *Hydrobiologia* 684, 1–14.

Singh, H.P., Batish, D.R., Kholi, R.K., Saxena, D.B. and Arora, V. (2003) Effect of parthenin – a sesquiterpene lactone from *Parthenium hysterophorus* L. on early growth and physiology of *Argeratum conyzoides*. *Journal of Chemical Ecology* 28, 2169–2179.

Sohal, S.K., Rup, P.J., Kaur, H., Kumari, N. and Kaur, J. (2002) Evaluation of the pesticidal potential of the congress grass, *Parthenium hysterophorus* Linn. on the mustard aphid, *Lipaphis erysimi* (Kalt.). *Environmental Biology* 23, 15–18.

Spottiswoode, C.N., Olsson, U., Mills, M.S.L., Cohen, C., Francis, J.E., *et al.* (2013) Rediscovery of a long-lost lark reveals the conspecificity of endangered *Heteromirafra* populations in the Horn of Africa. *Journal of Ornithology* 154, 813–825.

Stanley, R.G. and Linskens, F.H. (1974) *Pollen: Biology, Biochemistry and Management.* Springer-Verlag, Berlin.

Strehl, C.E. and White, J.A. (1986) Effects of superabundant food on breeding success and behaviour of the red-winged blackbird. *Oecologia* 70, 178–186.

Strong, D.R., Lawton, J.H. and Southwood, R. (1984) *Insects on Plants: Community Patterns and Mechanisms.* Blackwell Scientific Publications, Oxford.

Sukada, K. and Jayachandra (1980) Pollen allelopathy – a new phenomenon. *New Phytologist* 84, 739–746.

Sushilkumar, L. (2005) Economic benefits in biological control of parthenium by Mexican beetle, *Zygogramma bicolorata* Pallister (Coleoptera: Chrysomelidae) in India. *Annual Review of Entomology* 24(1–2), 75–78.

Tamado, T. and Milberg, P. (2000) Weed flora in arable fields of Eastern Ethiopia with emphasis on the occurrence of *Parthenium hysterophorus* L. *Journal of Weed Research* 40, 507–521.

Tanner, R.A., Varia, S., Eschen, R., Wood, S., Murphy, S. and Gange, A.C. (2013) Impacts of an invasive non-native annual weed, *Impatiens glandulifera*, on above- and below-ground invertebrate communities in the United Kingdom. *PloS One* 8(6), e67271. https://doi.org/10.1371/journal.pone.0067271.

Terfa, E.A. (2009) Impact of *Parthenium hysterophorus* L. (Asteraceae) on herbaceous plant biodiversity in the Awash National Park (ANP). MSc thesis, Addis Ababa University, Ethiopia.

Thomson, J.D., Andrews, B.J. and Plowright, R.C. (1982) The effect of a foreign pollen on ovule development in *Diervilla lonicera* (Caprifoliaceae). *New Phytologist* 90, 777–783.

Tilak, B.D. (1977) Pest control strategy in India. In: McFarlane, N.R. (ed.) *Crop Protection Agents – their Biological Evaluation.* Academic Press, London.

Timsina, B., Shrestha, B.B., Rokaya, M.B. and Münzbergová, Z. (2011) Impact of *Parthenium hysterophorus* L. invasion on plant species composition and soil properties of grassland communities in Nepal. *Flora* 206, 233–240.

Upadhyay, S.K., Ahmad, M. and Singh, A. (2013) Ecological impacts of weed (*Parthenium hysterophorus* L.) in saline soil. *International Journal of Scientific and Research Publications* 3, 1–4.

Urga, M., Tessema, T. and Yirefu, F. (2008) Impacts of *Parthenium hysterophorus* L. on herbaceous plant diversity in rangelands of Fentale District in the Central Rift Valley of Ethiopia. *Ethiopian Journal of Weed Management* 2, 25–41.

Valentine, E.L., Roberts, B. and Schwarzkopf, L. (2006) Mechanisms driving avoidance of non-native plants by lizards. *Journal of Applied Ecology* 44, 228–237.

Van der Putten, H.W., Klironomos, J.N.N.L. and Wardle, A.D. (2007) Microbial ecology of biological invasions. *International Society for Microbial Ecology* 1, 28–37.

Van Lill, W.S., Kruger, F.J. and Van Wyk, D.B. (1980) The effect of afforestation with *Eucalyptus grandis* Hill ex Maiden and *Pinus patula* Schlecht. et Cham. on streamflow from experimental catchments at Mokobulaan, Transvaal. *Journal of Hydrology* 48, 107–118. doi:10.1016/0022-1694(80)90069-4.

Van Nieuwenhuyse, D., Nollet, F. and Evans, A. (1999) The ecology and conservation of the red-backed shrike *Lanius collurio* breeding in Europe. *Aves* 36, 179–192.

Van Wingerden, W.K.R.E., Kreveld, A.R. and Bongers, W. (1992) Analysis of species composition and abundance of grasshoppers (Orth., Acrididae) in natural and fertilized grasslands. *Journal of Applied Entomology* 113, 138–152.

Vickery, P.D., Herkert, J.R., Knopf, F.L., Ruth, J. and Keller, C.E. (2000) Grassland birds: an overview of threats and recommended management strategies. In: Bonney, R., Pashley, D.N., Cooper, R.J. and Niles, L. (eds) *Strategies for Bird Conservation: The Partners in Flight Planning Process.* RMRS-P-16. US Department of Agriculture, Forest Service, Rocky Mountain Research Station, Fort Collins, Colorado, pp. 74–77.

Vilà, M., Basnou, C., Pyšek, P., Josefsson, M., Genovesi, P., *et al.* (2010) How well do we understand the impacts of alien species on ecosystem services? A pan-European, cross-taxa assessment. *Frontiers in Ecology and the Environment* 8, 135–144.

Vitousek, P.M. (1990) Biological invasions and ecosystem processes: Toward an integration of population biology and ecosystem studies. *Oikos* 57, 7–13.

Vitousek, P.M. and Walker, L.R. (1989) Biological invasion by *Myrica faya* in Hawaii: Plant demography, nitrogen fixation, ecosystem effects. *Ecological Monographs* 59, 247–265.

Wahab, S. (2005) Management of Parthenium through an integrated approach. Initiatives, achievements and research opportunities in India. In: *Proceedings of the Second International Conference on Parthenium Management.* University of Agricultural Sciences, Bangalore, India, pp. 36–43.

Wardle, D.A. (2002) *Communities and Ecosystems: Linking the Aboveground and Belowground Components.* Princeton University Press, Princeton, New Jersey.

Wardle, D.A. (2006) The influence of biotic interactions on soil biodiversity. *Ecology Letters* 9, 870–886.

Waser, N.M. (1978) Competition for hummingbird pollination and sequential flowering in two Colorado wildflowers. *Ecology* 59, 934–944.

Watling, I.J., Hickman, R.C. and Orrock, L.J. (2011) Invasive shrub alters native forest amphibian communities. *Biological Conservation* 144, 2597–2601.

Wegari, K.E. (2008) Perceived socio-economic impact of parthenium weed and its effect on crop production in Babile, Haramaya and Tulo Districts, East and West Hararge Zone, Ethiopia. MSc thesis, Haramaya University, Ethiopia.

Wilson, C.G., Flanagan, C.J. and Gillet, J.D. (1990) The phytophagous insect fauna of the introduced shrub *Mimosa pigra* in Northern Australia and its relevance to biological control. *Environmental Entomology* 19, 776–784.

Wissel, C. (1977) On the advantage of the specialization of flowers on particular pollinator species. *Journal of Theoretical Biology* 69, 11–22.

Yaduraju, N.T., Sushilkumar, L., Prasad Babu, M.B.B. and Gogoi, A.K. (2005) *Parthenium hysterophorus* – distribution, problems and management strategies in India. In: *Proceedings of the Second International Conference on Parthenium Management.* University of Agricultural Sciences, Bangalore, India, pp. 6–10.

Zavaleta, E. (2000) Valuing ecosystem services lost to *Tamarix* invasion in the United States. In: Mooney, H.A. and Hobbs, R.J. (eds) *Invasive Species in a Changing World.* Island Press, Washington, DC.

Zwolfer, H. (1965) Preliminary list of phytophagous insects attacking wild Cynareae (Compositae) in Europe. *Technical Bulletin of the Commonwealth Institute for Biological Control* 6, 81–154.

6

Impact of Parthenium Weed on Human and Animal Health

Sally Allan,[1]* Boyang Shi[2] and Steve W. Adkins[1]

[1]University of Queensland, Gatton, Queensland, Australia;
[2]Biosecurity Queensland, Department of Agriculture and Fisheries,
Brisbane, Queensland, Australia

6.1 Introduction

The annual, herbaceous plant parthenium weed (*Parthenium hysterophorus* L.) is a weed of global significance (Towers *et al.*, 1977; Dhileepan, 2009; Adkins and Shabbir, 2014). Originating from southern North America, Central America including the Gulf Coast, the West Indies and Caribbean Islands, and South America (Rollins, 1950, in Picman and Towers, 1982; Wedner *et al.*, 1989), parthenium weed has spread from its native range to over 40 tropical and subtropical countries around the world (Shabbir, 2012; Adkins and Shabbir, 2014; EPPO, 2014). The spread of parthenium weed has produced negative impacts on grazing animal production (Jayachandra, 1971; Chippendale and Panetta, 1994; Navie *et al.*, 2005), crop production (Tamado and Milburg, 2000; Khan, 2011; Adkins and Shabbir, 2014) and natural ecosystems (Dhileepan, 2009; Carnie, 2013; Belgeri *et al.*, 2014). It has been long recognized that parthenium weed also has detrimental effects on humans (Lonkar *et al.*, 1974; Towers *et al.*, 1977; Wedner *et al.*, 1987; McFadyen, 1995) and animal health (Narasimhan *et al.*, 1980; Tudor *et al.*, 1982; Ahmed *et al.*, 1988a; Shrestha *et al.*, 2015).

6.1.1 Issues and causes

The detrimental health effects of parthenium weed are attributed predominantly to the occurrence of sesquiterpene lactones in the plant, and in particular parthenin (Sharma and Verma, 2012), which has been shown to cause adverse responses in both humans and domesticated animals. Parthenin is the major sesquiterpene lactone (Rodriguez *et al.*, 1971) in the majority of global parthenium weed populations, including the northern American populations from which the majority of introduced populations are derived (Dale, 1981; Picman and Towers, 1982; Adkins and Shabbir, 2014). High levels of parthenin are found in the essential oil fraction extracted from the trichome hairs (Rodriguez *et al.*, 1976a; Picman *et al.*, 1979; McFadyen, 1995) present on the leaf, stem and achenes (Rodriguez *et al.*, 1976a; Picman *et al.*, 1979) or in the coating material of the pollen grains (Picman *et al.*, 1980; Roy and Shaik, 2013; Sharma *et al.*, 2013). The vast areas of parthenium weed infestation and the ability of the plant to produce large numbers of flowers and hence pollen grains (Towers and Mitchell, 1983; Dhileepan, 2012) have been major causal factors to the severity of the human

* sallyallan01@gmail.com

health problems. The largest impact on human health has been seen in India, where parthenium weed has been determined to be the major cause of contact dermatitis (Lonkar *et al.*, 1974; Khosla and Sobti, 1979; Sharma and Kaur, 1989; Kumar *et al.*, 2012; Sharma *et al.*, 2013) and has led to chronic disease and indirectly to several deaths (Lonkar *et al.*, 1974; Kololgi *et al.*, 1997).

Human disease occurs when a sensitized individual comes into contact with or inhales airborne particles of parthenium weed, such as its pollen or other fine plant debris. It is thought that disease is a response to one of two main immunological processes (Kumar *et al.*, 2012): a Type IV delayed hypersensitive reaction resulting in a common form of skin dermatitis caused by direct contact with airborne particles or a Type I reaction which is an immediate hypersensitive reaction from the inhalation or ingestion of plant parts, which becomes evident as allergic rhinitis, asthma and/or atopic dermatitis. It has been suggested that the induction and prolongation of parthenium weed disease may involve a combination of these two mechanisms (Lakshmi and Srinivas, 2007b). Medical treatment for these diseases is generally symptomatic (Sharma and Sethuraman, 2007) and can give some relief, although immediate side effects and health complications due to prolonged treatment are relevant issues. Medical treatments to reduce the sensitivity of people allergic to parthenium weed, such as hyposensitization therapy, have had variable results (Lakshmi and Srinivas, 2007a) and require further research. Avoiding or reducing contact with parthenium weed or airborne plant allergens is the best long-term management option for the disease (Lakshmi and Srinivas, 2007a; Schloemer *et al.*, 2015).

Although the majority of domestic animals avoid eating parthenium weed whenever possible, when it is grazed, animals can develop symptoms of contact dermatitis (Khosla and Sobti, 1979). When ingested in large amounts, animals can develop severe dermatitis, skin lesions, common erythematous eruptions, ulceration of the muzzle, gums and tongue (Narasimhan *et al.*, 1977, 1980; More *et al.*, 1982) and internal ulceration of the alimentary tract, followed by haemorrhage and rupture of internal organs and eventually death (Narasimhan *et al.*, 1977; Ahmed *et al.*, 1988a). Most animals recover when they are removed from grazing areas infested with parthenium weed.

6.1.2 Known extent of the health problems

Factors which affect the impact of parthenium weed on human health can be broadly divided into those that are human, plant or environmentally influenced. Human factors would include the body's response or reactiveness to parthenium weed allergens, behavioural traits such as lifestyle, work practices, clothing choices, personal hygiene and availability of medicines to treat the problem. Plant factors would include the amount and activity of the bioactive chemicals present in the weed, the size of the weed infestation, and their reproductive state at the time of contact. Environmental factors would include those that affect plant growth, weed spread, level of plant desiccation, the mode of conveyance used by the plant debris and pollen grains, including wind strength and direction, and the effect of the environment on the reactiveness of humans – for instance, strong sunlight or humid conditions are known to promote reactiveness.

The human health effects of this well-known allergenic plant have been recorded in almost all countries where the plant occurs and are considered a significant health concern in countries where the plant has been introduced (Lonkar *et al.*, 1974; Gupta and Chanda, 1991; Goldsworthy, 2005; Nadeem *et al.*, 2005; Table 6.1) and in areas where the plant is prolific (Wedner *et al.*, 1989).

While allergic contact dermatitis to ragweeds (*Ambrosia* spp.) in northern America is well documented, reports of allergies to parthenium weed are substantially fewer (Rao *et al.*, 1985). Allergic outbreaks, like those observed in India, have not been recorded in the USA, and there are

Table 6.1. Human disease impacts caused by parthenium weed in a number of countries.

Country	Diseases	Reference
Native range		
USA	Allergic disease	French (1930), Wedner *et al.* (1987, 1989), Warshaw and Zug (1996)
Cuba	Allergenic and irritant action	Ramos *et al.* (2002)
Introduced range		
India	Contact dermatitis, rhinitis, erythroderma, photo phyto dermatitis, oral hypersensitization	Lonkar *et al.* (1974), Towers *et al.* (1977), Rao *et al.* (1985), Sharma and Kaur (1989), Lakshmi and Srinivas (2007a), Agarwal *et al.* (2008), Kololgi *et al.* (1997), Handa *et al.* (2011), Kumar *et al.* (2012), Mahajan *et al.* (2014)
Australia	Dermatitis and respiratory symptoms, allergic eczematous contact dermatitis, allergic rhinitis	Haseler (1976), Bury and Kloot (1982), Chippendale and Panetta (1994), McFadyen (1995), Cheney (1998), Goldsworthy (2005)
Pakistan	Allergy and dermatitis	Nadeem *et al.* (2005), Khan (2011)
Nepal	Allergic dermatitis, eye redness, wounds, boils, cough and throat problems, asthma and fever	Karki (2009), Shrestha *et al.* (2015)
Ethiopia	Asthma, dermatitis, hay fever, rhinitis, bronchitis and high fever	Ayele (2007), Wiesner *et al.* (2007), Abdulkerim-Ute and Legesse (2016)
Bangladesh	Contact dermatitis	Roy and Shaik (2013)
Tanzania	Allergic dermatitis, asthma and bronchitis	ECHO East Africa Impact Centre (2016)
South Africa	Parthenium weed allergy	see Strathie and McConnachie (Chapter 14, this volume)

suggestions that people from the plant's native range may have a genetic or acquired tolerance, developed from early childhood exposure to parthenium weed and similar species such as ragweed (Lonkar *et al.*, 1974). Studies in Texas, USA, conducted on 18 patients thought to have a high sensitivity to western ragweed (*Ambrosia psilotachya* DC.), showed that there is significant cross-reactivity between western ragweed and parthenium weed pollen. However, this was not always the case and a few people were directly sensitive to the parthenium weed allergens (Wedner *et al.*, 1987).

6.1.3 Comparison with other Asteraceae

Parthenium weed is a member of the Asteraceae, which is responsible for the majority of phytogenic allergic contact dermatitis plants (McFadyen, 1995). Plants within this family, including various *Ambrosia, Helianthus, Solidago* and *Chrysanthemum* species, have been reported as the major cause of plant-related, airborne contact dermatitis (ABCD) within the USA (Swinnen and Goossens, 2013), while parthenium weed alone has been identified as the major cause of ABCD in India (Lakshmi and Srinivas, 2007a). In northern India, in addition to parthenium weed, other members of the Asteraceae causing ABCD include chrysanthemum (*C. morifolium* Ramat), dahlia (*Dahlia pinnata* Cav.) and Indian marigold (*Tagetes indica* L.; Sharma and Kaur, 1989). Cocklebur (*Xanthium strumarium* L.) has also been associated with ABCD in patients in India (Pasricha *et al.*, 1995). Sharma *et al.* (2005) reported that between 0.7 and 1.4% of the general population of India (*c.*9–18 million people) suffer dermatological effects from members of the Asteraceae family and that this value increases to 4.5% within the occupationally exposed groups, such as gardeners and farm labourers.

One characteristic of all Asteraceae is that they contain the most diverse range of sesquiterpene lactones, more than any other family, with between 3000 and 4000 different sesquiterpene lactones identified so far (McFadyen, 1995). These secondary plant metabolites have been identified as a common cause of Asteraceae allergic reactions (Rodriguez *et al.*, 1976a; Schloemer *et al.*, 2015). Many patients who report being allergic to species within the Asteraceae are probably sensitive to sesquiterpene lactones (Nandhakishore and Pasricha, 1994; Schloemer *et al.*, 2015). The Ambrosiinae subtribe of the Asteraceae produce three distinctive sesquiterpene lactone groups, the psilostachyanolides, ambrosanolides and the xanthanolides, with the latter two being identified within most *Parthenium* species (Miao *et al.*, 1995).

Cross-sensitivity can occur when patients who become sensitized to a primary allergen then react to other secondary allergens, which are often chemically closely related to the primary allergen (Warshaw and Zug, 1996). To investigate cross-sensitivity between allergens found in closely related genera within the Asteraceae, a trial in India tested the sensitivity of 63 patients who suffered from ABCD. Cross-sensitivity between parthenium weed and cocklebur was found in a high number of cases (74%), and between parthenium weed and edible chrysanthemum (*C. coronarium* L.) and sunflower (*Helianthus annuus* L.) in a significant but lower number of cases (21% and 11%, respectively; Nandhakishore and Pasricha, 1994).

The closely related *Ambrosia* genus, including the ragweeds, are known to be associated with significant human allergies (Mitchell *et al.*, 1971b; Huang *et al.*, 1991; Fisher, 1996b; Möller *et al.*, 2002; Ghiani *et al.*, 2012). Dermatitis can result from direct contact with these plants or their pollen (a Type IV allergic reaction) and the pollen alone can cause a Type I allergic rhinitis, bronchial asthma (Fisher, 1996b) or atopic dermatitis. It has been suggested that people sensitized to parthenium weed may develop a Type I hypersensitivity reaction when they come into contact with ragweed pollen, even though they have not been exposed to ragweed pollen before, and vice versa (Sriramarao and Rao, 1993).

Common ragweed (*A. artemisiifolia* L.) is a habitual cause of allergies in North America and typically results in allergic eczematous contact dermatitis (AECD) by direct contact with the plant or by contact with its pollen (Mitchell *et al.*, 1971b). Over the last few decades, there has been increasing concern in Europe about the health impacts of common ragweed as the weed continues to spread further northwards (Möller *et al.*, 2002; Ghiani *et al.*, 2012). Recent research has focused on how the environment can affect common ragweed pollen production and the impact that subsequent further spread will have on human health. On a local scale, work by Ghiani *et al.* (2012) showed that the allergenicity of common ragweed pollen can be increased by airborne motor traffic pollution. This work showed that the amount and number of allergens present in the pollen varied with the level of atmospheric pollution. This work may have implications for the closely related parthenium weed, because like common ragweed, parthenium weed often grows close to roads and in developing countries air pollution levels will continue to rise as an increasing number of vehicles use the road networks. This may also imply that parthenium weed allergenicity may become more potent in the future and in urban areas.

On a global scale, scientists are investigating the effect of climate change on plant growth and pollen production. Increasing temperatures have been shown to increase pollen production of western ragweed by 84% (Wan *et al.*, 2002). Apart from increasing temperatures, the early arrival of spring in the northern hemisphere and increasing atmospheric CO_2 concentrations are also impacts of global climate change. Studies looking at the effect of an early spring suggest that this would increase the growth and reproductive capacity of western ragweed, while increasing atmospheric CO_2 concentration would result in an increase in western ragweed growth and pollen production (Rogers *et al.*, 2006). Research on the effect of CO_2 alone showed that increasing atmospheric CO_2 concentrations, from pre-industrial levels (280 μmol/mol) to levels in 2000 (370 μmol/mol) and future levels

(600 μmol/mol), would result in a further increase in pollen production from 131% to 320% (Ziska and Caulfield, 2000). Results such as these have serious implications for human allergenic health in the future as similar increases in growth of parthenium weed under elevated levels of CO_2 have been reported (Navie *et al.*, 1996).

6.1.4 Statistics and economics

The invasion by parthenium weed has had a significant impact, particularly in India and Australia and now in Africa and South-east Asia. India is the country where parthenium weed has had the most devastating impact on agriculture, the environment and human health. Parthenium weed has affected approximately 35 million ha of land in India and has had an estimated proposed management cost of 300,000 million Indian rupees (INR; AUD$6 billion) per annum since introduction (Sushilkumar and Varshney, 2010). Allergens from parthenium weed are estimated to affect 4% (50 million) of the population (Sushilkumar and Varshney, 2010) and are the main cause of dermatitis in large regions of India (Lonkar *et al.*, 1974; Khosla and Sobti, 1979; Sharma and Kaur, 1989). It has been estimated that if each individual spends INR200 for parthenium weed-related health issues, this equates to INR880 crores (AUD$170 million) per annum (Sushilkumar and Varshney, 2010). Once patients are sensitized to parthenium weed, they often experience seasonal relapses of disease, often at peak flowering times, and the disease can become chronic (Handa *et al.*, 2001), leading to poor quality of life and in some instances induces death (Lonkar *et al.*, 1974; Kololgi *et al.*, 1997).

By 2000, 5.6% of the total area of Australia, *c.*43 million ha, had been affected by parthenium weed (Thorp and Lynch, 2000). In Queensland, at least 17 million ha have been significantly affected by parthenium weed (Australian Weeds Committee, 2012), with the plant having an occasional occurrence over approximately 30% of Queensland (*c.*60 million ha; Department of Agriculture

and Fisheries, 2015; Shi, 2016). Estimated costs of parthenium weed to the Queensland farming and grazing industries range from AUD$22 million per annum (Department of the Environment and Heritage and the CRC for Australian Weed Management, 2003) to AUD$109 million per annum (Adamson, 1996).

In infested areas of Australia, the cost of parthenium weed on human health has been estimated to be AUD$6.90 per person or AUD$19.90 per household (AEC group, 2002) or AUD$40.00 per month for people suffering significant allergies (Goldsworthy, 2005). In these infested areas, a survey conducted by graziers estimated that 10% of workers exposed to parthenium weed developed visible signs of allergic reactions (Chippendale and Panetta, 1994).

As parthenium weed infests more regions of Africa, research into the health implications has been initiated. Wiesner *et al.* (2007) interviewed 64 Ethiopian farmers about their health problems since parthenium weed had infested their land. Of these, 90% suffered from hay fever, 80% were fatigued, 62% had breathing or coughing difficulties, 30% had skin irritations and pustules, 21% had stretched or cracked skin and 22% had stomach pains, all believed to be caused by inhaling airborne parthenium weed particles and pollen.

6.2 Human Health

Parthenium weed can cause a variety of immuno-inflammatory diseases, both external and internal. External diseases include a variety of dermatitis forms (Fig. 6.1), with 11 listed by Lakshmi and Srinivas (2012), including the commonly occurring airborne contact dermatitis, and less commonly, chronic actinic dermatitis, mixed pattern dermatitis, photosensitivity and contact dermatitis (particularly on hands), seborrhoeic dermatitis and atopic dermatitis (Kololgi *et al.*, 1997; Kohli *et al.*, 2006; Sharma and Verma, 2012). Internal diseases include allergic asthma, bronchitis and rhinitis (hay fever). Both external and internal

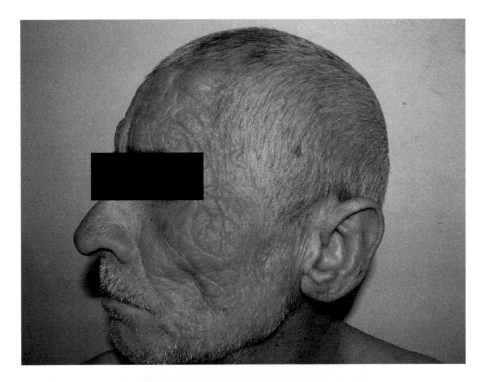

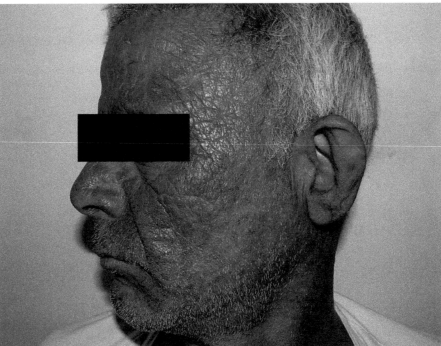

Fig. 6.1. Human disease occurs when a sensitized individual comes into contact with or inhales airborne particles of parthenium weed, such as its pollen or other fine plant debris. Airborne contact dermatitis is the most common form of parthenium weed dermatitis. (Photographs from India, K. Verma.)

conditions are predominantly the result of contact with, or the inhalation of, airborne particles of parthenium weed pollen or plant debris. Direct contact with the plant can also lead to dermatitis, but this is a less common form of the disease. Some nausea and giddiness have also been reported (Kohli *et al.*, 2006).

6.2.1 Types of disease

Airborne contact dermatitis (ABCD)

Airborne contact dermatitis (ABCD) is the most common form of dermatitis, with the first reported case attributed to parthenium weed occurring in the USA (French, 1930) and several decades later in India (Lonkar *et al.*, 1974; Towers *et al.*, 1977) and Australia (Haseler, 1976; Bury and Kloot, 1982). Since then, parthenium weed has been a moderate cause of ABCD in localized areas of Pakistan (Khan, 2011), Nepal (Karki, 2009; Shrestha *et al.*, 2015), Ethiopia (Wiesner *et al.*, 2007; Abdulkerim-Ute and Legesse, 2016), South Africa (see Strathie and McConnachie, Chapter 14, this volume), Tanzania (ECHO East Africa Impact Centre, 2016) and Australia (McFadyen, 1995; Goldsworthy, 2005). However, the widespread distribution of parthenium weed in both rural and urban areas of India has resulted in parthenium weed being a major cause of ABCD; in some areas, parthenium weed is the most common cause of ABCD (Sharma and Kaur, 1989; Kumar *et al.*, 2012). In India, cases of ABCD can occur throughout the year because the climate can create a perennial nature to the weed populations, but is often exacerbated during summer (Sharma and Verma, 2012) and in the rainy season (Akhtar *et al.*, 2010), both times of vigorous plant growth (Akhtar *et al.*, 2010; Sharma and Verma, 2012).

In sensitized people, dermatitis can appear within 24 h of exposure or be delayed for a few days (Sharma and Verma, 2012). ABCD often presents as a classical pattern of diffuse dermatitis on the face, particularly on the upper eyelids, neck and the 'V' line of the chest and flexural areas of the arms and legs, such as the inner elbow and behind the knee (Sharma *et al.*, 2005; Sharma and Verma, 2012). Mild ABCD cases present as erythema and itching; moderate cases as itching and burning, swelling, erythema, papules or papulovesicles; and severe cases present as extensive vesiculation, exudation and oedema. After repeated exposures, people may suffer chronic and extensive lichenified dermatitis (Sharma and Verma, 2012). Some studies have recorded a clinical progression of parthenium weed dermatitis over time, from ABCD to mixed pattern dermatitis or to chronic actinic dermatitis (Sharma *et al.*, 2005).

Chronic actinic dermatitis (CAD)

A considerably less common form of dermatitis is chronic actinic dermatitis (CAD). Patients with CAD present with localized lichenified papules and nodules/plaques on exposed areas such as the dorsal surfaces of the hands and forearms, the forehead, ear rims, cheeks, neck and in the 'V' line of the chest (Sharma *et al.*, 2005; Sharma and Verma, 2012). Usually the non-sun-exposed areas (eyelids, under the chin, flexural areas) are relatively free from dermatitis in CAD-only patients. However, these patients show positive phototests on normal non-exposed skin, whereas patients with ABCD show negative phototests from these same skin areas (Sharma *et al.*, 2005).

Mixed pattern dermatitis

Usually during the transition from ABCD to CAD, patients can present with a mixed pattern of dermatitis. This may involve scaly papules on exposed areas, similar to CAD and also in flexural areas of the arms and legs and on eyelids, such as seen in patients with ABCD (Sharma and Verma, 2012).

Photosensitive dermatitis

Photosensitive dermatitis (PD) has been reported by various researchers. PD exhibits as a distinct pattern of photosensitive lichenoid dermatitis of pruritic, flat, purple-coloured papules and plaques (Sharma and

Verma, 2012), typically not present on the upper eyelids, behind the ears and under the chin (Sharma *et al.*, 2005). The lack of dermatitis in these areas differentiates these patients from those suffering ABCD (Handa *et al.*, 2001; Schloemer *et al.*, 2015). In an Indian study by Sharma *et al.* (2005), of the 74 patients who had at least a 3-year history of parthenium weed-induced dermatitis, more than half had varying strengths of PD, from mild to severe.

Direct contact dermatitis

People who are exposed to parthenium weed can develop contact dermatitis. A classic study case was reported by Subba Rao *et al.* (1977), where 300 workers were employed to uproot parthenium weed plants from a local infestation. Of this group, 12 people (4%) developed contact dermatitis by presenting first with erythematous papulovesicular or maculopapular eruptions on exposed areas (face, forearms and the 'V' of neck/chest) and also upper eyelids, flexures and behind the ears. These patients produced a positive skin patch test, suggesting typical delayed hypersensitivity (Type IV), while 56% became sensitized to parthenium weed without exhibiting dermatitis and there was no report of internal allergic reactions.

Seborrhoeic dermatitis

Tiwari *et al.* (1979) describes seborrhoeic dermatitis as the appearance of dry, itchy pigmented lesions, possibly on the nasolabial folds, upper lip, chin, cheeks, eyebrows, skin folds at the base of the ear, chest and upper back. It is not an infection and is not transmitted from individual to individual. The cause of seborrhoeic dermatitis is still unknown, although many attempts have been made to relate it to infection by the *Malassezia* genus of skin fungi (Gupta *et al.*, 2004). Sethuraman *et al.* (2008) suggested that perhaps the airborne allergens of parthenium weed can accumulate in the nasolabial folds and then move into the eyelids in people who have deep and prominent skinfolds and there cause seborrhoeic dermatitis.

Atopic dermatitis

Atopic dermatitis is a chronic inflammatory, non-contagious and pruritic (an unpleasant sensation that provokes the desire to scratch) skin disease. This disease can develop in people with a genetic predisposition toward developing certain allergic hypersensitivity reactions (Type I). Different people may present with different symptoms, but in general symptoms often appear as red, inflamed and itchy areas which sometimes quickly develop into exacerbated bumps. The first signs of parthenium weed atopic dermatitis are the red to brownish-grey coloured patches on the face that may become itchy. Most atopic dermatitis starts in childhood before the age of 5 years (Chan and Burrows, 2009).

Rhinitis

Allergic rhinitis is an allergic inflammation of the nasal airways, usually associated with watery nasal discharge and itching of the nose and eyes (Bousquet *et al.*, 1997). It occurs when an allergen, such as pollen, dust or animal dander is inhaled by an individual with a sensitized immune system. Airborne pollen of parthenium weed is a potent atopic allergen that is a common cause of allergic rhinitis, particularly in India and in certain parts of the southern USA and north-eastern Australia (Rao *et al.*, 1985; Wedner *et al.*, 1989; McFadyen, 1995).

In Bangalore (now Bengaluru) in Southern India, a steady increase in cases of allergenic rhinitis occurred at the same time as parthenium weed spread into that area (Rao *et al.*, 1985). A survey found that 40% of 71 patients suffering seasonal or perennial rhinitis developed symptoms or aggravation of symptoms during the months of maximum growth of parthenium weed, July to October (Rao *et al.*, 1985). Skin-prick tests by Rao *et al.* (1985) found that 34% of 143 rhinitis sufferers produced positive results to a crude extract of parthenium weed pollen antigens and 12% of 33 bronchial asthma sufferers reacted positively to parthenium weed antigen tests.

It is thought that the capture of parthenium weed pollen in the nasal mucosa will more commonly lead to allergic rhinitis than bronchial asthma. The relatively large size of parthenium weed pollen grains (18 μm diameter; Rao *et al.*, 1985; McFadyen, 1995) and the clumping of pollen grains suggest that the pollen may not often pass beyond the nasal mucosa and instead produce localized allergic rhinitis (Rao *et al.*, 1985).

Asthma

Asthma or allergic bronchitis is a common chronic inflammatory disease of the airways characterized by variable and recurring symptoms and reversible airflow obstruction (Martinez, 2007). It can occur when small parthenium weed antigenic structures, particularly pollen, are inhaled into the lungs and there induce an immune response. This is more likely to occur if people breathe through their mouth rather than their nose, for example when doing hard work or when nasal passages are blocked or inflamed (McFadyen, 1995). It can also occur more often in stormy or rainy weather as it has been suggested that when pollen grains come into contact with water, very small starch granules are released from the pollen and can be breathed into the airways (The Australian Society of Clinical Immunology and Allergy, 2015), taking allergens with them.

6.2.2 Disease factors and mechanisms

The severity of the allergic reaction in a parthenium weed-sensitive patient is thought to depend upon the degree of their hypersensitivity at the time of contact, the quality and quantity of the antigen and the amount of antigen which penetrates the skin barrier (Verma *et al.*, 2000).

Quantity of antigen

The main bioactive chemical involved in parthenium weed-induced disease is the antigen parthenin (Sharma and Verma, 2012), which is commonly found in all plant parts, particularly the trichome hairs on the leaves and stems (Picman *et al.*, 1979) and in

the coating of pollen grains (Rodriguez *et al.*, 1976a). Airborne particles of parthenium weed plant debris and pollen grains are the main agents for most allergenic disease. Therefore, it is the distribution, the number of plants and the production of pollen grains that have a significant impact on the quantity of the antigen in the environment and therefore the potential to cause disease. The large number of healthy, flowering parthenium weed plants found close to public areas and dwellings have attributed to the devastating human health impact of this plant in areas of India (Fig. 6.1).

The vast global spread of parthenium weed is well documented, and each plant has the potential to produce a large amount of pollen. The number of capitula or flowers per plant defines the amount of pollen produced (Dhileepan, 2012), with a recorded 345,600 pollen grains produced per flower (Gupta and Chanda, 1991) or an estimated 624 million grains per plant (Towers and Mitchell, 1983). Parthenium weed pollen grains, which are 15–20 μm in diameter (Rao *et al.*, 1985; McFadyen, 1995), are then carried away as either single grains or in clusters of 600–800 grains (Towers and Mitchell, 1983).

As a result of the significant impact of parthenium weed in that country, India has been the focus of parthenium weed pollen count research. From past research, two main peaks of aerial pollen occur within a year, from January/February to March/April and from August/September to October/November, although some variation exists due to the study locations. These two peaks can be broadly related to springtime flowering trees and shrubs, and to weeds and grasses flowering towards the end of the monsoon and immediately after it (Chaubal and Kotmire, 1982; Gupta and Chanda, 1991; Singh *et al.*, 2003). It is the latter peak, from August/September to October/November, which corresponds with moderate temperatures and humidity, reducing rainfall and the peak of aerial parthenium weed pollen (Jain, 1993).

The reported proportion of parthenium weed pollen in aerial pollen counts varies considerably, between different locations, the annual time of sampling, year of sampling and varying sampling techniques (Gupta and Chandra, 1991; Medicinal Plants Archive, 2016). Significant proportions of parthenium weed pollen have been recorded from the southern city of Bengaluru, such as 66% (Agashe and Abraham, 1988), 44% (Mangala *et al.*, 1981) and 25% (Seetharamiah *et al.*, 1981), moderate proportions of 18% from Gwalior (Jain, 1993) and less significant proportions from the northern cities of Delhi (2.5% over a 7-year period; Singh *et al.*, 2003) and Rohtak (6.8% over a 2-year period; Ahlawat *et al.*, 2013).

Disease mechanism

Kumar *et al.* (2012) suggest that human allergic reactions to parthenium weed occur through two main immunological mechanisms. A Type IV mechanism is a delayed hypersensitive reaction resulting in dermatitis caused by airborne parthenium weed plant particles. This is the most common mechanism for a variety of dermatitis types. The second immunological mechanism, Type I, is an immediate hypersensitive reaction to inhaled or ingested parthenium weed plant particles, or direct contact, and is evident as the internal conditions of allergic rhinitis and asthma and the external condition of atopic dermatitis.

A typical Type IV hypersensitivity immune response takes the form of dermatitis, particularly on exposed parts of the body. It is thought when parthenium weed debris, containing the antigen(s), comes into contact with the dermal layers of the skin, epidermal Langerhans cells and cutaneous dendritic cells transport the allergen to regional lymph nodes. This causes proliferation of T-cells (a subtype of white blood cells) along with the production of effector and memory cells. The T-cells then move to areas of re-exposed skin sites resulting in cutaneous inflammation (Sharma and Verma, 2012). A Type IV response can be detected by patch testing.

A Type I immediate hypersensitive immune reaction occurs when a person with a genetic hyperallergic predisposition is re-exposed to the antigens in parthenium weed. When the body comes into contact with the antigen, B-cells are stimulated to produce immunoglobulin E (IgE) antibodies specific to the antigen, which then bind to the surface of mast cells and basophils. When re-exposure occurs, interaction with the bound IgE antibodies results in degranulation and secretion of histamine, leukotriene and prostaglandin. These secretions cause vasodilation, therefore increased blood flow, and the movement of white blood cells (leukocytes) out of the circulatory system and into body tissue (Wikipedia, 2017b). A Type I response can be detected by a skin-prick test.

6.2.3 Treatments

The best way to manage the detrimental health effects of parthenium weed is for people to avoid the allergens coming from the plant, remove themselves from regions where the plant grows (Rao *et al.*, 1985; Schloemer *et al.*, 2015) or to use medicinal treatments.

When people suffering allergic reactions to parthenium weed are transferred to areas without the weed, most patients experience a full recovery (Lonkar *et al.*, 1974; Khosla and Sobti, 1979; Sharma and Verma, 2012). However, this approach can be difficult due to the ubiquitous distribution of the plant in some regions of India (Rao *et al.*, 1985; Sharma and Sethuraman, 2007) and because, for some people, their only form of occupation involves direct or indirect contact with the plant. Where the option is available, people with contact dermatitis tend to change occupations more so than the general population (Bhatia *et al.*, 2015), if possible working indoors and, in the case of photo-dermatitis sufferers, avoiding sunlight as much as possible (Kololgi *et al.*, 1997).

Unfortunately, once someone develops a sensitivity to parthenium weed it is lifelong (Sharma and Sethuraman, 2007), and the chance of a spontaneous cure is rare (Handa *et al.*, 2001). To minimize contact with plant allergens and thereby reduce the incidence and severity of diseases, several physical or cultural strategies can be employed (Table 6.2). These strategies are particularly important where the option of medicinal treatments is not available or is prohibitive due to cost.

There are no established approaches to prevent sensitivity to parthenium weed, such as successful hypo- or de-sensitizing therapies (Handa et al., 2001; Adkins and Shabbir, 2014). Limited trials of hypo-sensitization or de-sensitization by orally administering extracts of parthenium weed to sensitive patients have had inconsistent results (Srinivas et al., 1988; Handa et al., 2001; Veien, 2011). Trials by Handa et al. (2001) resulted in 70% of parthenium weed sensitive patients experiencing disease improvement and in 30% aggravation of the allergic condition. Trials by Veien (2011) also resulted in temporary improvement for some patients, but the aggravation of dermatitis in others to the extent they had to withdraw from the trial.

The use of medical treatments is predominantly symptomatic (Rao et al., 1985; Sharma and Sethuraman, 2007), with better outcomes if patients are treated as early as possible after diagnosis. In general, topical steroid creams are used for localized and mild to moderate dermatitis (Narasimha et al., 2005; Sharma and Sethuraman, 2007; Bhatia et al., 2015) and systemic (oral) corticosteroids, immunosuppressant and antihistamine medications are used for more persistent and severe dermatitis, atopic dermatitis, asthma and bronchitis (Sharma and Sethuraman, 2007; Lakshmi et al., 2008; Adkins and Shabbir, 2014; Table 6.3).

The use of steroids and antihistamine medication is common for immune-mediated diseases, such as caused by parthenium weed. In general, steroids suppress an immune response by interfering with the production of IgE antibodies, the trademark of a Type I hypersensitivity reaction. Apart from immune suppression, many of these steroids, such as cyclosporine, have an anti-inflammatory action as well. In the case of cyclosporine, it is able to suppress the immune system by interfering with T-cell production and it can also inhibit the release of histamine from mast cells (Lakshmi et al., 2008). It is the mast cells, the white blood cells of the immune system, which contain both histamine and heparin. Antihistamine medication can also be used to block the action between histamines released from the mast cells and nerve endings (Wikipedia, 2017a), thereby suppressing the immediate reaction or Type I hypersensitivity (Smith et al., 1980).

Table 6.2. A variety of avoidance strategies to minimize interaction between people and parthenium weed antigens.

Avoidance strategy	Reference
As much parthenium weed as possible needs to be removed from the immediate living/working area	Lakshmi and Srinivas (2007a)
People exposed to parthenium weed need to cover as much of their skin as possible with clothing	Lakshmi and Srinivas (2007a)
Clothes need to be washed and dried indoors to avoid airborne parthenium weed allergens	Srinivas (2005); Goldsworthy (2005)
People exposed to parthenium weed need to shower or wash the uncovered areas of their skin with soap and water as often as possible, to wash off the antigen before it infects the skin	Lakshmi and Srinivas (2007a), Goldsworthy (2005)
People exposed to parthenium weed, particularly field workers, should wear disposable face masks that cover their nose and mouth and are rated for dust exclusion	McFadyen (1995)
People to wear gloves	Bhatia et al. (2015)
Use of barrier creams to slow down the skin penetration of antigens and to avoid the exposure to sunlight	Lakshmi and Srinivas (2007a)

Table 6.3. External disease caused by parthenium weed and suggested medical treatments, which should be taken as soon as possible.

Type of disease	Medication	References
Airborne contact dermatitis (ABCD)	Antihistamines	Smith *et al.* (1980)
	Corticosteroids	Gallant and Kenny (1986)
	Azathioprine	Verma *et al.* (2008)
	Betamethasone	Verma *et al.* (2008)
Hand dermatitis	Ahloroquine	Fisher (1996a)
Atopic dermatitis	Cyclosporine	Srinivas (2006)
Seborrhoeic dermatitis	Corticosteroids	Fisher (1996a)

Systemic immune-suppressant steroids have been effectively used for patients with parthenium weed-induced disease (Lakshmi *et al.*, 2008; Bhatia *et al.*, 2015). In particular, corticosteroids, such as prednisolone and betamethasone (Sharma and Sethuraman, 2007), have been successfully and widely used and several researchers (Gallant and Kenny, 1986; Andersson *et al.*, 1988) claim that the early and brief use of corticosteroids can reduce the immediate symptoms. Corticosteroids can also be applied topically for several days to diminish the mast cell activity in the affected skin, and as a result decrease the response to histamine-releasing agents (Verma and Pasricha, 1996). However, corticosteroids can produce unwanted side effects, such as sedation, which can pose safety risks to workers and the long-term use of corticosteroids for chronic conditions can lead to suppression of the adrenocortical axis (Gallant and Kenny, 1986; Lakshmi *et al.*, 2008), which can lead to debilitating conditions, e.g. Cushing's syndrome.

To avoid or reduce the use of corticosteroids, other medications, termed 'steroid-sparing', have been used as alternatives. Medications such as azathioprine, cyclosporine and methotrexate have been used effectively for parthenium weed-induced allergy. Although azathioprine is cheap, there is a time delay of 2–3 months before benefits can be seen. This produces the risk that some

people will stop treatment before seeing any benefit (Lakshmi *et al.*, 2008), however corticosteroids could be used in the initial phase (Sharma and Sethuraman, 2007). Cyclosporine, however, has been reported to produce benefits within 2 weeks (Lakshmi *et al.*, 2008). Cyclosporin is normally used in the acute situation of parthenium weed dermatitis as an immediate measure (Srinivas, 2006).

6.2.4 Biogeography

Native versus introduced range

Parthenium weed is a naturally occurring plant in Central and South America, around the Gulf Coast of the USA, the West Indies and Caribbean Islands (Rollins 1950, in Picman and Towers, 1982; Wedner *et al.*, 1989). Due to increased global transportation between areas, countries and continents, and because of its invasive attributes, parthenium weed has spread from the countries of its native range to successfully invade over 40 tropical and subtropical countries (Shabbir, 2012; Adkins and Shabbir, 2014; EPPO, 2014).

Differences in the plant from country to country

Research by Picman and Towers (1982) used 31 samples of parthenium weed to study the variations in sesquiterpene lactone chemistry between different global populations. From this work, seven chemical types of parthenium weed were identified, which can be broadly partitioned into a parthenin-dominant group and a more variable hymenin-dominated group.

In this work, all samples from North and Central America, India, Australia, South Africa, Venezuela, Brazil and one sample from Jamaica had parthenin as the major sesquiterpene lactone. Although the number of samples was limited, this chemical link suggests that these populations are closely related and had a similar origin, possibly from North America around the Gulf of Mexico (Picman and Towers, 1982). This is consistent with reports of parthenium weed being introduced into India and Australia by contaminated grain or pasture seed from Texas, USA (Lonkar et al., 1974; Haseler, 1976; McFadyen, 1995).

The second broad group within the 31 samples, the hymenin-dominated group from South America, was chemically more variable (Picman and Towers, 1982). A unique sample from Bolivia contained neither parthenin nor hymenin as the major sesquiterpene lactone, but only coronopilin. However, an earlier report on samples from Bolivia suggests that hymenin is the major sesquiterpene lactone found in these plants (Towers et al., 1977). Samples from Argentina and Jamaica all had hymenin as the major sesquiterpene lactone constituent, although early work by Towers et al. (1977) shows some samples from Jamaica having parthenin as the major sesquiterpene lactone. Parthenium weed with hymenin as the major sesquiterpene lactone was shown to also contain hysterin and dihydroisoparthenin (Picman et al., 1982). Later work by Dale (1981) noted morphological differences between these two broad groups and referred to them as: a white-flowered parthenium group (parthenin-dominant) and a cream to yellow-flowered South American group (hymenin-dominated). Other morphological differences between these two groups include axillary branch development and the size of capitula, disc flowers, petals and pollen.

Genetics

The impact of parthenium weed on human health varies within communities and between geographically separate communities. Globally, parthenium weed affects human health in most places where it occurs, but especially so where the weed has invaded and where it is most abundant (Wedner et al., 1989), in particular, India. No reports of negative health impacts have been found from South America, where the plant is both native and contains hymenin, instead of parthenin, as the dominant sesquiterpene lactone. The degree of impact on human health is also significantly less in the USA

and Australia. Lonkar et al. (1974) suggests that people in the USA may have a genetic or acquired tolerance, possibly from past exposure to parthenium weed or closely related Asteraceae plants, such as western ragweed.

Although it is prolific in many areas of India, the propensity and severity of the health impact on the human population is thought to result from additional factors, apart from the abundance of parthenium weed alone (Lonkar et al., 1974). There are differing research views on whether the health effects of parthenium weed are influenced by gender or age. In America (Towers et al., 1977) and India, it has been reported (Warshaw and Zug, 1996; Sharma and Sethuraman, 2007; Lakshmi and Srinivas, 2012) that men tend to be more affected by parthenium weed allergens than women or children. It has been hypothesized that men were more exposed to parthenium weed and that women and children may not be sensitized as often as men (Sharma and Kaur, 1989; Warshaw and Zug, 1996). However, this is not the case in India, as it is more common for the women and children to work in the field (Lonkar et al., 1974). In contrast, people surveyed in Nepal thought that children, the elderly and people new to the area were more susceptible to allergic reactions caused by parthenium weed (Karki, 2009). More recent observations of workers involved in parthenium weed research in Ethiopia (W. Mersie, Virginia State University Petersburg, 2016, personal communication) and South Africa (L. Strathie, Agricultural Research Council Pretoria, 2016, personal communication) have suggested that women workers were more affected by allergies than men and that the severity of the response and the exposure time required to produce an allergic response varied considerably.

In Australia, skin-prick testing of 207 individuals for atopic sensitivity to parthenium weed found that females were twice as likely to produce a positive test and be sensitized or allergic to parthenium weed (Goldsworthy, 2005). From this survey, there was also found to be no link between prevalence of allergy symptoms, a positive allergy status, gender, age, whether working indoors or outside and the amount of time residing in a parthenium weed area. There was an association between allergy symptoms and whether people reported contact with parthenium weed at home.

6.2.5 Indirect effects on human health: malaria management

Recent studies have shown that parthenium weed (or its extracts) may have a significant impact upon the behaviour and survival of certain kinds of insects (Ahmad *et al.*, 2011; Kumar *et al.*, 2011) and, in turn, since some of these are pathogenic, on human health. As explained by Nikbakhtzadeh *et al.* (2014), the effectiveness of a pathogen, and by association the occurrence of malaria, depends on the ability of the *Anopheles gambiae* mosquito to find a suitable sugar supply to prolong its survival and allow the pathogen to complete its life cycle. Nyasembe *et al.* (2014) have shown that the nectar-seeking habit of *A. gambiae* is influenced by the pathogen *Plasmodium falciparum* (William H. Welch, 1897), one of the pathogens that cause malaria in humans, and *A. gambiae* mosquitos have been reported to have a preference for parthenium weed as a sugar source when compared to several other common weed species (Nikbakhtzadeh *et al.* 2014; Nyasembe *et al.*, 2014).

Floral sugar or nectar availability impacts on the survival, reproduction and biting frequency of *A. gambiae* (Manda *et al.*, 2007). To find sugar, insects use floral scent and/or visual cues. The attraction of *A. gambiae* mosquitos to parthenium weed is thought to be facilitated by the volatile organic compounds, in particular, the terpenoids, released from the plant (Nikbakhtzadeh *et al.*, 2014). The ability of bioactive compounds of parthenium weed to modify the behaviour of *A. gambiae* highlights the potential for these chemicals, or synthetic equivalents, to be used as attractants (Nikbakhtzadeh *et al.*, 2014). Chemical attractants may form a powerful tool for surveillance and control of *A. gambiae*, thereby slowing the spread of malaria.

6.3 Animal Health

At the same time as research was being undertaken into the effects of parthenium weed on human health, work was also being conducted in India and Australia (Lonkar *et al.*, 1974; Haseler, 1976; Towers *et al.*, 1977; Bury and Kloot, 1982) on domesticated animal production systems. It was found that parthenium weed can cause diverse effects, both direct and indirect, on livestock. These effects include the indirect effects of reduced availability and quality of feed and the direct effects on livestock health and meat, milk and egg production and quality (Evans, 1997).

In India, it has been observed that parthenium weed is differentially grazed by livestock. Goats (*Capra aegagrus hircus* L.) were reported to graze parthenium weed freely, while cattle (Bos taurus L.) and buffalo (*Bubalus* sp.) tended to only graze it occasionally and other livestock tended to avoid the plant (Narasimhan *et al.*, 1977). Similar outcomes were observed in Nepal, where goats tended to graze parthenium weed and other animals preferred to avoid it (Karki, 2009). In Australia, sheep (*Ovis aries* L.) have been reported to graze parthenium weed more readily than cattle (Tudor *et al.*, 1982). Khosla and Sobti (1979) reported that cattle avoided eating parthenium weed and chickens (*Gallus gallus domesticus* L.) only peck at the flowers. The unpalatability of the weed to most domestic animals is thought to result from the odour, taste and occurrence of trichome hairs on the plants, leaves and stems (Gnanavel, 2013). It has been suggested that it is the sesquiterpene lactones in the trichome hairs that act as feed deterrent (Burnett *et al.*, 1974; Khosla and Sobti, 1979; Picman *et al.*, 1979; McFadyen, 1995).

Although cattle avoid eating parthenium weed as much as possible, when they do consume it, they develop symptoms of contact dermatitis, including rashes on their udders and over their bodies (Khosla and Sobti, 1979). Horses (*Equus ferus caballus* L.) grazing parthenium weed-infested areas have been reported to suffer from allergic skin reactions (Dhileepan, 2009) and black rhinoceros (*Diceros bicornis* L.), in the Pongola Nature Reserve in South Africa, developed swollen eyes and bright pink lips after grazing in areas dense with parthenium weed (Carnie, 2013).

In Nepal, it was noted that animals developed significant skin allergies and lesions when parthenium weed was used as bedding (Karki, 2009; Shrestha *et al.*, 2015) and skin allergies, boils, wounds (especially around the mouth), eye redness, loss of appetite and thirst resulted when cattle grazed parthenium weed. The use of injections, such as Avil and Flexona, have been effective treatments for these allergic responses in cattle (Karki, 2009). Other symptoms, such as coughing and diarrhoea, have been observed in cattle, sheep, goats and camels after grazing parthenium weed (Ayele, 2007). It has been suggested that diarrhoea is more commonly associated with stock eating the younger, green, immature plants (Ayele, 2007; Karki, 2009). In the survey undertaken by Karki (2009), local Nepalese farmers suspected that consumption of parthenium weed may have resulted in a reduction of fertility in cattle and the reproductive capacity of goats. Farmers also thought that new stock was more susceptible to parthenium weed and that allergic responses decreased over a few years.

Parthenium weed-infested areas in Australia have been associated with cattle that have a lower market value when compared with cattle from parthenium weed-free areas (Chippendale and Panetta, 1994). In cropping areas, parthenium weed infestations can lead to significant yield losses, by as much as 40% in certain areas of Pakistan (Khan, 2011), which then reduces the availability of feed for more intensive livestock production systems.

Ramappa *et al.* (1987) found the inclusion of parthenium weed meal into the diet of both layer and broiler chickens had no significant detrimental effect on egg or meat output. In layer chickens, the inclusion of 2–8% parthenium weed meal did not affect either egg number or weight. The birds remained normal and healthy and appeared unaffected when examined internally during post-mortem. Of some interest was the increased yellow colour of the egg yolk in chickens fed on parthenium weed meal, where the yellow intensity increased as the amount of parthenium weed meal increased (Ramappa *et al.* 1987). In some cultures, a more yellow egg yolk can cause a positive impact on demand (LeVaux, 2013).

Parthenium weed has been reported to taint milk and meat (Ayele, 2007; Karki, 2009; Lakshmi and Srinivas, 2012; Shrestha *et al.*, 2015). In Nepal, milk bitterness increased after cattle grazed parthenium weed and the local perception was that the quality of goat meat was reduced after consumption of parthenium weed. Goats were often underweight with a low fat content and had blistered intestines and dark spots on their throat and lungs after slaughter. The potential of parthenium weed to taint the meat of sheep was investigated by Tudor *et al.* (1982). In one experiment, where sheep were fed *c*.150 g of dried parthenium per day incorporated into a sorghum (*Sorghum bicolor* (L.) Moench) and grass mix for up to 28 days prior to slaughter, there was no reduction in meat quality or taint detected. However, in a second experiment, where sheep were grazed on parthenium weed-infested pastures for either 2 or 12 weeks before slaughter, mild tainting of the meat was detected. This taint was reduced by removing the sheep from the parthenium weed-infested area 14 days before slaughter. It was suggested that the level of tainting may be affected by the season and the growth stage of the parthenium weed and by the fat content of the meat, if this is where the chemicals responsible for tainting are deposited (Tudor *et al.*, 1982).

Due to the abundance of parthenium weed in some areas, parthenium weed has the potential to comprise a significant proportion of the diet of certain wild animals, if they will eat it. Even in areas where it is less abundant, bite-count observations by Meyer *et al.* (1984) found that parthenium weed could account for up to 9% of the diet of free-foraging deer (*Odocoileus virginianus* Zimmermann) in Texas during spring. The nutritional value of parthenium weed compared favourably with other forage sources, with a crude protein content of 22% and with a digestible energy of 2.53 kcal/g dry matter. This may also suggest that those animals which have evolved with parthenium weed may have more tolerance to the biochemical found within the weed.

6.3.1 Parthenium weed toxicology

Early livestock studies focused on feeding parthenium weed to cross-bred cattle and buffalo (Narasimhan *et al.*, 1977, 1980). When the animals were given a diet with a high content of parthenium weed shoots, all animals had an initial period of diarrhoea. The buffalo calves developed itching and severe dermatitis, such as erythematous papular eruptions, skin lesions, depigmentation and alopecia and later oedema around the eyelids and facial muscles. The cross-bred cattle did not suffer dermatitis, but all animals developed ulceration of the muzzle, nares, lips, gums, tongue and palate (Narasimhan *et al.*, 1977, 1980).

Apart from the external health effects reported in the studies above, consumption of parthenium weed caused serious internal damage to the cross-bred cattle and buffalo, including chronic or acute toxicity, a condition that has been termed partheniosis (Narasimhan *et al.*, 1980). Internal damage included ulceration of the alimentary tract and severe damage and necrosis to the liver, gall bladder and kidneys, with the majority of animals dying within 1–4 weeks and as early as 96 h after the onset of feeding. It was noted (Narasimhan *et al.*, 1980) that the buffalo calves had a lower mortality rate than cross-bred cattle, however all surviving animals recovered after the parthenium weed was withdrawn from their diet (Narasimhan *et al.*, 1977).

Additional detailed information on the internal damage caused by consumption of parthenium weed has been described by Rajkumar *et al.* (1988), citing inhibition of liver dehydrogenase in sheep, and by Ahmed *et al.* (1988a), who found that 'out-standing features of acute toxicity was congestion haemorrhages and rupture of elastic muscle

and connective tissue in most of the organs'. Haematological examinations revealed that buffalo calves suffering parthenium weed toxicity had significantly reduced red and white blood cell counts, haemoglobin and packed cell volume, and increased blood clotting time (Ahmed *et al.*, 1988b).

Additional toxicology information has come from studies of parthenium weed on the air-breathing giant gourami (*Colisa fasciatus* BL. and SCHN.), a local fish species in India (Prasad and Haider, 1986). The study found that exposure to aqueous extracts of parthenium weed seed caused increased metabolic activity, respiration rate and hyperactivity, including restlessness, darting movements, increased opercula function and gill ventilation. With continued exposure for more than 72 h, the higher extract concentrations resulted in neurotoxic-driven paralysis of the respiratory muscles and death. The fish exposed to the lower extract concentrations for extended periods adjusted to the extract and respiration rates slowly normalized. These authors suggest that chemicals in the parthenium weed extract inhibited two enzymes involved in respiratory muscle function and plasma-bicarbonate breakdown, thereby resulting in asphyxiation.

6.4 Chemicals Involved

Many chemical constituents of parthenium weed, identified by various researchers, have been tabulated in Roy and Shaik (2013) and include terpenoids, volatile oils, amino acids and sugars, phenolic derivatives and flavonoids (Table 6.4). Many of these secondary plant products have various bioactivities, including allelopathic effects on other plants (Kanchan and Jayachandra, 1980; Batish *et al.*, 2002; Shi *et al.*, 2015), negative health effects on animals (Narasimhan *et al.*, 1980; Tudor *et al.*, 1982; Ahmed *et al.*, 1988a; Shrestha *et al.*, 2015) and humans (Lonkar *et al.*, 1974; Towers *et al.*, 1977; Wedner *et al.*, 1987; McFadyen, 1995), and ethnobotanical uses (refer to Towers *et al.*, 1977).

Table 6.4. The known allergenic sesquiterpene lactones found in parthenium weed (*Parthenium hysterophorus* L.).

Compound	Activity on health	Reference of allergic activity
Parthenin	Identification as a sesquiterpene lactone in parthenium weed, positive patch test in sensitized patients (27/50), having parthenium weed dermatitis Allergic eczematous dermatitis Positive patch test in sensitized patients, having parthenium weed dermatitis	Mitchell and Dupuis (1971), Lonkar *et al.* (1974), Rodriguez *et al.* (1976a), Towers *et al.* (1977), Sohi *et al.* (1979), Khosla and Sobti (1979)
Hymenin	No positive reaction or very occasional weak positive patch test reaction in sensitized patients	Subba Rao *et al.* (1978)
Ambrosin	Positive patch test in patients with allergic eczematous contact dermatitis; positive patch test in sensitized patients (33/50), having parthenium weed dermatitis, positive patch test in people with contact allergy to *Chrysanthemum* sp.	Mitchell *et al.* (1970), Lonkar *et al.* (1974), Rodriguez *et al.* (1976a), Bleumink *et al.* (1976), Khosla and Sobti (1979)
Coronopilin	Positive patch test in patients with allergic eczematous contact dermatitis; strong allergen	Mitchell *et al.* (1970), Picman *et al.* (1980)
Tetraneurin A	Positive patch test in sensitized patients, having parthenium weed dermatitis	Sohi *et al.* (1979)
Hysterophorin	Positive patch test in sensitized patients, having parthenium weed dermatitis	Sohi *et al.* (1979)
Hysterin	Identification as a sesquiterpene lactone in parthenium weed and suspected as an allergen	Mitchell and Dupuis (1971), Picman *et al.* (1982)
Dihydroisoparthenin	Identification as a sesquiterpene lactone in parthenium weed and suspected as an allergen	Picman *et al.* (1982)

6.4.1 The main chemical culprits

Sesquiterpene lactones are reported to be responsible for the majority of deleterious impacts on humans and animals, with at least 45 sesquiterpene lactones being identified from parthenium weed (Roy and Shaik, 2013). Most plant aerial parts of parthenium weed are reported to contain bioactive sesquiterpene lactones (Picman and Towers, 1982; Picman, 1986; Roy and Shaik, 2013), the highest concentration occurring in the flowers and leaves, making up to 8% of the dry weight of the flower heads and 5% of the dry weight of the leaves, with nonsignificant amounts of sesquiterpene lactones in the stem and roots (Rodriguez *et al.*, 1976a). Sesquiterpene lactones are lipophilic and mainly occur in the plant oleoresin fraction (Sharma and Sethuraman, 2007). Some sesquiterpene lactones are biosynthesized in secretory structures, such as trichomes and capitate sessile glands, these structures being found to exude parthenin and ambrosin (Rodriguez *et al.*, 1976a). Glandular trichomes and the chemicals they produce are thought to be part of the plant's defence system to prevent grazing and deter pathogen attack (Burnett *et al.*, 1974; Rodriguez *et al.*, 1976a; Picman, 1986).

Most parthenium weed plants either contain parthenin or hymenin as the major sesquiterpene lactone (Towers *et al.*, 1977), with lower amounts of ambrosin, coronopilin, tetraneurin A and hysterophorin (Mitchell and Dupius, 1971; Sohi *et al.*, 1979; Picman *et al.*, 1982). Parthenium weed with hymenin as the major sesquiterpene lactone also contain hysterin (Mitchell and Dupius, 1971) and dihydroisoparthenin (Picman *et al.*, 1982).

Parthenin, identified as the main allergen of parthenium weed, is in the class of sesquiterpene lactones known as pseudoguinolides. It has an alpha methylene group exocyclic to the gamma lactone, which is thought to be primarily responsible for immune responses such as dermatitis (Mitchell *et al.*, 1971a; Rodriguez *et al.*, 1976b; Sharma and Sethuraman, 2007).

The pseudoguaianolide hymenin is a diastereomer of parthenin (Subba Rao *et al.*, 1978). Testing of individuals who are sensitized to parthenin has found them not to be cross-sensitive to hymenin (Subba Rao *et al.*, 1978), indicating some stereospecificity of these allergic reactions (Towers and Mitchell, 1983). Saxena *et al.* (1991) also demonstrated that parthenin is readily transformed by chemical or photochemical reactions into other derivatives, some of which have stronger bioactivity than parthenin itself.

6.4.2 Toxicity of the main allergen, parthenin

Following reports of animals developing chronic and acute toxicity after being fed various quantities of parthenium weed (Narasimhan *et al.*, 1977, 1980), research by Narasimhan *et al.* (1984) identified that the main toxic component of parthenium weed was parthenin. Parthenin was found to be the main chemical constituent responsible for vertebrate toxicity and when tested on rats (*Rattus* sp,) had an LD_{50} of 42.3 mg/kg body weight.

To investigate how animals metabolize parthenin, radioactively labelled parthenin was given both orally and directly to the heart of lactating guinea pigs (*Cavia porcellus* L.) and intravenously to a lactating cow. It was discovered that the excretion of parthenin took longer when given orally and that parthenin could be detected in the milk of both animals. Parthenin given directly to the heart had a much shorter interval before being detected in the milk, peaking just 1 h after being administered. While parthenin given orally (guinea pig) or intravenously (cow) produced maximum levels in the milk at *c.*5 h and could still be detected after 72 h, these results highlight that cows (and by inference, goats) feeding on parthenium weed would produce milk containing

parthenin. This is supported by the reports in Nepal of tainted milk coming from cows fed on parthenium weed (Shrestha *et al.*, 2015). This could possibly lead to health concerns for humans, particularly children being fed parthenium weed-tainted milk. Tanner and Mattocks (1987) have suggested that parthenin content in tainted cow's milk could be a contributing factor to Indian childhood cirrhosis.

Parthenin levels detected in the blood of the lactating cow were highest in its kidney, liver, mammary glands and bone marrow (Narasimhan *et al.*, 1984). These observations support the earlier observations that cattle and buffalo fed parthenium weed exhibited signs of liver and kidney damage (Narasimhan *et al.*, 1980). Ramos *et al.* (2002), investigating the mutagenic effects of parthenin, showed no significant effects when studied in bacterial reversion tests and a very limited effect in micronucleus induction tests. However, cytogenetic tests revealed parthenin to cause dose-dependent chromosomal aberrations, predominantly through the formation of chromatid breaks. Parthenin also produced nuclear morphological changes and was determined to be toxic at concentrations of 40–60 µM. These cytological aberrations are thought to have been caused by an indirect mechanism, possibly one that inhibits key metabolic enzymes, rather than having a direct effect on DNA or chromosome structure (Ramos *et al.*, 2002).

6.5 Summary

Parthenium weed is an aggressive, invasive weed of global significance. The successful survival strategies of this species have enabled it to spread and inhabit many tropical and subtropical locations, causing a negative impact on many agricultural industries, the environment and the health of humans, domesticated animals as well as other plants.

Some biological features of parthenium weed that enable it to be environmentally successful and significantly impact human and animal health include: (i) an ability to grow in a wide range of environmental conditions; (ii) a high reproductive capacity, allowing for a high rate of spread and involving the production of large quantities of pollen; and (iii) an arsenal of bioactive chemical compounds, in particular, the sesquiterpene lactones. Sesquiterpene lactones from many plants in the Asteraceae family, including parthenium weed, have been linked to allergenic diseases (Nandhakishore and Pasricha, 1994; Schloemer et al., 2015). Of the sesquiterpene lactones present, parthenin is the most abundant (Rodriguez et al., 1971) and has been identified as the most toxic (Narasimhan et al., 1984) and the main constituent responsible for the allergenic diseases associated with parthenium weed (Sharma and Verma, 2012).

Parthenium-induced allergenic diseases occur when a sensitized individual comes into contact with or inhales airborne particles of parthenium weed, such as pollen or other fine plant debris. These diseases include an array of dermatitis, most commonly ABCD, caused by a Type IV, delayed hypersensitive reaction, and internal conditions such as rhinitis, asthma and atopic dermatitis, as a result of Type I immediate hypersensitive reaction (Kumar et al., 2012). Similar diseases are reported in domesticated animals which have ingested or come into contact with parthenium weed, along with decreases in produce quality (Chippendale and Panetta, 1994; Karki, 2009).

The most devastating impact of parthenium weed on human and animal health is in India, where the plant is ubiquitous over vast land areas and most people and animals cannot completely avoid contact with the plant or its airborne parts. To reduce the impact of parthenium weed, avoidance strategies and planned medicinal interventions are recommended.

6.5.1 Recommendations and treatments

The long-term solution to prevent the detrimental effects attributed to parthenium weed is the reduction of plant numbers in the invaded landscape. The eradication of

the plant may not be possible once it is fully established, but management methods to reduce plant numbers and to reduce flowering, and therefore pollen production, can be implemented. These include cultural methods, such as plant removal or cultivation prior to flowering, the use of chemicals (Chapter 8), biological control or the use of competitive plants to displace parthenium weed (Chapter 7).

To reduce the detrimental effects of parthenium weed on human health, people should avoid coming into contact with the plant allergens and may need to remove themselves from the location where the plant is growing (Rao et al., 1985; Schloemer et al., 2015). To help reduce the incidence and severity of the diseases caused by the plant, several physical or cultural control methods can be employed, including: (i) wearing a full covering of clothes, gloves and face masks; (ii) regular washing of exposed skin; (iii) washing and drying of clothes indoors; and (iv) the application of barrier creams.

Quality of life can be improved by treating disease symptoms using topical steroid creams, antihistamines, systemic corticosteroids and steroid-sparing immune-suppressive drugs, such as cyclosporine.

Animals such as cattle, which are more susceptible to allergic reactions, should not be penned with parthenium weed alone but be able to forage for more preferable feed, where available. Some animals, such as goats, which may have or develop a tolerance for feeding on parthenium weed, could be used to reduce or control parthenium weed. Exclusion from parthenium weed for several weeks prior to slaughter or milking is required to ensure quality produce.

6.5.2 Gaps in knowledge

There have been relatively few reports on the impact of parthenium weed on human health in recent years. It would be beneficial to know if parthenium weed is still having a devastating impact on human health, especially in India, whether cultural methods and medical treatments are being successfully implemented and whether the human immune system is developing a tolerance or is becoming less responsive to parthenium weed. Work needs to continue on understanding how different biochemicals in parthenium weed affect the human body.

There is also limited data on the impact of parthenium weed on human health in the plant's native range countries. The impact on health is significantly less in native countries as opposed to the invaded countries. This may reflect tolerance developed by the immune system of people in its native countries over time, or it may reflect different biochemical profiles of the native parthenium weed population compared with the invasive populations (Shi, 2016). Further work is warranted to address these suppositions.

There is little data available on the impact of parthenium weed on native fauna. A number of wild animal species are becoming uncommon, which may be due to parthenium weed infestation. Research to investigate any possible impacts on native animals should be done in areas that have had a significant history of parthenium weed and also in areas where parthenium weed is proximal and information can be collected prior to any significant infestation.

As parthenium weed continues to spread through Africa and Asia, information needs to be collected from those countries which have recently been invaded and those that are proximal to infestations. There is a real opportunity to collect baseline data so that the true impact of parthenium weed on the environment, agricultural production systems and human and animal health can be ascertained. The collection of data and the monitoring of parthenium invasion may ultimately be used to manage both the spread and impact of parthenium weed.

Acknowledgements

We are grateful to Dr Asad Shabbir and K. Dhileepan for editing this chapter. Thanks to Dr K. Verma for contributing information on parthenium weed in India and for providing photographic images.

References

Abdulkerim-Ute, J. and Legesse, B. (2016) *Parthenium hysterophorus* L: distribution, impact and possible measures in Ethiopia. *Tropical and Subtropical Agroecosystems* 19, 61–72.

Adamson, D.C. (1996) Introducing dynamic considerations when economically evaluating weeds. MSc thesis, The University of Queensland, Brisbane, Australia.

Adkins, S. and Shabbir, A. (2014) Biology, ecology and management of the invasive parthenium weed (*Parthenium hysterophorus* L.). *Pest Management* 70, 1023–1029.

AEC group (2002) *Economic Impact of State and Local Government Expenditure on Weed and Pest Animal Management in Queensland*. Local Government Association of Queensland, Brisbane.

Agarwal, K.K., Kumar Nath, A., Jaisankar, T.J. and Souza, M.D. (2008) Parthenium dermatitis presenting as erythroderma. *Contact Dermatitis* 59, 182–183.

Agashe, S.N. and Abraham, J.N. (1988) Pollen calendar of Bangalore City, Part I. *Indian Journal of Aerobiology* 1, 35–38.

Ahlawat, M., Dahiya, P. and Chaudhary, D. (2013) Aeropalynological study in Rohtak city, Haryana, India: A 2-year survey. *Aerobiologia* 29, 121–129.

Ahmad, N., Fazal, H., Abbasi, B.H. and Iqbal, M. (2011) In vitro larvicidal potential against *Anopheles stephensi* and antioxidative enzyme activities of *Ginkgo biloba*, *Stevia rebaudiana* and *Parthenium hysterophorus*. *Asian Pacific Journal of Tropical Medicine* 4, 169–175.

Ahmed, M.N., Rao Rama, P. and Mahendar, M. (1988a) Experimental introduction of acute toxicity in buffalo calves by feeding *Parthenium hysterophorus* Linn. *Indian Journal of Animal Sciences* 58(6), 731–734.

Ahmed, M.N., Rao Rama, P. and Mahendar, M. (1988b) Hematological observations in experimental partheniosis in buffalo calves. *Indian Veterinary Journal* 65, 972–974.

Akhtar, N., Verma, K.K. and Sharma, A. (2010) Study of pro- and anti-inflammatory cytokine profile in the patients with parthenium dermatitis. *Contact Dermatitis* 63, 203–208.

Andersson, M., Andersson, P. and Pipkarull, U. (1988) Topical glucocorticosteroids and allergen induced increase in nasal reactivity: Relationship between treatment time and inhibitory effect. *Journal of Allergy and Clinical Immunology* 82, 1019–1026.

Australian Weeds Committee (2012) *Parthenium (Parthenium hysterophorus L.) Strategic Plan 2012–17.* Weeds of National Significance. Australian Government Department of Agriculture, Fisheries and Forestry, Canberra.

Ayele, S. (2007) The impact of parthenium (*Parthenium hysterophorus* L.) on the range ecosystem dynamics of the Jijiga Rangelands, Ethiopia. MSc thesis, Haramaya University, Dire Dawa, Ethiopia.

Batish, D.R., Singh, H.P., Kohli, R.K., Saxena, D.B. and Kaur, S. (2002) Allelopathic effects of parthenium against two weedy species, *Avena fatua* and *Bidens pilosa*. *Environmental and Experimental Botany* 47, 149–155.

Belgeri, A., Navie, S.C., Vivian-Smith, G. and Adkins, S.W. (2014) Early recovery signs of an Australian grassland following the management of *Parthenium hysterophorus* L. *Flora* 209, 587–596.

Bhatia, R.., Sharma, V.K., Ramam, M., Sethuraman, G. and Yadav, C.P. (2015) Clinical profile and quality of life of patients with occupational contact dermatitis from New Delhi, India. *Contact Dermatitis* 73, 171–181.

Bleumink, E., Mitchell, J.C., Geissman, T.A. and Towers, G.H.N. (1976) Contact hypersensitivity to sesquiterpene lactones in *Chrysanthemum* dermatitis. *Contact Dermatitis* 2, 81–88.

Bousquet, J., Reid, J. and Van Weel, C. (1997) Allergic rhinitis management pocket. *Allergy* 63, 990–996.

Burnett, W.C., Jones, S.B., Mabry, T.J. and Padolina, W.G. (1974) Sesquiterpene lactones – insect feeding deterrents in *Vernonia*. *Biochemical Systematics and Ecology* 2, 25–29.

Bury, J.N. and Kloot, P.M. (1982) The spread of composite (Compositae) weeds in Australia. *Contact Dermatitis* 8, 410–413.

Carnie, T. (2013) Invader spreading fast: Toxic plant threatens farms, health. *IOL News* 14 November. Available at: https://www.iol.co.za/news/toxic-plant-threatens-farms-health-1607181 (accessed 12 May 2018).

Chan, S.K. and Burrows, N.P. (2009) Atopic dermatitis. *Common Dermatitis* 37, 242–245.

Chaubal, P.D. and Kotmire, A.T. (1982) Aerobiological studies at Kolhapur. *Acta Biologica Indica* 10, 100–102.

Cheney, M. (1998) Determination of the prevalence of sensitivity to parthenium in areas of Queensland affected by the weed. Dissertation submitted in partial requirement for the Master of Public Health, Queensland University of Technology, Brisbane.

Chippendale, J.F. and Panetta, F.D. (1994) Cost of parthenium weed to the Queensland cattle industry. *Plant Protection Quarterly* 9, 73–76.

Dale, I.J. (1981) Parthenium weed in the Americas. *Australian Weeds* 1, 8–14.

Department of Agriculture and Fisheries (2015) Parthenium (*Parthenium hysterophorus*) Queensland distribution 2013–14. Available at: http://www.daf.qld.gov.au/_data/assests/pdf_file/0003/790491/Parthenium_2013.pdf (accessed 20 February 2017).

Department of the Environment and Heritage and the CRC for Australian Weed Management (2003) Parthenium weed (*Parthenium hysterophorus*) weed management guide. Available at: http://www.environment.gov.au/biodiversity/invasive/weeds/publications/guidelines/wons/p--hysterophorus.html (accessed 8 August 2016).

Dhileepan, K. (2009) Managing parthenium weed across diverse landscapes: prospects and limitations. In: Inderjit (ed.) *Management of Invasive Weeds*. Invading Nature – Springer Series in Invasion Ecology 5. Springer Science and Business Media, Amsterdam, pp. 227–259.

Dhileepan, K. (2012) Reproductive variation in naturally occurring populations of the weed *Parthenium hysterophorus* (Asteraceae) in Australia. *Weed Science* 60, 571–576.

ECHO East Africa Impact Centre (2016) ECHO EA Parthenium video. Available at: https://www.youtube.com/watch?v=9NlNKM4sCxc&feature=youtu.be (accessed 28 February 2017).

European and Mediterranean Plant Protection Organization (2014) Data sheets on invasive alien plants: *Parthenium hysterophorus* L. Asteraceae – Parthenium weed. *EPPO Bulletin* 44, 474–478.

Evans, H.C. (1997) *Parthenium hysterophorus*: A review of its weed status and the possibilities for biological control. *Biocontrol News Information* 18, 89–98.

Fisher, A.A. (1996a) Esoteric contact dermatitis. Part IV: Devastating contact dermatitis in India produced by American Parthenium weed. (The Scourge of India). *Cutis* 57, 297–298.

Fisher, A.A. (1996b) Esoteric contact dermatitis. Part III: Ragweed dermatitis. *Cutis* 57(4), 199–200.

French, S.W. (1930) A case of skin sensitivity to *Parthenium hysterophorus*. L. *Surgeon* 66, 673.

Gallant, C. and Kenny, P. (1986) Oral glucocorticoids and their complications. A review. *Journal of the American Academy of Dermatology* 14, 161–177.

Ghiani, A., Aina, R., Asero, R., Bellotto, E. and Citterio, S. (2012) Ragweed pollen collected along high-traffic roads shows a higher allergenicity than pollen sampled in vegetated areas. *Allergy* 67, 887–894.

Gnanavel, I. (2013) *Parthenium hysterophorus* L.: A major threat to natural and agro eco-systems in India. *Science International* 1(5), 124–131.

Goldsworthy, D. (2005) Parthenium weed, allergy symptoms and health status in a rural community. Conference paper, Queensland Health and Medical Scientific Meeting, Brisbane, November 2005.

Gupta, A.K., Madzia, S.E. and Batra, R. (2004) Etiology and management of seborrheic dermatitis. *Dermatology* 208, 89–93.

Gupta, S. and Chanda, S. (1991) Aerobiology and some chemical parameters of *Parthenium hysterophorus* pollen. *Grana* 30(2), 497–503.

Handa, S., De, D. and Mahajan, R. (2001) Airborne contact dermatitis – current perspectives in etiopathogenesis and management. *Indian Journal of Dermatology* 56, 700–706.

Haseler, W.H. (1976) *Parthenium hysterophorus* L. in Australia. *PANS* 22(4), 515–517.

Huang, S.-K., Zwollo, P. and Marsh, D.G. (1991) Class II major histocompatibility complex restriction of human T cell response to short ragweed allergen, *Amb a* V*. *European Journal of Immunology* 21, 1469–1473.

Jain, A.K. (1993) Allergenic pollen grains of *Parthenium hysterophorus* Linn. In the atmosphere of Gwalior (MP). *Proceedings of the National Academy of Sciences, India, Section B* 59, 55–58.

Jayachandra (1971) Parthenium weed in Mysore State and its control. *Current Science* 40, 568–569.

Kanchan, S.D. and Jayachandra (1980) Allelopathic effects of *Parthenium hysterophorus* L. Part II. Leaching of inhibitors from aerial vegetative parts. *Plant and Soil* 55, 61–66.

Karki, D. (2009) Ecological and socio-economic impact of *Parthenium hysterophorus* L. invasion in two urban cities of south-central Nepal. MSc thesis, Tribhuvan University, Kathmandu, Nepal.

Khan, H. (2011) Distribution and management of parthenium weed (*Parthenium hysterophorus* L.) in Peshawar Valley, Northwest-Pakistan. PhD thesis, Khyber Pakhtunkhwa Agricultural University, Peshawar, Pakistan.

Khosla, S.N. and Sobti, S.N. (1979) Parthenium – a national health hazard, its control and utility. A review. *Pesticides* 13, 121–127.

Kohli, R.K., Batish, D.R., Singh, H.P. and Dogra, K.S. (2006) Status, invasiveness and environmental threats of three tropical American invasive weeds (*Parthenium hysterophorus* L.,

Ageratum conyzoides L., *Lantana camara* L.) in India. *Biological Invasions* 8, 1501–1510.

Kololgi, P.D., Kololgi, S.D. and Kololgi, N.P. (1997) Dermatologic hazards of parthenium in human beings. In: Mahadevappa, M. and Patil, V.C. (eds) *First International Conference on Parthenium Management Vol. 1*, Dharwad, India, pp. 18–19.

Kumar, S., Singh, A.P., Nair, G., Batra, S., Seth, A., Wahab, N., and Warikoo, R. (2011) Impact of *Parthenium hysterophorus* leaf extracts on the fecundity, fertility and behavioural response of *Aedes aegypti* L. *Parasitology Research* 108, 853–859.

Kumar, S., Khandpu, S., Rao, D.N., Wahaab, S. and Khanna, N. (2012) Immunological response to *Parthenium hysterophorus* in Indian patients with parthenium sensitive atopic dermatitis. *Immunological Investigations* 41, 75–86.

Lakshmi, C. and Srinivas, C.R. (2007a) Parthenium: A wide angle view. *Indian Journal of Dermatology, Venereology and Leprology* 73, 296–306.

Lakshmi, C. and Srinivas, C.R. (2007b) Type I hypersensitivity to *Parthenium hysterophorus* in patients with parthenium dermatitis. *Indian Journal of Dermatology, Venereology and Leprology* 73, 103–105.

Lakshmi, C. and Srinivas, C.R. (2012) Parthenium the terminator: An update. *Indian Dermatology Online Journal* 3, 89–100.

Lakshmi, C., Srinivas, C.R. and Jayaraman, A. (2008) Ciclosporin in parthenium dermatitis – a report of 2 cases. *Contact Dermatitis* 59, 245–255.

LeVaux, A. (2013) Marketing the perfectly colored egg yolk. Available at: http://modernfarmer. com/2013/12/marketing-perfectly-colored-egg-yolk/ (accessed 20 February 2017).

Lonkar, A., Mitchell, J.C. and Calnan, C.D. (1974) Contact dermatitis from *Parthenium hysterophorus*. *Transactions of the St John's Hospital Dermatological Society* 60, 43–53.

Mahajan, V.K., Sharma, V., Gupta, M., Chauhan, P.S., Mehta, K.S. and Garg, S. (2014) Parthenium dermatitis: Is parthenolide an effective choice for patch testing? *Contact Dermatitis* 70, 340–343.

Manda, H., Gouagna, L.C., Nyandat, E., Kabiru, E.W., Jackson, R.R., *et al.* (2007) Discriminative feeding behaviour of *Anopheles gambiae* s.s. on endemic plants in western Kenya. *Medical Veterinary Entomology* 21, 103–111.

Mangala, A., Viswanath, B. and Subba Rao, P.V. (1981) Atmospheric survey of pollen of *Parthenium hysterophorus*. *Annals of Allergy, Asthma and Immunology* 47, 192–196.

Martinez, F.D. (2007) Genes, environments, development and asthma: A reappraisal. *European Respiratory Journal* 29(1), 179–184.

McFadyen, R.E. (1995) Parthenium weed and human health in Queensland. *Australian Family Physician* 24, 1455–1459.

Medicinal Plants Archive (2016) Pollen calendar of Bangalore City India. Available at: http://www.medicinalplantsarchive.us/pollen-grains/pollen-calendar-of-bangalore-city-india.html (accessed 8 February 2017).

Meyer, M.W., Brown, R.D. and Graham, M.W. (1984) Protein and energy content of white-tailed deer diets in the Texas Coastal Bend. *The Journal of Wildlife Management* 48, 527–534.

Miao, B., Turner, B.L. and Mabry, T.J. (1995) Systematics implications of chloroplast DNA variation in the subtribe Ambrosiinae (*Asteraceae*: Heliantheae). *American Journal of Botany* 82(7), 924–932.

Mitchell, J.C. and Dupuis, G. (1971) Allergic contact dermatitis from sesquiterpenoids of the Compositae family of plants. *British Journal of Dermatology* 84, 139–150.

Mitchell, J.C., Fritig, B., Singh, B. and Towers, G.H.N. (1970) Allergic contact dermatitis from *Frullania* and Compositae: The role of sesquiterpene lactones. *Journal of Investigative Dermatology* 54, 233–239.

Mitchell, J.C., Geissman, T.A., Dupuis, G. and Towers, G.H.N. (1971a) Allergic contact dermatitis caused by *Artemisia* and *Chrysanthemum* species. *Journal of Investigative Dermatology* 56(2), 98–101.

Mitchell, J.C., Roy, A.K., Dupuis, G.D. and Towers, G.H.N. (1971b) Allergic contact dermatitis from ragweeds (*Ambrosia* species): The role of sesquiterpene lactones. *Contact Dermatitis* 104, 73–76.

Möller, H., Spiren. A., Svensson, Å., Gruvberger, B., Hindsen, M. and Bruze, M. (2002) Contact allergy to the Asteraceae plant *Ambrosia artemisiifolia* L. (ragweed) in sesquiterpene lactone-sensitive patients in southern Sweden. *Contact Dermatitis* 47, 157–160.

More, P.R., Vadlamudi, V.P. and Qureshi, M.I. (1982) Note on the toxicity of *Parthenium hysterophorus* in livestock. *Indian Journal of Animal Science* 52, 456–457.

Nadeem, M., Rani, Z., Aman, S. and Shabbir, A. (2005) Parthenium weed: A growing concern in Pakistan. *Journal of Pakistan Association of Dermatologists* 15(1), 4–8.

Nandhakishore, T. and Pasricha, J.S. (1994) Pattern of cross-sensitivity between 4 Compositae plants, *Parthenium hysterophorus*, *Xanthium strumarium*, *Helianthus annuus* and *Chrysanthemum coronarium*, in Indian patients. *Contact Dermatitis* 30(3), 162–167.

Narasimha, S.K., Srinivas, C.R. and Mathew, A.C. (2005) Effect of topical corticosteroid frequency on histamine induced wheals. *International Journal of Dermatology* 44, 425–427.

Narasimhan, T.R., Ananth, M., Naryana Swamy, M., Rajendra Babu, M., Mangala, A. and Subba Rao, P.V. (1977) Toxicity of *Parthenium hysterophorus* L. *Current Science* 46, 15–16.

Narasimhan, T.R., Ananth, M., Naryana Swamy, M., Rajendra Babu, M., Mangala, A. and Subba Rao, P.V. (1980) Toxicity of *Parthenium hysterophorus* L.: parthenoiosis in cattle and buffaloes. *Indian Journal of Animal Science* 50, 173–178.

Narasimhan, T.R., Keshava Murthy, S., Harindranath, N. and Subba Rao, P.V. (1984) Characterization of a toxin from *Parthenium hysterophorus* and its mode of excretion in animals. *Journal of Bioscience* 6(5), 729–738.

Navie, S.C., McFadyen, R.E., Panetta, F.D. and Adkins, S.W. (1996) The biology of Australian weeds. 27. *Parthenium hysterophorus* L. *Plant Protection Quarterly* 11, 76–88.

Navie, S.C., McFadyen, R.E., Panetta, F.D. and Adkins, S.W. (2005) The effect of CO_2 enrichment on the growth of a C3 weed (*Parthenium hysterophorus* L.) and its competitive interaction with a C4 grass (*Cenchrus ciliaris* L.). *Plant Protection Quarterly* 20, 61–66.

Nikbakhtzadeh, M.R., Terbot II, J.W., Otienoburu, P.E. and Foster, W.A. (2014) Olfactory basis of floral preference of the malaria vector *Anopheles gambiae* (Diptera: Culicidae) among common African plants. *Journal of Vector Ecology* 39, 372–383.

Nyasembe, V.O., Teal, P.E.A., Sawa, P., Tumlinson, J.H., Borgemeister, C. and Torto, B. (2014) *Plasmodium falciparum* infection increases *Anopheles gambiae* attraction to nectar sources and sugar uptake. *Current Biology* 24, 1–5.

Pasricha, J.S., Verma, K.K. and D'Souza, P. (1995) Air-borne contact dermatitis caused exclusively by *Xanthium strumarium*. *Indian Journal of Dermatology, Venereology and Leprology* 61, 354–355.

Picman, A.K. (1986) Biological activities of sesquiterpene lactones. *Biochemical Systematics and Ecology* 14(3), 255–281.

Picman, A.K. and Towers, G.H.N. (1982) Sesquiterpene lactones in various populations of *Parthenium hysterophorus*. *Biochemical Systematics and Ecology* 10, 145–153.

Picman, A.K., Rodriguez, E. and Towers, G.H.N. (1979) Formation of adducts of parthenin and related sesquiterpene lactones with cysteine and glutathione. *Chemico-Biological Interactions* 28, 83–89.

Picman, A.K., Towers, G.H.N. and Subba Rao, P.V. (1980) Coronopilin – another major sesquiterpene lactone in *Parthenium hysterophorus*. *Phytochemistry* 19, 2206–2207.

Picman, A.K., Balza, F. and Towers, G.H.N. (1982) Occurrence of hysterin and dihydroisoparthenin in *Parthenium hysterophorus*. *Phytochemistry* 21, 1801–1802.

Prasad, M.S. and Haider, Z.A. (1986) Effect of parthenium extract on the metabolism of young *Colisa fasciatus* (BL. and SCHN.) (Pisces, Perciformes). *Clean: Soil Air Water* 14, 661–666.

Rajkumar, E.D.M., Kumar, N.V.N. and Haran, N.V.H. (1988) Antagonistic effect of *Parthenium hysterophorus* on succinate-dehydrogenase of sheep liver. *Journal of Environmental Biology* 9, 231–237.

Ramappa, B.S., Devegowda, G., Umakantha, B. and Jain, A.K. (1987) Effect of inclusion of meal of parthenium (*Parthenium hysterophorus*) in layer diets. *Indian Journal of Animal Sciences* 57, 1016–1018.

Ramos, A., Rivero, R., Visozo, A., Piloto, J. and García, A. (2002) Parthenin, a sesquiterpene lactone of *Parthenium hysterophorus* L. is a high toxicity clastogen. *Mutation Research* 514, 19–27.

Rao, M., Prakash, O. and Subba Rao, P.V. (1985) Reaginic allergy to parthenium pollen evaluation by skin test and RAST. *Clinical Allergy* 15, 449–454.

Rodriguez, E., Hirosuke, Y. and Mabry, T.J. (1971) The sesquiterpene lactone chemistry of the genus *Parthenium* (compositae). *Phytochemistry* 10(5), 1145–1154.

Rodriguez, E., Dillon, M.O., Mabry, T.J., Mitchell, J.C. and Towers, G.H.N. (1976a) Dermatologically active sesquiterpene lactones in trichomes of *Parthenium hysterophorus* L. (Compositae). *Experimentia* 32, 236–238.

Rodriguez, E., Towers, G.H.N. and Mitchell, J.C. (1976b) Review: Biological activities of sesquiterpene lactones. *Phytochemistry* 15, 1573–1580.

Rogers, C.A., Wayne, P.M., Macklin, E.A., Mullenberg, M.L., Wagner, C.J., Epstein, P.R. and Bazzaz, F.A. (2006) Interaction of the onset of spring and elevated atmospheric CO_2 on ragweed (*Ambrosia artemisiifolia* L.) pollen production. *Environmental Health Perspectives* 114, 865–869.

Roy, D.C. and Shaik, M. (2013) Toxicology, phytochemistry, bioactive compounds and pharmacology of *Parthenium hysterophorus*. *Journal of Medicinal Plant Studies* 1(3), 126–141.

Saxena, D.B., Dureja, P., Kumar, B., Rani, D. and Kohli, R.K. (1991) Modification of parthenin. *Indian Journal of Chemistry* 30(9), 849–852.

Schloemer, J.A., Zirwas, M.J. and Burkhart, C.G. (2015) Airborne contact dermatitis: Common causes in the USA. *International Journal of Dermatology* 54(3), 271–274.

Seetharamiah, A.M., Viswanath, B. and Subba Rao, P.V. (1981) Atmospheric survey of pollen of *Parthenium hysterophorus*. *Annals of Allergy* 47, 192–196.

Sethuraman, G., Bansal, A., Sharma, V.K. and Verma, K.K. (2008) Seborrhoeic pattern of parthenium dermatitis. *Contact Dermatitis* 58, 372–374.

Shabbir, A. (2012) Towards the improved management of parthenium weed: Complementing biological control with plant suppression. PhD thesis, The University of Queensland, Brisbane, Australia.

Sharma, S.C. and Kaur, S. (1989) Airborne contact dermatitis from composite plants in Northern India. *Contact Dermatitis* 21, 1–5.

Sharma, S.C. and Sethuraman, G. (2007) Parthenium dermatitis. *Dermatitis* 18, 183–190.

Sharma, V.K. and Verma, P. (2012) Parthenium dermatitis in India: past, present and future. *Contact Dermatitis* 78(5), 560–568.

Sharma, S.C., Sethuraman, G. and Bhat, R. (2005) Evolution of clinical pattern of parthenium dermatitis: a study of 74 cases. *Contact Dermatitis* 53, 84–88.

Sharma, V.K., Verma, P. and Maharaja (2013) Parthenium dermatitis in India. *Photochemical & Photobiological Sciences* 12, 85–94.

Shi, B. (2016) Invasive potential of the weed *Parthenium hysterophorus*: the role of allelopathy. PhD thesis, the University of Queensland, Brisbane, Australia.

Shi, B., Aslam, Z. and Adkins, S.W. (2015) The invasive potential of parthenium weed: a role for allelopathy. In: Price, J.E. (ed.) *New Developments in Allelopathy Research*. Nova Science Publishers, New York.

Shrestha, B.B., Shabbir, A. and Adkins, S.W. (2015) *Parthenium hysterophorus* in Nepal: A review of its weed status and possibilities for management. *Weed Research* 55, 132–144.

Singh, A.B., Pandit, T. and Dahiya, P. (2003) Changes in airborne pollen concentrations in Dehli, India. *Grana* 42, 168–177.

Smith, J.A., Mansfield, L.E., DeShazo, R.D. and Nelson, H.S. (1980) An evaluation of the pharmacologic inhibition of the immediate and late cutaneous reactions to allergen. *Journal of Allergy and Clinical Immunology* 65, 185.

Sohi, A., Tiwari, V., Lonkar, A., Rangachar, S. and Nagasampagi, B. (1979) Allergenic nature of *Parthenium hysterophorus. Contact Dermatitis* 5, 133–136.

Srinivas, C.R. (2005) Transmission of parthenium dermatitis by clothing. *Archives of Dermatological Research* 141, 1605.

Srinivas, C.R. (2006) Parthenium dermatitis treated with azathioprine weekly pulse doses. *Indian Journal of Dermatology, Venereology and Leprology* 72, 234.

Srinivas, C.R., Krupashankar, D.S., Singh, K.K., Balachandran, C. and Shenoi, S.D. (1988) Oral hyposensitization in parthenium dermatitis. *Contact Dermatitis* 18, 242–243.

Sriramarao, P. and Rao, P.V. (1993) Allergenic cross-infectivity between parthenium and ragweed pollen allergens. *International Archives of Allergy Immunology* 100, 74–85.

Subba Rao, P.V., Mangala, A., Subba Rao, B.S. and Prakash, K.M. (1977) Clinical and immunological studies on persons exposed to *Parthenium hysterophorus* L. *Specialia Experientia* 33, 1387–1388.

Subba Rao, P.V., Mangala, A., Towers, G.H.N. and Rodriguez, E. (1978) Immunological activity of parthenium and its diasteriomer in persons sensitized by *Parthenium hysterophorus* L. *Contact Dermatitis* 4, 199–203

Sushilkumar and Varshney, J.G. (2010) Parthenium infestation and its estimated cost management in India. *Indian Journal of Weed Science* 42(1&2), 73–77.

Swinnen, I. and Goossens, A. (2013) An update on airborne contact dermatitis: 2007–2011. *Contact Dermatitis* 68, 232–238.

Tamado, T. and Milberg, P. (2000) Weed flora in arable fields of eastern Ethiopia with emphasis on the occurrence of *Parthenium hysterophorus. Weed Research* 40, 507–521.

Tanner, M.S. and Mattocks, A.R. (1987) Hypothesis: Plant and fungal biocides, copper and Indian childhood liver disease. *Annals of Tropical Paediatrics* 7, 264–269.

The Australian Society of Clinical Immunology and Allergy (2015) Pollen allergy. Available at: http://www.allergy.org.au/images/pcc/ASCIA_PCC_Pollen_allergy_2015.pdf (accessed 10 August 2016).

Thorp, J.R. and Lynch, R. (2000) The determination of weeds of national significance, National Weeds Strategy Executive Committee, Launceston.

Tiwari, V.D., Sohi, A.S. and Chopra, T.R. (1979) Allergic contact dermatitis due to *Parthenium hysterophorus. Indian Journal of Dermatology, Venereology and Leprology* 45, 392–400.

Towers, G.H. and Mitchell, J.C. (1983) The current status of the weed *Parthenium hysterophorus* L. as a cause of allergic contact dermatitis. *Contact Dermatitis* 15, 465–469.

Towers, G.H.N., Mitchell, T.C., Rodriguez, E., Bennett, F.D. and Subba Rao, P.V. (1977) Biology and chemistry of *Parthenium hysterophorus* L, a problem weed in India. *Journal of Scientific and Industrial Research* 36, 672–684.

Tudor, G.D., Ford, A.L., Armstrong, T.R. and Bromage, E.K. (1982) Taints in meat from sheep grazing *Parthenium hysterophorus. Australian Journal of Experimental Agriculture and Animal Husbandry* 22, 43–46.

Veien, N.K. (2011) Systemic contact dermatitis. *International Journal of Dermatology* 50, 1445–1456.

Verma, K.K. and Pasricha, J.S. (1996) Azathioprine as a corticosteroid sparing agent in air borne contact dermatitis. *Indian Journal of Dermatology, Venereology and Leprology* 62, 30–32.

Verma, K.K., Manchanda, Y. and Pasricha, J.S. (2000) Azathioprine as a corticosteroid sparing agent for the treatment of dermatitis caused by the weed parthenium. *Acta Dermato Venereologica* 80, 31–32.

Verma, K.K., Mahesh, R., Srivastava, P., Ramam, M. and Mukhopadhyaya, A.K. (2008) Azathioprine versus betamethasone for the treatment of parthenium dermatitis: a randomized controlled study. *Indian Journal of Dermatology, Venereology and Leprology* 74, 453–457.

Wan, S., Yuan, T., Bowdish, S., Wallace, L., Russell, S.D. and Luo, Y. (2002) Response of an allergenic species, *Ambrosia psilostachya* (Asteraceae), to experimental warming and clipping: Implications for public health. *American Journal of Botany* 89, 1843–1846.

Warshaw, E.M. and Zug, K.A. (1996) Sesquiterpene lactone allergy. *American Journal of Contact Dermatology* 7, 1–23.

Wedner, H.J., Zenger, V.E. and Lewis, W.H. (1987) Allergic reactivity of *Parthenium hysterophorus* (Santa Maria Feverfew) pollen: An unrecognized allergen. *International Archives of Allergy and Applied Immunology* 84, 116–122.

Wedner, H.J., Wilson, P. and Lewis, W.H. (1989) Allergic reactivity to *Parthenium hysterophorus* pollen: An ELISA study of 582 sera from the United States Gulf Coast. *Journal of Allergy and Clinical Immunology* 84, 263–271.

Wiesner, M., Tessema, T., Hoffmann, A., Wilfried, P., Buttner, C., Mewis, I. and Ulrichs, C. (2007). Impact of the pan-tropical weed *Parthenium hysterophorus* L. on human health in Ethiopia. Institute of Horticultural Science, Urban Horticulture, Berlin, Germany.

Wikipedia (2017a) Mast cell. Available at: http://www.wikipedia.org/wiki/Type_I_hypersensitivity (accessed 20 February 2017).

Wikipedia (2017b) Type I hypersensitivity. Available at http://www.wikipedia.org/wiki/Type_I_-hypersensitivity (accessed 20 February 2017).

Ziska, L.H. and Caulfield, F.A. (2000) Rising carbon dioxide and pollen production of common ragweed, a known allergy-inducing species: Implications for public health. *Australian Journal of Plant Physiology* 27, 893–898.

7 Biological Control

Kunjithapatham Dhileepan,[1]* Rachel McFadyen,[2] Lorraine Strathie[3] and Naeem Khan[4]

[1]Biosecurity Queensland, Department of Agriculture and Fisheries, Brisbane, Queensland, Australia; [2]PO Box 88, Mt Ommaney, Queensland, Australia; [3]Agricultural Research Council – Plant Health and Protection, Hilton, South Africa; [4]Department of Weed Sciences, University of Agriculture, Peshawar, Pakistan

7.1 Introduction

Management options for parthenium weed (*Parthenium hysterophorus* L.) include chemical, physical, grazing management and biological methods (Dhileepan, 2009). Chemical control is often the first line of defence, but high costs of herbicides preclude their long-term use for parthenium weed management in grazing areas, public and uncultivated areas, forests and woodlands. Physical methods such as grading, slashing and ploughing can provide some relief over the short term, but they may exacerbate the associated health risk due to exposure and are not effective in long-term management. Management of parthenium weed can be achieved by maintaining sufficient levels of grass cover to maximize competition against the weed. However, biological control is regarded as the most effective and economic method for parthenium weed in many situations. Biological control is broadly defined as the use of a biological agent, a complex of biological agents or a biological process to bring about weed suppression. Various biological control options that include classical biological control, mycoherbicides and suppressive plants have been reviewed extensively (e.g. Dhileepan and Strathie, 2009; Dhileepan, 2009; Strathie *et al.*, 2011; Dhileepan and McFadyen, 2012; Khan *et al.*, 2014).

7.2 Classical Biological Control

Parthenium weed is susceptible to herbivory (Raghu and Dhileepan, 2005), but there are no herbivores or pathogens in the introduced ranges that are known to exert any critical impact on parthenium weed (e.g. Dhileepan and Strathie, 2009). Classical biological control, using host-specific (highly selective) natural enemies introduced from the native range of the plant, to cause vegetative or reproductive suppression, is the most cost-effective, long-term management option for parthenium weed (e.g. Page and Lacey, 2006). The intention is that the natural enemies will form self-sustaining populations that disperse naturally (or are deliberately dispersed) through the invaded range, within their range of suitability, with limited further interventions required. The benefits of this process accumulate with time.

7.2.1 History

Biological control of parthenium weed was initiated in Australia in 1976. Testing and release of biological control agents continued there until 2004 (Table 7.1), after which agents continued to be evaluated and redistributed. Based on the successful

---

* k.dhileepan@qld.gov.au

Table 7.1. Introduction history and current status of parthenium weed biological control agents in various countries. (Data from Dhileepan and Strathie 2009; Dhileepan and McFadyen 2012; K. Dhileepan unpublished data; L. Strathie unpublished data; S. Adkins unpublished data; S. Tang, unpublished data.)

Biological control agents	Country of introduction	Source country	Year imported/ reported	Year release approved/ commenced	Establishment status
Lepidoptera: Tortricidae					
Epiblema strenuana Walker	Australia	Mexico	1982	1982	Widespread and abundant
	India	Australia	1985	Not released	Unknown
	Sri Lanka	Australia	2003	2004	Localised
	Papua New Guinea	Australia	2004	Colony failed	
	Vanuatu	Unknown	2014	No deliberate release	
	South Africa	Australia	2010 and 2018	Testing in progress	
	China	Australia	1990	Released on ragweed	Established on parthenium in Guangxi Province
Platphalonidia mystica (Razowski & Becker)	Australia	Argentina	1991	1992	Localised
Lepidoptera: Sessidae					
Carmenta nr *ithacae* (Beutenmüller)	Australia	Mexico	1996	1998	Widespread and abundant
	South Africa	Australia	2014	Testing in progress	
Lepidoptera: Bucculatricidae					
Bucculatrix parthenica Bradley	Australia	Mexico	1983	1985	Widespread
Coleoptera: Chrysomelidae					
Zygogramma bicolorata Pallister	Australia	Mexico	1980	1981	Widespread and abundant
	India	Mexico	1983	1984	Widespread and abundant
	Pakistan	India	2003	No deliberate release	Widespread
	Nepal	India	2009	No deliberate release	Widespread and abundant
	South Africa	Australia	2005	2013	Localised establishment
	Ethiopia	South Africa	2007	2013	No establishment
	Tanzania	South Africa	2013	2013	No establishment
	Uganda	South Africa	2018	2018	Unknown

Taxon	Country	Origin	Year		Status
Coleoptera: Curculionidae					
Listronotus setosipennis Hustache	Australia	Argentina & Brazil	1981	1982	Abundant and widespread
	South Africa	Australia	2003	2013	Widespread establishment
	Ethiopia	South Africa	2007	2013	No large scale releases.
	Uganda	South Africa	2018	2018	Unknown
Smicronyx lutulentus Dietz	Australia	Mexico	1980	1981	Abundant and widespread
	India	India	1985 and 2018	Colony failed in 1985. Testing in progress	
	South Africa	Australia	2010	Testing in progress	Widespread establishment
	Ethiopia	South Africa	2015		
Conotrachelus albocinereus Fiedler	Australia	Argentina	1992	1995	Localised
Homoptera: Delphacidae					
Stobaera concinna (Stål)	Australia	Mexico	1982	1983	Localised
Basidiomycotina: Uredinales					
Puccinia abrupta partheniicola Parmelee	Australia	Mexico	1991	1991	Localised and abundant
	Ethiopia	Unknown	1997	No deliberate release	Widespread and abundant
	India	Unknown	1980	No deliberate release	Localised
	South Africa	Unknown	1995	No deliberate release	Localised
	Nepal	Unknown	2011	No deliberate release	Localised
	China	Unknown	2007	No deliberate release	Localised
Puccinia xanthii var. *parthenii-hysterophorae* Seier, H.C. Evans & Á. Romero	Australia	Mexico	1999	1999	Widespread
	Sri Lanka	Australia	2003	Unknown	Unknown
	South Africa	Australia	2004	2010	Localised

outcomes in Australia, biological control was subsequently initiated in India, then Sri Lanka, South Africa, Ethiopia, Tanzania and Oceania (Table 7.1). Attempts to broaden the use of biological control (e.g. in East Africa) are in progress.

7.2.2 Native range surveys

Native range surveys for biological control agents were first conducted by the Queensland Government (Australia) from 1976 to 1992. Based on preliminary surveys in Mexico, southern USA, Brazil, Argentina and the Caribbean islands, more intensive studies were continued in both North and South America (Dhileepan and McFadyen, 2012). The Gulf of Mexico was considered to be the origin of the invasive parthenium weed biotype in Australia, so survey efforts were focused in Mexico. Surveys there were conducted from 1978 to 1994 (McClay *et al.*, 1995). Explorations for pathogens were conducted in Mexico from 1995 to 1997 (Evans, 1997). In South America, explorations were conducted in Bolivia, Argentina and southern Brazil from 1977 to 1980 and in Argentina from 1987 to 1992.

Around 262 phytophagous arthropod species and several fungal pathogens were catalogued from parthenium weed in the southern USA and Mexico (McClay *et al.*, 1995; Evans, 1997). In Brazil and Argentina, around 100 insect species were recorded, of which only three were prioritized for further host-specificity tests (Dhileepan and McFadyen, 2012). Nine insect species (six from Mexico and three from South America) and two rust fungi underwent host-range testing and were released in Australia from 1980 onwards (Table 7.1). None of the 15 insect species recorded on parthenium weed in the Caribbean islands was selected for host-specificity tests as none seemed to be adequately host-specific. No surveys have been conducted in Bolivia as yet, although the lowlands of this country may yield additional agents.

7.2.3 Key biological control agents

Zygogramma bicolorata Pallister (Coleoptera: Chrysomelidae)

This leaf-feeding beetle from Mexico feeds and develops on only one or two species in the genera *Parthenium* and *Ambrosia* (McFadyen and McClay, 1981) and was not recorded feeding on sunflower (*Helianthus annuus* L.) in its native range. Both adults (Fig. 7.1a) and larvae (Fig. 7.1b) feed on parthenium weed leaves, preferentially on younger leaves, and can be highly damaging under some conditions, entirely defoliating stands of parthenium weed. Adults lay eggs singly or in groups on the leaves, flower heads, stems and on terminal and axillary buds. Newly hatched larvae feed voraciously on young leaves. Fully grown larvae burrow into the soil below the plant to pupate inside an earthen cell, the pupal stage lasting 2 weeks. The life-cycle is completed in 6–8 weeks, with up to four generations per year, depending on rainfall and food availability. Adult beetles can spend up to six months in diapause in the soil in autumn and winter in Australia (McFadyen, 1992).

Smicronyx lutulentus Dietz (Coleoptera: Curculionidae)

The seed-feeding weevil occurs commonly on parthenium weed in Mexico and Texas. It can also survive and reproduce on Gray's feverfew (*Parthenium confertum* A. Gray), but its incidence on Gray's feverfew was less than on parthenium weed (McFadyen and McClay, 1981). Adults are 1.5–2.0 mm long, black/grey in colour (Fig. 7.1c) and feed on flower buds and young leaves, causing negligible damage (Strathie, 2014). They live for up to 3 months, and females lay an average of 237 eggs, in buds or freshly opened capitula (Fig. 7.1d) but not in open flowers. Emerging larvae feed within the developing seeds, usually with a single larva per capitulum (Fig. 7.1d). The attacked seed swells greatly and other seeds in the same flower head often do not develop. When the mature flower dehisces and seeds fall to the ground, larvae leave the seeds and burrow into the

Fig. 7.1. Biological control agents of parthenium weed: leaf-feeding beetle *Zygogramma bicolorata* adult (a) and larva (b); seed-feeding weevil *Smicronyx lutulentus* adult (c) and larva (d); stem-galling moth *Epiblema strenuana* adult (e) and larva (f); stem-boring weevil *Listronotus setosipennis* adult (g) and larvae in shoot tip (h) and root (i). (K. Dhileepan, Biosecurity Queensland.)

soil to pupate in earthen cells 2–6 cm below the surface. An extended prepupal stage of 7–8 weeks occurs in the soil (McFadyen and McClay, 1981). In Mexico, *S. lutulentus* appears to have two generations per year, coinciding with autumn and summer rains, and can cause up to 30% seed destruction.

Epiblema strenuana Walker
(Lepidoptera: Tortricidae)

The stem-galling moth (Fig. 7.1e) is widely distributed in North America, including Mexico, Virgin Islands and Antigua in the Caribbean. *Ambrosia* and *Xanthium* species are host plants in the north and parthenium weed in the southern native range (McClay, 1987). Females lay 100–1000 eggs singly on the leaves and the emerging larvae feed initially on the leaf, before boring into the stem through the terminal or axillary meristem. Once inside the stem, larval feeding induces a hollow fusiform gall in which the larva feeds (Fig. 7.1f) and pupates. The moth emerges through a hole in the gall 4–6 weeks later, leaving the pupal skin extruded from the exit hole. Gall development can occur in all stages of plant growth; usually there is one larva per gall, but with multiple galls per stem.

Listronotus setosipennis (Hustache)
(Coleoptera: Curculionidae)

The stem-boring weevil has a wide distribution in its native range in northern Argentina and southern Brazil, where the only known host plant is parthenium weed. Adult weevils (Fig. 7.1g) are 5 mm long, nocturnal and feed on leaves and flowers. Adults are winged, but are not known to fly. Adults are long-lived, up to at least a year recorded in the laboratory, and very tolerant of extended dry periods. Females insert eggs singly into holes chewed in flowers, preferably, covered by a frass cap, but will also utilize leaf bases or stem surfaces. Newly emerged larvae tunnel directly from the peduncle into the stem to feed (Fig. 7.1h). Mature larvae move to the roots (Fig. 7.1i), where they feed prior to pupation. Adult feeding and oviposition damage is negligible. Larval feeding has the

ability to kill or prevent further development of parthenium weed seedlings and rosettes (Wild *et al.*, 1992). Pupation occurs in a cell constructed in the soil and the life-cycle takes 4–8 weeks. Adult emergence from the soil is triggered by rain; prolonged dry conditions delay emergence.

Stobaera concinna (Stål)
(Homoptera: Delphacidae)

In Mexico, this sap-feeding plant-hopper was found only on parthenium weed (McClay, 1983). In the southern USA and in the West Indies, *S. concinna* feeds and reproduces primarily on *Ambrosia* spp. (Calvert *et al.*, 1987). Adults (Fig. 7.2a) are 3–4 mm long, with males smaller and more strongly pigmented than females. The eggs are inserted singly into stems and hatch after 2 weeks. There are five nymphal instars and development from egg to adult takes 30–56 days. Nymphs (Fig. 7.2b) feed on leaves and new shoots, while adults feed on main and axillary shoots. At high population densities, *S. concinna* causes yellowing of the leaves and spindly plant growth (McClay, 1983).

Bucculatrix parthenica Bradley
(Lepidoptera: Bucculatricidae)

The leaf-mining moth is native to Mexico, where it is rare and causes no major damage to parthenium weed, probably due to specialist natural enemies (McClay *et al.*, 1990). Adults (3–4 mm long) lay eggs singly on the leaves. The first and second instar larvae are leaf-miners and later instars feed externally on the leaves. Pupation takes place within a characteristic ribbed silk cocoon usually attached to the leaf midrib. Adults emerge within 7–11 days and survive for up to 14 days. The life-cycle is completed in about 25 days under field conditions. In Mexico, larval feeding is evident on all growth stages of parthenium weed.

Conotrachelus albocinereus Fiedler
(Coleoptera: Curculionidae)

The stem-galling weevil is native to South America, where it occurs in Colombia,

Fig. 7.2. Biological control agents of parthenium weed: sap-feeding planthopper *Stobaera concinna* adult (a) and nymph (b); stem-galling weevil *Conotrachelus albocinereus* adult (c) and larva (d); the root-feeding moth *Carmenta* sp. near *ithacae* adults (e) and larvae (f); the summer rust *Puccinia xanthii* var. *parthenii-hysterophorae* (g); the winter-rust *Puccinia abrupta* var. *partheniicola* (h). (K. Dhileepan, Biosecurity Queensland.)

Argentina and Brazil, feeding on parthenium weed and annual ragweed (*Ambrosia artemisiifolia* L.). There are three generations per year in northern Argentina and the adults overwinter in the leaf litter or the soil. Adults (Fig. 7.2c) are 4–5 mm long, nocturnal and live for up to 3 months (McFadyen, 2000). They feed on the stem tips and leaves, although adult feeding is not particularly damaging (McFadyen, 2000). Eggs are laid singly on the stem, mainly in the leaf axils, and hatch within 10–14 days. Newly emerged larvae feed on epidermal cells and burrow vertically into the stem to feed on the nutrient-rich parenchyma cells, fracturing the vascular tissues, thereby disrupting the host plant's overall metabolism (Florentine *et al.*, 2002). Larval feeding induces elliptical galls on the main shoot axes (Fig. 7.2d). Larvae remain within the galls for about 2 months before exiting to pupate in earthen cells constructed in the soil. Adults emerge from the soil after 30 days, triggered by rainfall. Gall induction usually destroys axillary shoots, but the main stem remains unaffected.

Platphalonidia mystica (Razowski and Becker) *(Lepidoptera: Tortricidae)*

The stem-boring moth is native to northern Argentina, Bolivia and southern Brazil, where the only known host plant is parthenium weed. Adults are light grey to white, 3–8 mm long and partly diurnal, with oviposition occurring in the daytime. Females lay up to 119 eggs. Newly emerged larvae feed on the leaf surface initially then enter the main stem via axillary shoots. Larvae generally feed close to the surface of the phloem tissue of the stem and pupate within the stem. The life-cycle from egg to adult takes 8–10 weeks. Larval feeding results in the death of shoot tips and weakening of the main stem, and high larval densities can kill the plant (Griffiths and McFadyen, 1993).

Carmenta sp. nr ithacae (Beutenmüller) *(Lepidoptera: Sesiidae)*

The root-feeding clearwing moth (Fig. 7.2e) has a wide native range distribution in the USA, where larvae have been recorded feeding in common sneeze weed (*Helenium autumnale* L.) and false sunflower (*Heliopsis helianthoides* (L.) Sweet). The only records of the moth on parthenium weed are from Mexico (Withers *et al.*, 1999). The population of *C. ithacae* from *P. hysterophorus* in Mexico may be a different species from the USA species, and hence is referred to as *Carmenta* sp. nr *ithacae* (Withers *et al.*, 1999). Moths are 10 mm long and diurnal. Emergence occurs soon after sunrise, mating between 8am and 11am, and oviposition after 11am. Females lay eggs on any part of the plant, but particularly the leaves and stems, and the eggs hatch within 10–14 days. Larvae migrate to the stem where they feed just below the outer layer, then move down the stem to feed in the cortical tissue of the tap root and crown (Fig. 7.2f). After 5–6 weeks, the fully developed larvae pupate in the root or stem base in silk cocoons. After 10–12 days, adults emerge from the clearly visible silk tube pupal cases protruding from the plant, at or above the soil level. Adults live for 3–12 days and females lay 64–235 eggs in their lifetime. Larvae are found on all growth stages of parthenium weed, and heavily infested plants often die.

Thecesternus hirsutus Pierce *(Coleoptera: Curculionidae)*

In Mexico, the root-feeding weevil was collected on both parthenium weed and Gray's feverfew (McClay and Anderson, 1985). Although developing larvae were collected from the roots of parthenium weed, it is suspected that its primary host is Gray's feverfew. The large (7–10 mm long) apterous adults are long-lived and field-collected adults lived for more than 10 months. Adults feed on leaves, and eggs are laid singly onto the soil surface, either at the base of the host plant or within the root zone. Newly emerged larvae feed externally on the roots, inducing a superficial gall at the feeding site. Mature larvae form an earthen cell around each larva, attaching to the feeding site on the gall. Though no larvae have ever been reared from eggs to adults in the laboratory, field observations suggest that larval development may last up to 6 months. Pupation occurs within the earthen cell attached to the gall and takes 16–20 days. In Mexico, up to 12 larvae have been found feeding on the roots of single parthenium weed plants.

Puccinia abrupta var. *partheniicola* (H.S. Jackson) Parmelee *(Uredinales)*

The winter rust occurs naturally in Argentina, Bolivia, Brazil and Central America (Evans, 1997). The winter rust is a macrocyclic and autoecious rust, having all spore stages on parthenium weed (Parker *et al.*, 1994). It forms brown powdery pustules (uredinia) on the upper leaf surface (Fig. 7.2g) and stems throughout the year, and black non-powdery pustules on leaves, stems and inflorescences (telia) seasonally (Parker *et al.*, 1994). The teliospores are not dispersed and remain on the host plant. The rust appears to overwinter in the urediniospore stage. The rust requires surface moisture for spore germination and infection, and the highest rate of infection occurs at 15°C with a dew period of 8–10 hours. Spore germination decreases with increasing temperature, with no germination at 30°C (Fauzi *et al.*, 1999).

Puccinia xanthii var. *parthenii-hysterophorae* Seier, H.C. Evans and Á. Romero *(Uredinales)*

The summer rust (previously known as *Puccinia melampodii* Dietel & Holw.) is native to Central America and Mexico (Evans, 1997). It is a microcyclic rust species which cycles exclusively through the telial stage, and completes its reduced life-cycle on only one host, parthenium weed. It forms telia predominantly on the lower leaf surface (Fig. 7.2h) as well as on the stems, initially as distinct sori, which constitute aggregates of individual telia that coalesce over time (Seier, 1999). Sporulation can occur over the entire leaf surface, leading to necrosis and eventual dieback of affected leaves. Successive infection cycles can cause severe stunting and premature plant death (Seier, 1999).

7.2.4 Biological control implementation

Australia

Nine species of insects and two rust fungi were introduced from 1980 to 2004 (Table 7.1; Dhileepan and McFadyen, 2012). All have established (Table 7.1), although the time taken for establishment, their geographic range, abundance and impact varies greatly (Dhileepan and McFadyen, 2012). Since 2004, efforts have focused on monitoring the establishment and spread of agents, and redistributing effective agents into new areas.

The leaf-feeding beetle *Z. bicolorata* was imported into Australia from Mexico in 1980, after preliminary host-specificity testing testing in Mexico (McFadyen and McClay, 1981). Field release in 1981 followed detailed host-specificity tests in Australia (McFadyen, 1980). More than 80,000 adult beetles (200–18,000 per site) were released across 40 sites in Central and North Queensland between 1981 and 1983. They became established on annual ragweed in south-eastern Queensland and north New South Wales (NSW) within 2 years of first releases, but establishment on parthenium weed took longer (about 10 years) (Dhileepan and McFadyen, 2012). In Central Queensland, field establishment of *Z. bicolorata* on parthenium weed was first observed in January 1990. Since 1993, widespread establishment and large outbreaks of the beetle have occurred in Central Queensland, resulting in significant reductions in plant vigour, flower production, weed density and soil seed-bank (Dhileepan *et al.*, 2000; Dhileepan, 2003b). Due to natural spread by the beetle and deliberate spread by farmers, *Z. bicolorata* is now widespread throughout Central, Southern and South East Queensland. However, there is no indication of establishment north of 23°S.

The seed-feeding weevil *S. lutulentus* was approved for field release in 1980 after host-specificity testing was conducted in Mexico and Australia (McFadyen and McClay, 1981). Over 15,000 adults were released between 1981 and 1983 across 11 sites in Central and North Queensland. The weevil was initially thought to have failed to establish (McFadyen, 1992), but its widespread establishment was observed in Central Queensland in 1996. Currently, *S. lutulentus* is widespread and abundant in almost all parthenium weed-infested areas in Central and North Queensland. It is now being redistributed to South East Queensland.

The stem-galling moth *E. strenuana*, sourced from parthenium weed in Mexico, was approved for field release in Australia in 1982 after host-specificity testing was conducted in Mexico and Australia (McClay, 1987). Field releases of pupae (700–1500 per site) were made at 64 sites in Queensland from 1982 to 1984. Subsequent field releases were made using galls collected from initial establishment sites. The moth became established on *P. hysterophorus*, annual ragweed and Noogoora burr (*Xanthium occidentale* Bertol.) soon after field releases. Within 3 years the moth was widespread in all areas where host plants occur, utilizing the alternative host plants when parthenium weed is seasonally not available. The larvae induce hollow, fusiform galls on both rosette and flowering stages of parthenium weed throughout the growing season (McFadyen, 1992). Galling reduced the photosynthesis, transpiration and stomatal conductance rates significantly (Florentine *et al.*, 2005), resulting in reduced plant growth and reproductive output (Dhileepan and McFadyen, 2001; Dhileepan, 2003b). The impact became significant when young plants were attacked (Dhileepan and McFadyen, 2001) and in the presence of pasture competition (Navie *et al.*, 1998). The moth can have more than six generations each year (McFadyen, 1992). However, levels of infestation by *E. strenuana* in Australia vary considerably (Dhileepan, 2003b) due to lack of synchrony between parthenium weed germination and moth emergence in spring. Despite this, *E. strenuana* is considered to be the most damaging agent on parthenium weed in Australia, being more abundant and widespread than the other agents.

The stem-boring weevil *L. setosipennis* was imported into Australia from northern Argentina and southern Brazil during 1982–1986 (Wild *et al.*, 1992). Field release followed host-specificity tests confirming that the insect is highly host-specific (Wild *et al.*, 1992). A total of 23,800 adult weevils were released (1000–3000 per site, per release) between December 1982 and March 1985 at 30 sites in Central and North Queensland. Additional field releases were undertaken in North Queensland from December 1991 to May 1992. The stem-boring weevil became established in 1983, soon after the first releases, but its field incidence remained low and sporadic (McFadyen, 1992; Dhileepan, 2003a). The weevil is now widespread throughout Central and North Queensland, and is being redistributed to South and South East Queensland.

The planthopper *S. concinna* was imported from Mexico into Australia in 1982 and approved for field release in 1983, after testing in Mexico and Australia confirmed that it could feed and reproduce only on parthenium weed and *Ambrosia* spp. (McClay, 1983). Releases of over 500,000 adults and nymphs (5000–10,000 adults and nymphs per site, per release) were made at 60 sites across Queensland until 1986. The planthopper initially did not establish on parthenium weed, although it has since become established at low levels on parthenium weed and annual ragweed in South East Queensland (McFadyen, 1992). The planthopper has recently been found on parthenium weed in North Queensland, but in very low numbers. The causes for its failure to establish in Central Queensland and very low population levels in South East and North Queensland are unknown.

The leaf-feeding moth *B. parthenica* from Mexico was imported into quarantine in Australia in 1983. Host-specificity testing demonstrated that parthenium weed was the only suitable host (McClay *et al.*, 1990), and the moth was released in Australia from 1984 to 1985. Releases of over 50,000 adults, as plant material infested with pupal cocoons, as well as adult moths (1200–3500 adults per site, per visit) were undertaken at 14 sites in Central and North Queensland. Field establishment in Central Queensland was confirmed in 1987 (McClay *et al.*, 1990). The moth spread very readily and is now widespread and seasonally abundant in all parthenium weed-infested areas in Queensland, although it does not appear to have a major impact.

The gall weevil *C. albocinereus* sourced from Argentina was imported into quarantine in Australia in 1992. In host-specificity tests, no feeding or oviposition was recorded on any non-target plant except annual ragweed (McFadyen, 2000). The agent was approved for field release in 1995, and more than 15,000 weevils were subsequently released (from 40 to 900 weevils per site, per release) at 22 sites in Central and North Queensland between 1995 and 2000. Although recovered from a few release sites in Central Queensland, there is no evidence of widespread field establishment of this weevil.

Host-specificity tests for the stem-boring moth *P. mystica* were conducted in Argentina and Australia and the moth was approved for field release in 1992 (Griffiths and McFadyen, 1993). Field releases were undertaken using potted parthenium weed infested with larvae and pupae (20–80 plants per site, per visit) at 12 sites in Central Queensland from 1992 to 1996. Occasionally, newly emerged adults were also field released. Monitoring for field establishment was complicated by the widespread presence of *E. strenuana*, which occupies the same niche on plants, and the difficulty of distinguishing larvae in the field. However, adults of *P. mystica* have been reared from field-collected stems on several occasions and the species is believed to be established at a low level.

The root-feeding moth *Carmenta* sp. nr *ithacae* sourced from parthenium weed in Mexico was approved for field release in Australia in 1998 (Withers *et al.*, 1999). From 1998 to 2002, 2816 larval-infested plants and 387 adults were released at 31 sites across Queensland, Australia (Dhileepan *et al.*, 2012). Evidence of field establishment was first observed in two release sites in Central Queensland in 2004. The moth is now widespread and abundant in Central and North Queensland, and is being redistributed to South East Queensland.

The winter rust *P. abrupta* var. *partheni-icola* from higher altitude regions of Mexico

(Parker *et al.*, 1994) was approved for release in Australia in 1991 after tests demonstrated that the host range was limited to parthenium weed and Gray's feverfew (Parker *et al.*, 1994). The winter rust was field-released across 51 sites (60–150 infected plants per site, per visit) in South East and Central Queensland from 1991 to 1996. It established widely in South East Queensland, but in Central Queensland it established only in a few localized areas with wet winters (Dhileepan and McFadyen, 2012). The rust has not established in North and Western Queensland, where conditions are hotter and drier. Failure to establish more widely was due to the inability of the urediniospores to survive in the summer (Fauzi *et al.*, 1999). Under laboratory conditions the winter rust had a negative impact on parthenium weed (Fauzi, 2009), but its field impact appears minor (Dhileepan, 2003b).

The summer rust *P. xanthii* var. *parthenii-hysterophorae*, sourced from low-altitude regions of Mexico (Evans, 1997), is highly host-specific, virulent and adapted to areas with high temperatures and limited periods of humidity (Seier, 1999). Host-specificity tests confirmed that the summer rust has a sufficiently restricted host range (Seier, 1999) and the rust was approved for release in 1999. Field releases (12–60 rust infected plants per site, per visit) in Queensland commenced in 2000 and continued until 2004, at more than 50 sites in North and Central Queensland (Dhileepan *et al.*, 2006). Field establishment of summer rust was evident in most release sites soon after release (Dhileepan *et al.*, 2006). Currently the summer rust is widely established in North and Central Queensland. However, its prevalence and damage levels vary widely between years and sites, due to variation in rainfall. Its impact on parthenium weed is yet to be studied. The summer rust is currently being redistributed within South and South East Queensland.

Parthenium weed is primarily a weed of grazing areas in Queensland, and hence all biological control efforts have focused on grazing areas. In rangelands in Queensland, biological control has had a significant negative impact on parthenium weed and its soil seed banks, but the impact varied widely between years and locations (Dhileepan, 2003a, 2003b). Overall, the impact was greater in Central than in North Queensland, likely due to lower temperatures and higher, more consistent rainfall. A significant increase in grass biomass production due to biological control was observed more often in Central Queensland than in North Queensland (Dhileepan, 2007). In 1995, the benefit from increased grass production due to biological control was estimated to be AUD$1.25/ha/year in Central Queensland and AUD$1.19/ha/year in North Queensland (Adamson and Bray, 1999). Biological control saved AUD$8 million annually in health costs for the treatment of allergic dermatitis and asthma in workers in parthenium weed-infested areas, giving an overall benefit/cost ratio for the programme of 7.2 and a net present value (in 2005) of AUD$33.3 million for a total cost of AUD$11.0 million (Page and Lacey, 2006). The above cost-benefit analysis was based on field evaluations conducted during 1996–2000, when the stem-galling moth *E. strenuana*, the stem-boring weevil *L. setosipennis* and the leaf-feeding beetle *Z. bicolorata* were the primary agents active in the field. Since then, other agents including the seed-feeding weevil *S. lutulentus* and the root-feeding moth *Carmenta* nr *ithacae* have also become widely established, increasing the suite of active agents. An evaluation based on the current state of agent establishment and abundance would show an even higher economic benefit to the Queensland beef industry. The economic benefit is expected to be much higher if classical biological control is combined with grazing management as well.

In recent years, parthenium weed has continued spreading from the core infested areas in Central Queensland to new areas in South and South East Queensland. Recent surveys in South and South East Queensland suggest that the leaf-feeding beetle (*Z. bicolorata*), the stem-galling moth (*E. strenuana*) and the winter rust are the only agents that occur there. Currently, the other effective agents, such as *S. lutulentus*, *L. setosipennis*, *Carmenta* sp. nr *ithacae* and *P. xanthii*

var. *parthenii-hysterophorae*, are being redistributed from Central and North Queensland to South and South East Queensland (Callander and Dhileepan, 2016).

India

India commenced a biological control programme against parthenium weed in 1983 based on the Australian programme. Three agents, the leaf-feeding beetle *Z. bicolorata*, the seed-feeding weevil *S. lutulentus* and the stem-galling moth *E. strenuana*, already released in Australia, were imported into quarantine in Bangalore, India between 1983 and 1987 (Table 7.1). However, only *Z. bicolorata* has been field-released in India.

The beetle *Z. bicolorata*, sourced from Mexico in 1983, was approved for field release in 1984 (Jayanth and Nagarkatti, 1987) and released in the Bangalore region where it became established in the same year (Jayanth, 1987a). It attained damaging levels after three years (Jayanth and Visalakshy, 1994), causing extensive defoliation and significant reduction in the parthenium weed density (Jayanth and Bali, 1994; Jayanth and Visalakshy, 1996). The beetle spread widely in the neighbouring states by natural dispersal as well as by deliberate introductions. However, reports of non-target feeding by *Z. bicolorata* on sunflower (*H. annuus* L.) crops (e.g. Kumar, 1992; Chakravarthy and Bhat, 1994) halted the further release and redistribution of this agent and the testing of other agents in India. Later studies (Jayanth *et al.*, 1997) indicated that the risk of *Z. bicolorata* becoming a pest of sunflower was negligible and there was no economic loss due to *Z. bicolorata* feeding on sunflower (Kulkarani *et al.*, 2000). Once this controversy was resolved, field releases of *Z. bicolorata* continued in 15 States in India, resulting in defoliation outbreaks (e.g. Sushilkumar, 2009). Defoliation by *Z. bicolorata* also resulted in the re-establishment of native vegetation (Jayanth and Visalakshy, 1996; Sridhara *et al.*, 2005). However, information on the long-term impact of defoliation by *Z. bicolorata* in India is lacking. Thirty years after the initial release, the beetle occurs in most areas in India with parthenium weed infestations, apart from the hot, dry north-west region (e.g. Dhileepan and Wilmot Senaratne, 2009). Spread of Z. bicolorata from India into neighbouring Pakistan and Nepal has been recorded (discussed below).

The moth E. strenuana was imported from Australia into India in 1985. In host-specificity tests in India, non-target oviposition and larval feeding occurred on niger (Guizotia abyssinica (L.f.) Cass.) and sunflower crops (Jayanth, 1987b). Hence, E. strenuana was not released there. The weevil S. lutulentus was imported from Australia into India in 1985, but could not be reared in the quarantine laboratory so no further work was undertaken. The summer rust P. xanthii var. parthenii-hysterophorae, although not tested in India, is regarded as unsuitable due to acceptance of marigold (Calendula officinalis L.), a species of cultural importance (Seier, 2015). The winter rust P. abrupta var. partheniicola, although not intentionally released as a biological control agent (Table 7.1), has been reported in India (e.g. Parker et al., 1994). However, the winter rust strain in India does not appear to be widespread or aggressive (e.g. Kumar and Evans, 2005), and the introduction of a highly virulent isolate from Mexico or Australia would be beneficial.

Despite the widespread establishment of Z. bicolorata in India, parthenium weed continues to be a major problem, so the introduction of additional agents warrants future consideration.

Sri Lanka

Sri Lanka commenced a biological control programme against parthenium weed in 2003 with the importation of the stem-galling moth E. strenuana and the summer rust P. xanthii var. parthenii-hysterophorae from Australia (Table 7.1). Host-specificity tests confirmed the suitability of E. strenuana as a biological control agent for parthenium weed in Sri Lanka, and the moth was field released in 2004 (Jayasuriya, 2005), but nothing is known of the release efforts or its establishment. The summer rust was

imported into Sri Lanka in 2003, and host-specificity tests indicated that only parthenium weed was susceptible to the rust (Kelaniyangoda and Ekanayake, 2008). However, subsequent approval or release efforts are unknown. Attempts to culture Z. bicolorata in quarantine in Sri Lanka have not been successful. Recent surveys in Sri Lanka revealed no indication of field establishment of any agent (Dhileepan, personal observations, 2014 and 2015).

South Africa

South Africa commenced a biological control programme against parthenium weed in late 2003. Based on Australian experience, the leaf-feeding beetle (Z. bicolorata), the stem-boring weevil (L. setosipennis) and the summer rust (P. xanthii var. parthenii-hysterophorae), followed by the stem-galling moth (E. strenuana) and the root-feeding moth (Carmenta sp. nr ithacae), were prioritized for assessment in South Africa (Table 7.1; Strathie et al. 2005, 2011).

The winter rust P. abrupta partheniicola, although not intentionally released (Table 7.1), was first observed in South Africa in 1995 in the North-West Province (Wood and Scholler, 2002), and now also occurs in Mpumalanga and KwaZulu-Natal Provinces. It does not appear to be particularly aggressive (e.g. Kumar and Evans, 2005), although its impact is still to be evaluated.

The summer rust P. xanthii var. parthenii-hysterophorae was imported into South Africa from Queensland, Australia in 2004. Based on the pathogenicity and high degree of host specificity (Retief et al., 2013), its release in South Africa was approved in 2010 (Strathie et al., 2011). Subsequently, infected potted plants were field released at 18 sites in KwaZulu-Natal and Mpumalanga Provinces and release efforts are continuing (A. den Breeyen, South Africa, 2016, personal communication). Evidence of field establishment of P. xanthii var. parthenii-hysterophorae was first observed in 2012, with natural spread of up to 50km from release sites observed by 2015 (A. den Breeyen, South Africa, 2016, personal communication).

A colony of *Z. bicolorata* was established in quarantine in South Africa in 2005 from adults collected in Queensland, Australia (McConnachie, 2015a). A risk analysis based on host-specificity tests in South Africa determined that the risk of attack on sunflower was negligible and, as other non-target plants were not at risk, the beetle was deemed safe for release (McConnachie, 2015a). Field release was approved in 2013. More than 42,000 *Z. bicolorata* beetles (mostly adults) have been released at 162 sites in climatically suitable KwaZulu-Natal and Mpumalanga Provinces (L. Strathie, 2017, unpublished data), with mass-rearing and releases still continuing. Although *Z. bicolorata* has since established in both provinces, incidence is very low, with absence at many sites and low abundance at the few established sites, but with occasional outbreaks observed under suitable conditions at some sites (L. Strathie, 2017, unpublished data).

A colony of *L. setosipennis* collected from north-western Argentina was established in quarantine in South Africa in 2003 (Strathie *et al.*, 2005). Based on host-specificity tests which demonstrated no risk to most non-target species tested, and a risk analysis that demonstrated negligible risk to sunflower (Strathie and McConnachie, 2012), *L. setosipennis* was approved for field release in South Africa in 2013. More than 25,000 *L. setosipennis* adults have been field released at more than 108 sites in KwaZulu-Natal and Mpumalanga Provinces, with mass-rearing and field releases still continuing. Establishment was confirmed in 2014, and the weevil has established more readily than *Z. bicolorata*. However, spread from point of release after two years is limited to 500 m or less (L. Strathie, 2016, unpublished data).

Epiblema strenuana was sourced from Queensland, Australia in 2010 and a colony was established in quarantine in South Africa (Strathie *et al.*, 2011). In view of the ability of *E. strenuana* to complete development on niger in laboratory tests in India, an oil seed crop native to East Africa (Jayanth, 1987b), evaluation in South Africa initially focused on niger cultivars most commonly cultivated in Ethiopia (Strathie

et al., 2011). Although niger is not cultivated in South Africa, due to the rapid and wide dispersal of *E. strenuana* following its release in Australia, its use of *Ambrosia* and *Xanthium* species as alternative host plants, and the presence of parthenium weed and the alternative host plants in countries between the south and east of Africa, it was postulated that if released in South Africa the moth may reach Ethiopia, where niger is an important crop (McConnachie, 2015b). In laboratory studies in South Africa, significant larval feeding damage occurred on all five niger cultivars assessed under no-choice conditions, although development was completed on one cultivar only. In choice tests, no development and reduced larval feeding damage on niger occurred when parthenium weed was also available. However, complete development, with the formation of galls, occurred on four niger cultivars during larval development trials. For these reasons, *E. strenuana* was rejected as a potential agent for parthenium weed in South Africa until the impact of the moth on niger cultivars is investigated under open-field choice trials in Australia (McConnachie, 2015b).

A colony of *S. lutulentus* was established in quarantine in South Africa in 2010 using field-collected adults from Queensland, Australia (Strathie *et al.*, 2011). No eggs were laid on any of the non-target plants species tested in South Africa, indicating an extremely high degree of host-specificity (Strathie, 2014). The weevil was approved for field release in late 2014 and releases commenced in early 2015. Over 26,500 *S. lutulentus* adults have been released at 52 sites in KwaZulu-Natal and Mpumalanga Provinces. A year after release, establishment of *S. lutulentus* was confirmed at several sites, although in low abundance, possibly due to severe drought conditions that prevailed during the early release period. Mass-rearing and widespread releases are continuing, and the extent of establishment is broadening. A colony of *Carmenta* sp. nr *ithacae* was successfully established in quarantine in South Africa in 2014 from field collected larvae from Queensland, Australia. Host-specificity tests are still in progress.

Although the releases of four biological control agents are well underway in South Africa, larger scale rearing and release efforts would improve the rate and level of control achieved. Assessment of the spread and impact of established agents is underway. Although still relatively early in terms of the level of control achieved, the need for additional agents is apparent, so that parthenium weed can be managed throughout its invaded range. New agents such as *Carmenta* sp. nr *ithacae* will complement existing agents, however further investigation of *E. strenuana* and additional future agents is necessary.

Ethiopia

Ethiopia commenced a biological control programme against parthenium weed in 2005, in collaboration with the USA and African partner countries. Following the establishment of a basic quarantine facility, the leaf-feeding beetle *Z. bicolorata* and the stem-boring weevil *L. setosipennis* were prioritized, based on their host-specificity, rearing ease and consideration in South Africa.

The beetle *Z. bicolorata* was imported from South Africa into quarantine in Ethiopia in 2007. Screening on cultivars of niger and tef (*Eragrostis tef* (Zucc.) Trotter), a staple food for Ethiopia, was conducted beforehand in South Africa. Based on host-specificity test results in South Africa and Ethiopia (Mersie, 2010), *Z. bicolorata* was approved for release in Ethiopia and field releases commenced in 2013 (M. Abebe, Ethiopia, 2016, personal communication). More than 25,000 adults and larvae of *Z. bicolorata* have been released at 10 sites in and around Adama in central Ethiopia (M. Abebe and W. Mersie, Ethiopia, 2016, personal communication). Extensive localized defoliation by *Z. bicolorata* was observed near Adama in 2016 (M. Abebe and W. Mersie, Ethiopia, 2017, personal communication). Mass-rearing and releases are continuing.

The weevil *L. setosipennis* was imported from South Africa into quarantine in Ethiopia in 2007, after cultivars of niger and tef were screened in South Africa. Based on host-specificity test results in Ethiopia (Beyera, 2011), approval to release *L. setosipennis* in Ethiopia was granted in 2013. Field releases were initiated in 2016. So far, 2300 *L. setosipennis* adults have been field released in Ethiopia (W. Mersie, Ethiopia, 2017, personal communication). Mass-rearing efforts are continuing and have been broadened also to the north and east of Ethiopia.

A founder colony of *S. lutulentus* from laboratory culture in South Africa was imported into quarantine in Ethiopia in late 2015, to assess suitability for release, but rearing difficulties were experienced and a culture could not be established. The moth *Carmenta* sp. nr *ithacae* may be considered for importation and host-specificity tests in due course.

The stem-galling moth *E. strenuana* is not being considered for introduction into Ethiopia, in view of its potential risk of feeding on the economically important niger crop.

The winter rust *P. abrupta* var. *partheniicola* was first reported in Ethiopia in 1997. Although not deliberately introduced there (Table 7.1), it now occurs commonly in cool and humid areas at high altitudes (1500–2500 m above sea level) where rainfall varies from 400 to 700 mm (Taye *et al.*, 2004a). In Ethiopia, the rust fungus significantly reduced the plant height, number of leaves, number of branches and total biomass of parthenium weed (Taye *et al.*, 2004a; Bekeko *et al.*, 2012).

Pakistan

Currently there is no active biological control programme on parthenium weed in Pakistan. Although not introduced intentionally, the leaf-feeding beetle *Z. bicolorata* was first reported in the Punjab region in Pakistan in 2003 (Javaid and Shabbir, 2007), where it is believed to have spread from India (Table 7.1). Due to natural spread and deliberate distribution, the agent occurs widely in the Punjab Province in Pakistan (Dhileepan and Wilmot Senaratne, 2009; Shabbir *et al.*, 2012), but there is no information on population levels or impact.

Nepal

Although not deliberately introduced, *Z. bicolorata* was first reported in the Hetaunda District, Central Nepal in 2009, causing severe defoliation to parthenium weed (Shrestha *et al.*, 2010). Spread occurred and currently the beetle is widespread throughout the parthenium weed-infested areas in Kathmandu Valley. Complete defoliation was reported at a site south of Kathmandu Valley (Shrestha *et al.*, 2011). The presence of *P. abrupta* var. *partheniicola* in Kathmandu Valley was confirmed in 2011, but its impact appeared to be insignificant (Shrestha *et al.*, 2011).

Tanzania

Tanzania commenced a biological control programme against parthenium weed recently, using agents that had been released in South Africa (A. Witt, Kenya, 2015, personal communication). About 1000 *Z. bicolorata* adults and 1000 larvae from lab-reared colonies in South Africa were released at five field sites in and around Arusha Province, Tanzania in 2013, followed by 200 adults and 1000 larvae released into field cages in early 2014, and 1000 adults released in early 2015 (ARC-PHP, South Africa and Centre for Agriculture and Biosciences International (CABI), UK, unpublished data). There was no evidence of field establishment, but sites were not monitored regularly (A. Witt, Kenya, 2016, personal communication). Efforts to mass-rear agents in Arusha have been initiated, with *Z. bicolorata* the first agent to be reared for release. Subsequent field releases have led to early signs of localised establishment.

Papua New Guinea

Currently there is no active biological control programme against parthenium weed in Papua New Guinea, although parthenium weed is an eradication target. Earlier attempts to establish colonies of *Z. bicolorata* and *E. strenuana* imported from Australia into quarantine in Papua New Guinea in 2003 failed.

Vanuatu

Currently there is no active biological control programme against parthenium weed in Vanuatu. An unknown number of field-collected *Z. bicolorata* adults from Australia were released in Vanuatu in 2014, but no evidence of establishment has been observed yet (M. Day, Australia, 2016, personal communication). Though not deliberately released, *E. strenuana* galling was observed on parthenium weed in Vanuatu in 2016, but its mode of introduction is not known. There are no active biological control programmes against parthenium weed in other Pacific Islands.

China

Currently there is no active biological control programme against parthenium weed in China. However, the stem-galling moth *E. strenuana* was introduced into China from Australia in 1990 as a biological control agent for annual ragweed (Wan *et al.*, 1995). Galling by *E. strenuana* was observed on parthenium weed in southern China (A. Shabbir, Pakistan, 2012, personal communication), but no published information is available on the incidence and impact of the moth in China. Though not deliberately released, the winter rust (*P. abrupta* var. *partheniicola*) has been found in Yunnan Province.

Uganda

Uganda commenced a biological control programme against parthenium weed in 2018. So far, *Z. bicolorata* and *L. setosipennis* have been released from South Africa.

7.3 Mycoherbicides

Mycoherbicides are fungal-based bioherbicides. Information is widely available on the mycoflora associated with parthenium weed in the native range (e.g. Evans, 1997), India (e.g. Dhileepan, 2009), South Africa (Wood and Scholler, 2002) and Ethiopia (Taye *et al.*, 2004a, 2004b). Though there are currently no approved or registered mycoherbicides available for the control of parthenium weed in any country, mycoherbicides have the potential to play an important role in managing parthenium weed in cropping areas.

7.4 Suppressive Plants

Management approaches, such as maintaining natural plant biodiversity and sowing of selected suppressive plant species into parthenium weed-infested areas, are useful weed management tools (Wahab, 2005). Several beneficial pasture plants have the potential to displace parthenium weed (Table 7.2). Different plant species have varying suppressive abilities on the growth of parthenium weed due to their differing attributes, such as allelopathy, rapid establishment, faster growth rates and physical competition (e.g. Khan et al., 2013, 2014). Most research on beneficial plants that competitively displace parthenium weed has been restricted to India and Australia, with limited studies in South Africa, Ethiopia and Pakistan (Dhileepan, 2009).

In Australia, the suppressive ability of 30 plant species for the management of parthenium weed was studied, focusing mainly on cattle grazing areas where only pasture species would be acceptable (Table 7.2; O'Donnell and Adkins, 2005; Bowen et al., 2007; Khan et al., 2010, 2013, 2014). Most of these species can suppress the growth of parthenium weed under glasshouse conditions, however some of them (e.g. Setaria incrassata (Hochst.) Hack., Cenchrus ciliaris (L.), Themeda triandra Forssk., Bothriochloa insculpta (Hochst. ex A. Rich.) A. Camus, B. pertusa (L.) A. Camus, Clitoria ternatea L. and Astrebla squarrosa C.E. Hubb.) have shown good suppressive ability against parthenium weed under simulated grazing field conditions as well (Khan, 2011). However, their suppressive ability under natural cattle grazing conditions is yet to be demonstrated.

In India, there is extensive literature on using various plants to competitively suppress parthenium weed (e.g. Dhileepan, 2009; Knox et al., 2011). Based on laboratory and field research, a number of plant species have been identified as prospective suppressive plants to manage parthenium weed (Table 7.2). However, some species have no economic benefit other than parthenium weed suppression. Unlike in Australia, where use of competitive plants focused only on pasture species, in India the management focus has been for their use in wastelands, roadsides and fallow lands.

In South Africa, guinea grass (Panicum maximum Jacq. Nees) significantly reduced the growth of parthenium weed in field conditions (van der Laan et al., 2006), while two other native grasses, African love grass (Eragrostis curvula Schrad.) and digit grass (Digitaria eriantha Steud.) were less suppressive and less productive while competing with parthenium weed. The use of native suppressive species is preferable to exotic species, particularly for natural area. Palatability to herbivores should also be considered during the selection of suppressive plant species. The use of such native species to suppress parthenium weed may present challenges in terms of their palatability to native herbivores (L. Strathie and A. McConnachie, South Africa, 2010, personal observation).

In Pakistan, several palatable grass species like Dichanthium annulatum (Forssk.) Stapf. and Cenchrus pennisetiformis Hochst. & Steud. and major weed species like Imperata cylindrica (L.) Beauv. and Sorghum halepense (L.) Pers. in natural and cultivated areas were found to suppress the growth of parthenium weed (Table 7.2; Anjum and Bajwa, 2005; Javaid et al., 2005). More recently, several other species (Table 7.2; Khan et al., 2014) were also found to greatly suppress the growth of parthenium weed under experimental and natural field conditions, as well as to produce sufficient biomass for livestock consumption in north-western Pakistan.

Parthenium weed is a C_3/C_4 intermediate plant which is expected to benefit from the anticipated anthropogenic climate change (i.e. elevated CO_2 concentration), while pasture species in grazing areas are C_4 species which are unlikely to benefit from

Table 7.2. Plants investigated for their potential to suppress parthenium weed in various countries. (Dhileepan, McFadyen, Strathie and Khan.)

Country	Suppressive plants	Plant traits	References
India	**Acanthaceae**		Dhileepan
	Andrographis paniculata Nees.	Native and medicinal	(2009)
	Amaranthaceae		Knox *et al.*
	Achyranths aspera L.	Native, medicinal and religious use	(2011)
	Alternanthera sessilis (L.) DC.	Native and vegetable	
	Amaranthus spinosus L.	Non-native and vegetable	
	Aerva javanica Juss.	Non-native and fodder	
	Asteraceae		
	Tagetus erecta L.	Non-native and ornamental	
	T. patula L.	Non-native and ornamental	
	Capparaceae		
	Cleome gynandra L.	Medicinal and edible	
	Chenopodiaceae		
	Chenopodium album L.	Vegetable and poultry food	
	Euphorbiaceae		
	Croton bomplandianum Bill	Native and medicinal	
	C. sparsiflorus L	Native and medicinal	
	Fabaceae		
	Senna (*Cassia*) *auriculata* L.	Native, ornamental and medicinal	
	S. occidentalis (L.) Link.	Non-native, toxic and invasive	
	S. sericea (Symon) Albr., & Symon	Non-native and medicinal	
	S. tora L.	Native and medicinal	
	Stylosanthes scabra Vogel	Non-native pasture	
	Tephrosia purpurea L.	Native and medicinal	
	Lamiaceae		
	Hyptis suaveolens (L.) Poit.	Non-native and medicinal	
	Ocimum canum Sim.	Native and medicinal	
	Malvaceae		
	Malva pusilla Sm.	Non-native, food and medicinal	
	Sida acuta Burm.f.	Non-native and invasive	
	S. rhombifolia L.	Non-native, fibre and medicinal	
	S. spinosa L.	Non-native and invasive	
	Abutilon indicum (Link) Sweet.	Native and medicinal	
	Nyctaginaceae		
	Mirabilis jalapa L.	Non-native and ornamental	
	Poaceae		
	Cenchurus ciliaris L.	Non-native pasture	
	Cymbopogon coloratus (Hook.f.) Stapf.	Native and medicinal	
	Sorghum bicolor (L.) Moench	Crop and forage	
	Zea mays L.	Crop and forage	
Australia	**Fabaceae**		O'Donnell and
	Clitoria ternatea L.	Non-native legume and cattle feed	Adkins (2005), Bowen *et al.*
	Glycine latifolia Newell & Hymowitz	Native pasture legume	(2007), Khan
	Macroptilium bracheatum Marechal & Baudet	Non-native pasture legume	*et al.* (2013)
	Stylosanthes seabrana B.L.	Non-native pasture legume	
	S. scabra Vogel	Non-native pasture legume	

continued

Table 7.2. *Continued.*

Country	Suppressive plants	Plant traits	References
	Poaceae		
	Bothriochloa insculpta (R.Br.) A. Camus	Non-native pasture	
	B. pertusa (L.) cv. Keppel	Non-native pasture	
	B. ewartiana (Domin.) C.E. Hubb.	Native pasture	
	B. macra (Steud.) S.T. Blake	Native pasture	
	B. decipiens (Hack.) C.E. Hubb.	Native pasture	
	B. bladhii (Retz.) S.T. Blake.	Native pasture	
	Dicanthium aristatum (Poir.) C.E. Hubb	Non-native pasture	
	D. sericeum (R. Br.) A. Camus	Native pasture	
	Cenchrus ciliaris L.	Non-native pasture	
	Heteropogon contortus L. Roem. & Schult.	Native pasture	
	Chloris ventricosa R.Br.	Native pasture	
	Panicum maximum Jacq. var. maximum	Non-native pasture	
	Setaria incrassata (Hochst.) Hack cv. inverell	Non-native pasture	
	Digitaria brownii Roem & Schult.	Native pasture	
	Microlaena stipoides (Labill.) R.Br.	Native pasture	
	Themeda triandra Forssk.	Native pasture	
	Astrebla elymoides F. Muell. ex F.M. Bailey	Native pasture	
	A. squarrosa C.E. Hubb.	Native pasture	
	A. lappacea (Lindl.) Domin	Native pasture	
	Enteropogon acicularis (Lindl.) Lazarides	Native pasture	
	Eulalia aurea (Bory) Kunth.	Native pasture	
	Austrodanthonia racemosa (R.Br.) H.P. Lind.	Native pasture	
South Africa	**Poaceae**		van der Laan *et al.* (2006)
	Eragrostis curvula Nox.	Native pasture	
	Digitaria eriantha Steud.	Native pasture	
	P. maximum	Native pasture	
Pakistan	**Poaceae**		Anjum and Bajwa (2005), Javaid *et al.* (2005), Khan *et al.* (2014)
	Imperata cylindrica (L.) Beauv.	Native pasture	
	Desmostachya bipinnata (L.) Stapf.	Native perineal pasture	
	Dichanthium annulatum (Forssk.) Stapf.	Native pasture	
	Cenchrus pennisetiformis Hochst. & Steud.	Native perineal pasture	
	Sorghum halepense (L.) Pers.	Non-native pasture	
	Sorghum almum Parodi.	Non-native forage	
	Fabaceae		
	Vigna umbellata (Thunb.) Ohwi & H. Ohashi	Non-native forage legume	
Ethiopia	**Fabaceae**		Tamado and Milberg (2004)
	Vigna unguiculata (L.) Walp.	Crop and forage legume	
	Poaceae		
	Sorghum bicolor (L.) Moench	Crop	

the atmospheric CO_2 enrichment (e.g. Navie *et al.*, 2005), as C_3 and C_4 plants show differential growth responses under elevated atmospheric CO_2 (e.g. Khan *et al.*, 2015). In Australia, under experimental conditions, two C_4 pasture species (*S. incrassata* and *A. squarrosa*) were shown to decrease their suppressive ability in terms of biomass production, while one C_4 species (*Bothriochloa decipiens*) and one C_3 legume species (*C. ternatea*) were found to increase their growth and suppressive ability against parthenium weed. Future studies are needed to understand the interactions between the C_3 or C_3/C_4 intermediate parthenium weed and C_4 pasture species under natural field conditions under elevated CO_2 conditions.

Competitive plants can enhance the effectiveness of classical biological control agents under laboratory (Navie *et al.*, 1998) and experimental field conditions (Shabbir *et al.*, 2013, 2015). To date, all studies involving competitive plants on their own and in combination with biological control agents to manage parthenium weed have been either laboratory based or under experimental conditions in the field, and efficacy is yet to be demonstrated under large-scale field conditions.

7.5 Discussion

Classical biological control, using a suite of natural enemies, should certainly be considered as a primary method of managing parthenium weed in affected countries. The economic benefits of biological control programmes on invasive alien plants around the globe are widely recognized. Biological control will not eradicate parthenium weed, but it is essential for its significant reduction in extent and spread.

Classical biological control using a suite of introduced natural enemies has had considerable impact on parthenium weed populations under different seasons and habitats in introduced ranges such as Australia (Dhileepan and McFadyen, 2012). Classical biological control, together with grazing management and roadside chemical management, has reduced parthenium weed from the most invasive species in Queensland to a far lower rank, over a few decades. Other countries are in earlier stages of implementation so have not achieved the same degree of success as yet, although with continued effort this should be possible. Despite the broad and increasing invasion of parthenium weed on three continents and associated islands, the deliberate use of biological control has only been considered seriously in Australia, India, and more recently in South Africa and Ethiopia, with limited adoption elsewhere. However, concern over the increasing invasion and impact

of parthenium weed around the globe is growing, so biological control, with its economic benefits, is being increasingly considered critical for long-term, sustainable management of the weed. Indeed, in resource-poor countries where the feasibility of extensive chemical control operations is limited or impossible, biological control presents one of the most cost-effective and viable management options. However, impediments to adoption of the practice need to be addressed. These include the lack of support and financial backing by national governments, relevant legislation, resources, appropriate facilities, experienced capacity and knowledge to facilitate decision making on legislation and assessment of scientific data.

Countries where biological control of parthenium weed is being implemented have benefited considerably from the foundational programme in Australia, saving costs on native range surveys, agent selection, gaining ready access to starter colonies from field populations, and benefiting from the expertise available. Broader acceptance of existing host-range data would benefit new countries embarking on biological control and avoid duplicated efforts, particularly as agents may spread beyond country borders. New countries embarking on biological control programmes need to develop appropriate facilities and expertise for agent mass-rearing, to improve the rate and level of control achieved, rather than relying on limited introductions or natural dispersal. Site disturbance in areas with pastoral grazing practices, small-scale agriculture and rural settlements is challenging for the early phases of field establishment of biological control agents, particularly for those agents that disperse slowly. Hence, understanding of agent requirements and limitations and greater public awareness of biological control practice are required. Prioritization of areas best suited for biological control, in terms of infestation extent and land use may be useful, with selective use of biological control away from areas destined for broad-scale chemical operations.

For new countries considering biological control of parthenium weed, a combination of the agents *L. setosipennis*, *Carmenta* sp. nr *ithacae*, *S. lutulentus* and *P. xanthii* var. *parthenii-hysterophorae* is recommended. Although *E. strenuana* is one of the most damaging agents, resolution of host-range issues will be required for regions where niger is a valued crop. Some difficulties have been experienced in widely establishing *Z. bicolorata* in Australia, South Africa and Ethiopia. The beetle may be best suited for areas with reliable annual rainfall of more than 800 mm and moisture-retaining soils.

Agent establishment has been achieved with releases of relatively low numbers in some cases, but information on optimal release strategies and impact is desirable. Larger or inundative releases may achieve more successful establishment when factors such as predation or climatic conditions hamper agent establishment. It is likely that some agents will be more effective under anticipated anthropogenic climate change (e.g. Shabbir *et al.*, 2014).

Additional agents, beyond the nine established in Australia, may be required to achieve improved control over a wider range of habitats and seasons. Countries that have introduced only one or two agents should consider additional species to control the weed under the varying conditions affected. Broader adoption of biological control of parthenium weed in Africa and Asia is recommended.

Though not currently used to manage parthenium weed in any countries, mycoherbicides have the potential to be used in cropping areas where other methods, including chemical and classical biological control, have limitations.

Maintaining natural plant biodiversity and sowing selected suppressive plant species, preferably native, into infested areas, are useful tools in the management of parthenium weed. In Australia, under certain situations (e.g. grazing areas under no grazing or under experimental grazing conditions), the use of suppressive plants has enhanced the effectiveness of classical biological control, resulting in improved pasture production and reducing the vigour of parthenium weed (e.g. Shabbir *et al.*, 2015).

However, large-scale field studies at multiple locations under natural grazing conditions are needed. There is a need for the adoption of a similar approach in other countries where biological control agents are already present.

For effective management of parthenium weed, integrating various biological control options that include classical biological control, mycoherbicides and suppressive plants with management tools such as chemical, physical, grazing management and cultural management is desirable.

Acknowledgements

The authors wish to acknowledge contributions on the current status of parthenium weed biological control agents by Sushilkumar (India), Asad Shabbir (Pakistan), Arne Witt (Tanzania), Michael Day (South Pacific Islands), Jason Callander (Australia), Boyang Shi (China), Bharat Babu Shrestha (Nepal), Million Abebe and Wondi Mersie (Ethiopia), Alana den Breeyen (South Africa) and Wilmot Senaratne (Sri Lanka). Financial support from the Queensland Government in Australia, Meat and Livestock Australia, the Department of Environmental Affairs: Natural Resources Management Programme in South Africa and the United States Agency for International Development (USAID)-funded Feed the Future: Integrated Pest Management Innovation Lab is gratefully acknowledged.

References

Adamson, D.C. and Bray, S. (1999) The economic benefit from investing in insect biological control of parthenium weed (*Parthenium hysterophorus*). RDE Connections, NRSM, University of Queensland, Australia.

Anjum, T. and Bajwa, R. (2005) Biocontrol potential of grasses against parthenium weed. In: Prasad, T.V.R., Nanjappa, H.V., Devendra, R., Manjunath, A., Subramanya, S.C., *et al.* (eds), *Proceedings of the Second International Conference on Parthenium Management*. University of Agricultural Sciences, Bangalore, India, pp. 143–146.

Bekeko, Z., Hussien, T. and Tessema, T. (2012) Distribution, incidence, severity and effect of the rust (*Puccinia abrupta* var. *partheniicola*) on *Parthenium hysterophorus* L. in Western Hararghe Zone, Ethiopia. *African Journal of Plant Science* 6, 337–345.

Beyera, S. (2011) Life cycle, host specificity and effectiveness of the stem-boring weevil, *Listronotus setosipennis* (Hustache) against *Parthenium hysterophorus* L. under quarantine conditions. MSc thesis, Ambo University, Ethiopia.

Bowen, D., Ji, J. and Adkins, S.W. (2007) Management of parthenium weed through competitive displacement with beneficial plants: A field study, Brisbane. A report submitted to the Queensland Murray Darling Committee. The University of Queensland, Brisbane, Australia.

Callander, J.T. and Dhileepan, K. (2016) Biological control of parthenium weed: Field collection and redistribution of established biological control agents. In: Randall, R., Lloyd, S. and Borger, C. (eds), *Proceedings of the Twentieth Australasian Weeds Conference*. Weeds Society of Western Australia, Perth, pp. 242–245.

Calvert, P.D., Wilson, S.W. and Tsai, J.H. (1987) *Stobaera concinna* (Homoptera: Delphacidae): field biology laboratory rearing and descriptions of immature stages. *Journal of New York Entomological Society* 95, 91–98.

Chakravarthy, A.K. and Bhat, N.S. (1994) The beetle (*Zygogramma conjuncta* Rogers), an agent for the biological control of weed, *Parthenium hysterophorus* L. in India feeds on sunflower (*Helianthus annuus* L.). *Journal of Oilseeds Research* 11, 122–125.

Dhileepan, K. (2003a) Current status of the stem-boring weevil *Listronotus setosipennis* (Coleoptera: Curculionidae) introduced against the weed *Parthenium hysterophorus* (Asteraceae) in Australia. *Biocontrol Science and Technology* 13, 3–12.

Dhileepan, K. (2003b) Seasonal variation in the effectiveness of the leaf-feeding beetle *Zygogramma bicolorata* (Coleoptera: Chrysomelidae) and stem-galling moth *Epiblema strenuana* (Lepidoptera: Tortricidae) as biocontrol agents on the weed *Parthenium hysterophorus* (Asteraceae). *Bulletin of Entomological Research* 93, 393–401.

Dhileepan, K. (2007) Biological control of parthenium (*Parthenium hysterophorus*) in Australian rangeland translates to improved grass production. *Weed Science* 55, 497–501.

Dhileepan, K. (2009) Managing parthenium weed across diverse landscapes: prospects and limitations. In: Inderjit (ed.), *Management of Invasive Weeds*. Invading Nature-Springer Series in Invasion Ecology, Volume 5. Springer Science, Dordrecht, the Netherlands, pp. 227–259.

Dhileepan, K. and McFadyen, R.E.C. (2001) Effects of gall damage by the introduced biocontrol agent *Epiblema strenuana* (Lepidoptera; Tortricidae) on the weed *Parthenium hysterophorus* (Asteraceae). *Journal of Applied Entomology* 125, 1–8.

Dhileepan, K. and McFadyen, R.E. (2012) *Parthenium hysterophorus* L. – parthenium. In: Julien, M., McFadyen, R.E. and Cullen, J. (eds), *Biological Control of Weeds in Australia: 1960 to 2010*. CSIRO Publishing, Melbourne, Australia, pp. 448–462.

Dhileepan, K. and Strathie, L. (2009) *Parthenium hysterophorus* L. (Asteraceae). In: Muniappan, R., Reddy, D.V.P. and Raman, A. (eds), *Biological Control of Tropical Weeds with Arthropods in the Tropics: Towards Sustainability*, Cambridge University Press, Cambridge, pp. 274–318.

Dhileepan, K. and Wilmot Senaratne, K.A.D. (2009) How widespread is *Parthenium hysterophorus* and its biological control agent *Zygogramma bicolorata* in South Asia? *Weed Research* 49, 557–562.

Dhileepan, K., Setter, S.D. and McFadyen, R.E. (2000) Impact of defoliation by the biocontrol agent *Zygogramma bicolorata* on the weed *Parthenium hysterophorus* in Australia. *BioControl* 45, 501–512.

Dhileepan, K., Florentine, S.K. and Lockett, C.J. (2006) Establishment, initial impact and persistence of parthenium summer rust *Puccinia melampodii* in north Queensland. In: Preston, C., Watts, J.H. and Crossman, N.D. (eds), *Proceedings of the Fifteenth Australian Weeds Conference*. Weed Management Society of South Australia, Adelaide, Australia, pp. 577–580.

Dhileepan, K., Trevino, M., Vitelli, M.P., Senaratne, K.W., McClay, A.S. and McFadyen, R.E. (2012) Introduction, establishment, and potential geographic range of *Carmenta* sp. nr *ithacae* (Lepidoptera: Sesiidae), a biological control agent for *Parthenium hysterophorus* (Asteraceae) in Australia. *Environmental Entomology* 41, 317–325.

Evans, H.C. (1997) The potential of neotropical fungal pathogens as classical biological control agents for management of *Parthenium hysterophorus* L. In: Mahadeveppa, M. and Patil, V.C. (eds), *First International Conference on Parthenium Management*. University of Agricultural Sciences, Dharwad, India, pp. 55–62.

Fauzi, M.T. (2009) Biocontrol ability of *Puccinia abrupta* var. *partheniicola* on different growth stages of parthenium weed (*Parthenium hysterophorus* L.). *HAYATI Journal of Biosciences* 16, 83.

Fauzi, M.T., Tomley, A.J., Dart, P.J., Ogle, H.J. and Adkins, S.W. (1999) The rust *Puccinia abrupta* var. *partheniicola*, a potential biocontrol agent of parthenium weed: Environmental requirements for disease progress. *Biological Control* 14, 141–145.

Florentine, S.K., Raman, A. and Dhileepan, K. (2002) Responses of the weed *Parthenium hysterophorus* (Asteraceae) to the stem gall-inducing weevil *Conotrachelus albocinereus* (Coleoptera: Curculionidae). *Entomologia Generalis* 26, 195–206.

Florentine, S.K., Raman, A. and Dhileepan, K. (2005) Effects of gall induction by *Epiblema strenuana* on gas exchange, nutrients, and energetics in *Parthenium hysterophorus*. *BioControl* 50, 787–801.

Griffiths, M.W. and McFadyen, R.E. (1993) Biology and host-specificity of *Platphalonidia mystica* (Lep, Cochylidae) introduced into Queensland to biologically control *Parthenium hysterophorus* (Asteraceae). *Entomophaga* 38, 131–137.

Javaid, A. and Shabbir, A. (2007) First report of biological control of *Parthenium hysterophorus* by *Zygogramma bicolorata* in Pakistan. *Pakistan Journal of Phytopathology* 18, 199–200.

Javaid, A., Anjum, T. and Bajwa, R. (2005) Biological control of parthenium II: Allelopathic effect of *Desmostachya bipinnata* on distribution, and early seedling growth of *Parthenium hysterophorus* L. *International Journal of Biology and Biotechnology* 2, 459–463.

Jayanth, K.P. (1987a) Introduction and establishment of *Zygogramma bicolorata* on *Parthenium hysterophorus* at Bangalore, India. *Current Science* 56, 310–311.

Jayanth, K.P. (1987b) Investigations on the host-specificity of *Epiblema strenuana* (Walker) (Lepidoptera: Tortricidae). Introduced for biological control trials against *Parthenium hysterophorus* in India. *Journal of Biological Control* 1, 133–137.

Jayanth, K.P. and Bali, G. (1994) Biological control of *Parthenium hysterophorus* by the beetle *Zygogramma bicolorata* in India. *FAO Plant Protection Bulletin* 42, 207–213.

Jayanth, K.P. and Nagarkatti, S. (1987) Investigations on the host-specificity and damage potential of *Zygogramma bicolorata* Pallister (Coleoptera: Chrysomelidae) introduced into India for the biological control of *Parthenium hysterophorus*. *Entomon* 12, 141–145.

Jayanth, K.P. and Visalakshy, P.G. (1994) Dispersal of the parthenium beetle *Zygogramma bicolorata* (Chrysomelidae) in India. *Biocontrol Science and Technology* 4, 363–365.

Jayanth, K.P. and Visalakshy, G.P.N. (1996) Succession of vegetation after suppression of parthenium weed by *Zygogramma bicolorata* in Bangalore, India. *Biological Agriculture and Horticulture* 12, 303–309.

Jayanth, K.P., Visalakshy, P.N.G., Ghosh, S.K. and Chaudhary, M. (1997) Feasibility of biological control of *Parthenium hysterophorus* by *Zygogramma bicolorata* in the light of the controversy due to its feeding on sunflower. In: Mahadeveppa, M. and Patil, V.C. (eds), *First International Conference on Parthenium Management*. University of Agricultural Sciences, Dharwad, India, pp. 45–51.

Jayasuriya, A.H.M. (2005) Parthenium weed – status and management in Sri Lanka. In: Ramachandra Prasad, T.V., Nanjappa, H.V., Devendra, R., Manjunath, A., Subramanya, S.C., *et al.* (eds), *Proceedings of the Second International Conference on Parthenium Management*. University of Agricultural Sciences, Bangalore, India, pp. 36–43.

Kelaniyangoda, D.B. and Ekanayake, H.M.R.K. (2008) *Puccinia melampodii* Diet. & Hollow as a biological control agent of *Parthenium hysterophorus*. *Journal of Food and Agriculture* 1, 13–17.

Khan, N. (2011) Long term, sustainable management of parthenium weed (*Parthenium hysterophorus* L.) using suppressive plants. PhD thesis, University of Queensland, Brisbane, Australia.

Khan, N., O'Donnell, C., Shabbir, A. and Steve, W.A. (2010) Competitive displacement of parthenium weed with beneficial native and introduced pasture plants in central Queensland, Australia. In: Zydenbos, S.M. (ed.), *Proceedings of the 17th Australasian Weeds Conference*. New Zealand Plant Protection, Christchurch, New Zealand, pp. 131–134.

Khan, N., O'Donnell, C., George, D. and Adkins, S.W. (2013) Suppressive ability of selected fodder plants on the growth of *Parthenium hysterophorus*. *Weed Research* 53, 61–68.

Khan, N., Shabbir, A., George, D., Hassan, G. and Adkins, S.W. (2014) Suppressive fodder plants as part of an integrated management program for *Parthenium hysterophorus* L. *Field Crops Research* 156, 172–179.

Khan, N., George, D., Shabbir, A., Hanif, Z. and Adkins, S.W. (2015) Rising CO_2 can alter fodder–weed interactions and suppression of *Parthenium hysterophorus*. *Weed Research* 55, 113–117.

Knox, J., Jaggi, D. and Paul, M.S. (2011) Population dynamics of *Parthenium hysterophorus* (Asteraceae) and its biological suppression through

Cassia occidentalis (Caesalpiniaceae). *Turkey Journal of Botany* 35, 111–119.

Kulkarani, K.A., Kulkarani, N.S. and Santosh Kumar, G.H. (2000) Loss estimation in sunflower due to *Zygogramma bicolorata* Pallister. *Insect Environment* 6, 10–11.

Kumar, A.R.V. (1992) Is the Mexican beetle *Zygogramma bicolorata* (Coleoptera: Chrysomelidae) expanding its host range? *Current Science* 63, 729–730.

Kumar, P.S. and Evans, H.C. (2005) The mycobiota of *Parthenium hysterophorus* in its native and exotic ranges: opportunities for biological control in India. In: Prasad, T.V.R., Nanjappa, H.V., Devendra, R., Manjunath, A., Subramanya, S.C., *et al.* (eds), *Proceedings of the Second International Conference on Parthenium Management.* University of Agricultural Sciences, Bangalore, India, pp. 107–113.

Kumar Sushil (2009) Biological control of parthenium in India: status and prospects. *Indian Journal of Weed of Weed Science* 41, 1–18.

McClay, A.S. (1983) Biology and host-specificity of *Stobaera concinna* (Stal) (Homoptera: Delphacidae), a potential biocontrol agent for *Parthenium hysterophorus* L. (Compositae). *Folia Entomologica Mexicana* 56, 21–30.

McClay, A.S. (1987) Observations on the biology and host specificity of *Epiblema strenuana* (Lepidoptera, Tortricidae), a potential biocontrol agent for *Parthenium hysterophorus* (Compositae). *Entomophaga* 32, 23–34.

McClay, A.S. and Anderson, D.M. (1985) Biology and immature stages of *Thecesternus hirsutus* Pierce (Coleoptera: Curculionidae) in north-eastern Mexico. *Proceedings of the Entomological Society of Washington* 87, 207–215.

McClay, A.S., McFadyen, R.E. and Bradley, J.D. (1990) Biology of *Bucculatrix parthenica* Bradley sp. nova (Lepidoptera: Bucculatricidae) and its establishment in Australia as a biological control agent for *Parthenium hysterophorus* (Asteraceae). *Bulletin of Entomological Research* 80, 427–432.

McClay, A.S., Palmer, W.A., Bennett, F.D. and Pullen, K.R. (1995) Phytophagous arthropods associated with *Parthenium hysterophorus* (Asteraceae) in North America. *Environmental Entomology* 24, 796–809.

McConnachie, A.J. (2015a) Host range and risk assessment of *Zygogramma bicolorata*, a defoliating agent released in South Africa for the biological control of *Parthenium hysterophorus. Biocontrol Science and Technology* 25, 975–991.

McConnachie, A.J. (2015b) Host range tests cast doubt on the suitability of *Epiblema strenuana*

as a biological control agent for *Parthenium hysterophorus* in Africa. *BioControl* 60, 1–9.

McFadyen, R.E. (1980) Host specificity of the Parthenium leaf beetle *Zygogramma* sp. nr. *malvae* Stall. from Mexico, a potential biocontrol agent against *Parthenium hysterophorus* in Queensland. Internal report. Alan Fletcher Research Station, Sherwood, Queensland, Australia.

McFadyen, R.C. (1992) Biological control against parthenium weed in Australia. *Crop Protection* 11, 400–407.

McFadyen, R.C. (2000) Biology and host specificity of the stem galling weevil *Conotrachelus albocinereus* Fiedler (Col.: Curculionidae), a biocontrol agent for parthenium weed *Parthenium hysterophorus* L. (Asteraceae) in Queensland, Australia. *Biocontrol Science and Technology* 10, 195–200.

McFadyen, R.E. and McClay, A.R. (1981) Two new insects for the biological control of parthenium weed in Queensland. In: Wilson, B.J. and Swarbrick, J.T. (eds), *Sixth Australian Weeds Conference.* Weed Science Society of Queensland, Australia, pp. 145–149.

Mersie, W. (2010) Initial Environmental Examination (IEE) for Release of the Biological Agent, *Zygogramma (Zygogramma bicolorata* L.) to Control the Invasive Weed, Parthenium (*Parthenium hysterophorus* L.) in Ethiopia. Report submitted to the United States Agency for International Development.

Navie, S.C., Priest, T.E., McFadyen, R.E. and Adkins, S.W. (1998) Efficacy of the stem-galling moth *Epiblema strenuana* Walk. (Lepidoptera: Tortricidae) as a biological control agent for ragweed Parthenium (*Parthenium hysterophorus* L.). *Biological Control* 13, 1–8.

Navie, S.C., McFadyen, R.E., Panetta, F.D. and Adkins, S.W. (2005) The effect of CO_2 enrichment on the growth of a C_3 weed (*Parthenium hysterophorus*) and its competitive interaction with a C_4 grass (*Cenchrus ciliaris*). *Plant Protection Quarterly* 20, 61–66.

O'Donnell, C. and Adkins, S.W. (2005) Management of parthenium weed through competitive displacement with beneficial plants. *Weed Biology and Management* 5, 77–79.

Page, A.R. and Lacey, K.L. (2006) *Economic Impact Assessment of Australian Weed Biological Control.* Technical Series No. 10. CRC for Australian Weed Management, Adelaide, Australia.

Parker, A., Holden, A.N.G. and Tomley, A.J. (1994) Host specificity testing and assessment of the pathogenicity of the rust, *Puccinia abrupta* var. *partheniicola*, as a biological control agent of

Parthenium weed (*Parthenium hysterophorus*). *Plant Pathology* 43, 1–16.

Raghu, S. and Dhileepan, K. (2005) The value of simulating herbivory in selecting effective weed biological control agents. *Biological Control* 34, 265–273.

Retief, E., Ntushelo, K. and Wood, A.R. (2013) Host-specificity testing of *Puccinia xanthii* var. *parthenii-hysterophorae*, a potential biocontrol agent for *Parthenium hysterophorus* in South Africa. *South African Journal of Plant and Soil* 30, 7–12.

Seier, M.K. (1999) Studies on the rust *Puccinia melampodii* Diet. & Holw., a potential biological control agent for parthenium weed (*Parthenium hysterophorus* L.) in Australia. Report submitted to the Queensland Department of Natural Resources & Mines. CABI Bioscience UK Centre, Silwood Park, UK.

Seier, M.K. (2015) The potential of fungal pathogens for classical biological control of invasive alien weeds in the Asia Pacific region (Abstract). *25th Asian Pacific Weed Science Society Conference*, Hyderabad, India, 13–16 October 2016.

Shabbir, A., Dhileepan, K. and Adkins, S.W. (2012) Spread of parthenium weed and its biological control agent in the Punjab, Pakistan. *Pakistan Journal of Weed Science Research* 18(581), e588.

Shabbir, A., Dhileepan, K., O'Donnell, C. and Adkins, S.W. (2013) Complementing biological control with plant suppression: implications for improved management of parthenium weed (*Parthenium hysterophorus* L.). *Biological Control* 64, 270–275.

Shabbir, A., Dhileepan, K., Khan, N. and Adkins, S.W. (2014) Weed–pathogen interactions and elevated CO_2: Growth changes in favour of the biological control agent. *Weed Research* 54, 217–222.

Shabbir, A., Dhileepan, K., Zalucki, M.P., O'Donnell, C., Khan, N., Hanif, Z. and Adkins, S.W. (2015) The combined effect of biological control with plant competition on the management of parthenium weed (*Parthenium hysterophorus* L.). *Pakistan Journal of Botany* 47, 157–159.

Shrestha, B.B., Poudel, A., Khattri-Chettri, J., Karki, D., Gautam, R.D. and Jha, P.K. (2010) Fortuitous biological control of *Parthenium hysterophorus* by *Zygogramma bicolorata* in Nepal. *Journal of Natural History Museum* 25, 332–337.

Shrestha, B.B., Thapa-Magar, K.B., Paudel, A. and Shrestha, U.B. (2011) Beetle on the battle: Defoliation of *Parthenium hysterophorus* by *Zygogramma bicolorata* in Kathmandu valley, Nepal. *Botanica Orientalis: Journal of Plant Science* 8, 100–104.

Sridhara, S., Basavaraja, B.K. and Ganeshaiah, K.N. (2005) Temporal variation in relative dominance of *Parthenium hysterophorus* and its effect on native biodiversity. In: Prasad, T.V.R., Nanjappa, H.V., Devendra, R., Manjunath, A., Subramanya, S.C., *et al.* (eds), *Proceedings of the Second International Conference on Parthenium Management*. University of Agricultural Sciences, Bangalore, India, pp. 240–242.

Strathie, L.W. (2014) Application for permission to release the seed-feeding weevil, *Smicronyx lutulentus* (Coleoptera; Curculionidae) from quarantine at ARC-PPRI Cedara, for the biological control of *Parthenium hysterophorus* (Asteraceae) in South Africa. Report to the Department of Agriculture, Forestry and Fisheries, Republic of South Africa.

Strathie, L.W. and McConnachie, A. (2012) Application for permission to release the stem-boring weevil, *Listronotus setosipennis* (Coleopetera: Curculionidae) from quarantine at ARC-PPRI Cedara, for the biological control of *Parthenium hysterophorus* (Asteraceae) in South Africa. Report to the Department of Agriculture, Forestry and Fisheries, Republic of South Africa.

Strathie, L.W., Wood, A.R., van Rooi, C. and McConnachie, A. (2005) *Parthenium hysterophorus* (Asteraceae) in southern Africa, and initiation of biological control against it in South Africa. In: Prasad, T.V.R., Nanjappa, H.V., Devendra, R., Manjunath, A., Subramanya, S.C., *et al.* (eds), *Proceedings of the Second International Conference on Parthenium Management*. University of Agricultural Sciences, Bangalore, India, pp. 127–133.

Strathie, L.W., McConnachie, A.J. and Retief, E. (2011) Initiation of biological control against *Parthenium hysterophorus* L. (Asteraceae) in South Africa. *African Entomology: Biological Control of Invasive Alien Plants in South Africa (1999–2010): Special Issue 2* 19, 378–392.

Tamado, T. and Milberg, P. (2004) Control of parthenium (*Parthenium hysterophorus*) in grain sorghum (*Sorghum bicolor*) in the smallholder farming system in eastern Ethiopia. *Weed Technology* 18, 100–105.

Taye, T., Einhorn, G., Gossmann, M., Büttner, C. and Metz, R. (2004a) The potential of parthenium rust as biological control of parthenium weed in Ethiopia. *Pest Management Journal of Ethiopia* 8, 39–50.

Taye, T., Einhorn, G. and Metz, R. (2004b) *Parthenium hysterophorus*, an invasive species in Ethiopia – investigations on the occurrence and on its pathogens. *Journal of Plant Diseases and Protection* 19, 271–278.

van der Laan, M. (2006) Allelopathic interference potential of the alien invader plant *Parthenium hysterophorus*. MSc thesis, University of Pretoria.

Wahab, S. (2005) Management of parthenium weed through an integrated approach initiatives, achievements and research opportunities in India. In: Prasad, T.V.R., Nanjappa, H.V., Devendra, R., Manjunath, A., Subramanya, S.C., *et al.* (eds), *Proceedings of the Second International Conference on Parthenium Management*. University of Agricultural Sciences, Bangalore, India, pp. 127–333.

Wan, F., Wang, R. and Ding, J. (1995) Biological control of *Ambrosia artemisiifolia* with introduced agents, *Zygogramma suturalis* and *Epiblema strenuana* in China. In: Delfosse, E.S. and Scott, R.R. (eds), *Proceedings of the Eighth International Symposium on Biological control of Weeds*. Lincoln University, Canterbury, New Zealand, pp. 193–200.

Wild, C.H., McFadyen, R.E., Tomley, A.J. and Willson, B.W. (1992) The biology and host specificity of the stem-boring weevil *Listronotus setosipennis* (Col.: Curculionidae). A potential biocontrol agent for *Parthenium hysterophorus* (Asteraceae). *Entomophaga* 37, 591–598.

Withers, T.M., McFadyen, R.E. and Marohasy, J. (1999) Importation protocols and risk assessment of weed biological control agents in Australia: the example of *Carmenta* nr. *ithacae*. In: Follett, P.A. and Duan, J.J. (eds), *Non-target Effects of Biological Control*. Kluwer, Boston, Massachusetts, pp. 195–214.

Wood, A.R. and Scholler, M. (2002) *Puccinia abrupta* var. *partheniicola* on *Parthenium hysterophorus* in southern Africa. *Plant Disease* 86, 327.

8 Management: Physical, Cultural, Chemical

Shane D. Campbell,[1,2]* Wayne D. Vogler[1] and Tamado Tana[3]

[1]*Biosecurity Queensland, Tropical Weeds Research Centre, Charters Towers, Queensland, Australia;* [2]*Current affiliation: the University of Queensland, Gatton, Australia;* [3]*College of Agriculture and Environmental Sciences, Haramaya University, Ethiopia*

8.1 Introduction

There are many different options available to manage problematic weeds, with research continually advancing new technologies and products to enable development of more cost-effective and environmentally sustainable strategies. This applies to parthenium weed (*Parthenium hysterophorus* L.), which has been and continues to be the focus of a significant research effort aimed at improving management options within affected countries.

Although parthenium weed is a relatively easy plant to kill, its biology/ecology, widespread distribution in many areas and ability to grow in a diverse range of habitats/ situations (e.g. wastelands, cleared land, pastures, crops, roadsides, along streams and rivers) (Navie *et al.*, 1996, 1998a; Gnanavel, 2013) makes it a difficult plant to eliminate.

As previously discussed (see Adkins *et al.*, Chapter 2, this volume), parthenium weed has a long-lived seed bank (Butler, 1984; Navie *et al.*, 1998b), an ability to rapidly reach reproductive maturity (Parsons and Cuthbertson, 2001; Adkins and Shabbir, 2014) and multiple dispersal mechanisms (Auld *et al.*, 1983; Navie *et al.*, 1996, 1998a). This means that land managers may have to implement management on a regular basis for many years to achieve a desired level of control and even then there is a risk of reintroduction from external sources. Furthermore, at any given time several life stages can be present within an infestation (e.g. young seedlings, plants in the vegetative stage, mature plants with seeds) which may make management more difficult (Kohli *et al.*, 2006; Batish *et al.*, 2012). For example, some herbicides can provide high mortality of young plants but they may not be effective at controlling mature plants (see Section 8.4). The adverse health effects caused by parthenium weed to humans also pose another level of complexity to its control, with some people hesitant to come into contact with it (Batish *et al.*, 2012; Gnanavel, 2013).

Before commencing a management programme, land managers/owners need to decide what objective they want to achieve. Is it to completely eradicate parthenium weed, to contain it or to undertake a level of control that will minimize its impacts? A range of variables may influence this decision, including potential legislative requirements, the area infested, the prevailing habitat, anticipated economic returns and the resources available. The capacity to manage parthenium weed in terms of financial

* shane.campbell@uq.edu.au

and physical resources varies greatly between and even within affected countries. It is therefore not possible to develop standardized management practices applicable to all situations. The merits of available management options and the resources available should be compared so that informed decisions can be made by those responsible for managing parthenium weed.

Irrespective of the objective, in most instances an integrated weed management approach will be needed, using a range of individual techniques that have been found effective on parthenium weed (Kohli et al., 2006; Mahadevappa, 2009; Adkins and Shabbir, 2014). Several articles have highlighted the effectiveness and limitations of some of the control options and management strategies that have been tested on parthenium weed (e.g. Navie et al., 1996, 1998a; Parsons and Cuthbertson, 2001; Dhileepan, 2009; Mahadevappa, 2009; Batish et al., 2012; Gnanavel, 2013; Masum et al., 2013; Adkins and Shabbir, 2014; Kaur et al., 2014; Kumari, 2014; Tanveer et al., 2015). This chapter will expand on this collated information, provide a current update on the physical, cultural and chemical techniques that are available and discuss some of the integrated strategies that have proven effective in certain situations.

8.2 Cultural Management

8.2.1 Legislative measures

Prevention plays an important role in the management of most weeds. An initial prevention strategy is to put in place legislation that defines the objectives for management of the weed. When a weed is recognized in the legislation of a jurisdiction (e.g. State, Country, Province), this provides formal acknowledgment of the risk it poses, and responsibilities and actions are usually assigned based on its potential and current distribution and likely impacts. Key actions tend to include preventing introduction, minimizing spread into new areas, eradication at the early stages of invasion, or

containment and ongoing management of larger infestations to reduce impacts.

In Australia, parthenium weed has been declared a weed under the respective legislation of all states and territories, but the level of declaration and associated requirements varies greatly depending on the perceived level of risk (Table 8.1; Australian Weeds Committee, 2012). In Queensland, where the largest infestations of parthenium weed are to be found in Australia, it is classified as restricted matter under the Biosecurity Act 2014. This means that it must not be given away, sold or released into the environment without a permit. The Act also requires everyone to take all reasonable and practical steps to minimize the risks associated with invasive plants and animals under their control, including parthenium weed. This is called a general biosecurity obligation. At a local level, each local government must have a biosecurity plan that covers invasive plants and animals in its area. This plan may include actions to be taken on certain species and some of these actions may be required under local laws (DAF, 2016). In other Australian states and territories where parthenium weed is considered a high-risk weed but it has not become well established (i.e. in the Northern Territory, Western Australia, New South Wales), the legislation focuses on reporting and eradicating outbreaks and preventing further introductions (Table 8.1; Australian Weeds Committee, 2012). The New South Wales (NSW) legislation (NSW Noxious Weeds Act 1993) also has some specific requirements regarding movement of agricultural machines (crop harvesters and associated grain-handling equipment) from interstate. They must be thoroughly cleaned by the person in charge of the machine before entering NSW and the machinery must then be checked by a state border inspector. This process has greatly reduced the risk of further introductions of parthenium weed into NSW from Queensland. To help individuals and organizations meet their general biosecurity obligations (Biosecurity Act 2014) in Queensland, numerous wash-down facilities have been strategically established for removal of weed seeds from

Table 8.1. Declaration status of parthenium weed in various countries.

Region	Category	Declaration	Requirements
Australia			
Australian Capital Territory	Notifiable/ prohibited	Pest Plants and Animals Act 2005	Presence must be notified to the chief executive of the relevant Government agency. Importation, propagation, commercial supply and disposal is controlled.
New South Wales	Class 1	Noxious Weeds Act 1993 No.11 Noxious Weeds Amendment Act 2012 No. 25	The plant must be eradicated and the land must be kept free of the plant. It is a notifiable weed and all outbreaks must be reported to the local council within three days. The plant must not be sold anywhere within NSW. Restrictions on movement.
Northern Territory	Class A, C	Weeds Management Act 2001	Class A: To be eradicated. Class C: Not to be introduced into the Northern Territory.
Queensland	Restricted invasive plant (Category 3)	Biosecurity Act 2014	It must not be given away, sold or released into the environment without a permit. The Act requires everyone to take all reasonable and practical steps to minimize the risks associated with invasive plants and animals under their control.
South Australia	Class 1, Category 1, state wide	Natural Resource Management Act 2004	Movement and sale restricted. Notification of presence required and owner of land to take action to destroy or control.
Tasmania	Declared weed	Weed Management Act 1999	The importation, sale and distribution of parthenium weed is prohibited. Legal responsibilities of landholders and others stakeholders in dealing with parthenium weed are laid out in the parthenium weed Statutory Weed Management Plan.
Victoria	State prohibited weed	Catchment and Land Protection Act 1994	To be eradicated if possible or excluded from the state.
Western Australia	declared pest, prohibited, control category C1 exclusion	Biosecurity and Agriculture Management Act 2007	Prohibited from entry into Western Australia.
India			
Karnataka	Noxious weed	Agricultural Pests and Diseases Act, 1968	Occupiers of land could be required to carry out preventative or remedial measures if a notice under Section 4 of the Act is issued.
South Africa			
All	Category 1b	National Environmental Management: Biodiversity Act 2004 (Act No. 10 of 2004)	Persons must take steps to control the weed.
Sri Lanka			
All	Noxious weed	Plant Protection Ordinance No. 35 of 1999	
Kenya			
All	Noxious weed	Suppression of Noxious Weeds Act (CAP 325)	Its presence must be reported and persons responsible for an area of land should clear the noxious weed, or cause it to be cleared, from that land.

Sources: Dhileepan (2009); Kenya Law (2010); Australian Weeds Committee (2012); Adkins and Shabbir (2014); NSW Department of Primary Industries (2014); Republic of South Africa Government Gazette (2014); Environment, Planning and Sustainable Development Directorate (2015); Primary Industries and Regions SA (2015); DAF (2016); Department of Agriculture and Food (2016); Department of Primary Industries, Parks, Water and Environment (2016).

vehicles and machinery. Formal guidelines have also been produced by the state government on clean-down and inspection procedures (DAFF, 2013, 2014).

Parthenium weed is a declared weed in a few other affected countries, including South Africa, Sri Lanka and Kenya, with the legislation largely focused on preventing its movement into uninfested areas. In India, the State of Karnataka placed parthenium weed into the category of an agricultural pest in 1969, but this legislation could not be enforced and therefore was unable to prevent its spread to other states (Mahadevappa, 2009; Adkins and Shabbir, 2014). In spite of parthenium weed being a serious problem throughout India, there are no legislative statutes specifically for parthenium weed in other states. In other countries where parthenium weed has become a problem (e.g. Nepal, Bangladesh, Ethiopia, Pakistan, China), legislative measures should be considered, but only if authorities are committed to following/enforcing the laws/acts that are put in place.

8.2.2 Hygiene practices

Preventing the spread of parthenium weed is the most cost-effective management strategy (Dhileepan, 2009). While it can be spread into new areas via many dispersal mechanisms, human-related activities are generally associated with its long-distance dispersal (see Shabbir et al., Chapter 3, this volume). Where available, legislation can provide a framework to minimize spread and impose penalties if breaches occur, but practical measures can also be implemented to greatly reduce associated risks.

Contaminated feed for animals (e.g. fodder, grain) or seed for planting pastures and crops have been responsible for numerous outbreaks of parthenium weed in new areas around the world, emphasizing the importance of sourcing clean materials. To help landholders in Queensland (Australia), the state government developed weed hygiene declaration forms, which can be requested from providers to outline the weed status of any seed or feed material or animals being purchased (Chamberlain and Gittens, 2004). If the status cannot be confirmed, where possible newly purchased animals should be quarantined in yards or small paddocks for a period of time and animals should be fed in the same areas. This will minimize the area that needs to be monitored and controlled if parthenium weed plants are found. Regular checking of new areas planted with pastures or crops should also occur, particularly if the seed has been sourced from areas infested with parthenium weed.

Vehicles, machinery and equipment can also spread parthenium weed and should be cleaned thoroughly if being moved from infested to uninfested areas. In Australia, sampling of public wash-down facilities in or near parthenium weed-infested areas has found large quantities of parthenium weed seed in the remaining sludge, highlighting the potential risk of spread by vehicles (Nguyen, 2011; Khan et al., 2013). Furthermore, in an evaluation of the probable sources of new outbreaks of parthenium weed on private properties in NSW (Australia) between 1982 and 2004, 59% were attributed to contaminated crop harvesters while vehicles/trucks/other machinery accounted for another 14% (Blackmore and Johnson, 2010).

The use of high-pressure and/or low-pressure, high-volume water hoses is recommended to ensure thorough cleaning of vehicles, machinery and equipment (Parsons and Cuthbertson, 2001; Khan, 2012), and in Australia there are numerous roadside wash-down facilities that are either manually or automatically operated (Fig. 8.1). If public wash-down facilities are not available, a designated area should be assigned as a wash-down area so that any seeds that are removed and germinate in the immediate area can be destroyed promptly (Holman and Dale, 1981; Parsons and Cuthbertson, 2001; DAFF, 2014).

8.2.3 Competition and suppression

In native and improved pastures used for grazing by domestic livestock, pasture

Fig. 8.1. An automated wash-down facility being used in Queensland, Australia. (Queensland Department of Agriculture and Fisheries.)

management is one of the most important tools to reduce the impact of parthenium weed. Parthenium weed has a preference for disturbed areas, such as pastures in poor condition, and any steps that can be implemented to favour the pasture will be advantageous. A number of pasture species have been found to be highly suppressive against parthenium weed (O'Donnell and Adkins, 2005; Khan *et al.*, 2012b) and it is usually only through animals preferentially selecting them that their vigour is reduced, providing parthenium weed with the opportunity to proliferate. Several grass species, including bisset bluegrass (*Bothriochloa insculpta* Hochst. ex A. Rich.), floren bluegrass (*Dichanthium aristatum* (Poir.) C.E. Hubb), buffel grass (*Cenchrus ciliaris* L.), purple pigeon grass (*Setaria incrassata* Hochst. Hack.), green panic (*Panicum maximum* Jacq. var. *trichoglume* Robyns); Indian bluegrass (*Bothriochloa pertusa* L.), bull Mitchell grass (*Astrebla squarrosa* C.E. Hubb), Kangaroo grass (*Themeda triandra* Forssk.), pitted

bluegrass (*Bothriochloa decipiens* Hack. C.E. Hubb.), hoop Mitchell grass (*Astrebla elymoides* F. Muell ex. F.M. Bailey) and a legume species (butterfly pea, *Clitoria ternatea* L.), studied under glasshouse and field conditions, have been shown to suppress the growth of parthenium weed (O'Donnell and Adkins, 2005; Khan *et al.*, 2012b).

Domesticated animal stocking rates should be set at a density that allows for the maintenance of a uniform coverage of pasture, but these rates may need to be adjusted according to season and frequency of rainfall events to maintain dominant grass cover (Holman and Dale, 1981; Navie *et al.*, 1996, 1998a).

In Australia, heavily infested areas can be fenced off and destocked to help prevent seed being spread by domesticated animals to parthenium weed-free areas (Holman and Dale, 1981). To restore heavily infested areas, they should be spelled (free of domesticated animals) for at least one full growing season and also grazed very lightly in the following season (Holman and Dale, 1981; Parsons and Cuthbertson, 2001). Even then, this may not be sufficient and the area may need to be sown with suitable pasture species, especially if the soil seed bank of the desirable grasses has been depleted. Any newly sown pastures should be spelled to allow the pasture to establish and then initially grazed lightly (Parsons and Cuthbertson, 2001). Rotational grazing (i.e. use of resting periods) has also been shown to enhance the control of annual broadleaf species, such as parthenium weed, within perennial grass-based grasslands (DAF, 2016).

In cropping situations, selection of the most competitive crops and varieties and manipulation of plant densities may help reduce parthenium weed. The challenge is to establish a density of a particular crop that maximizes competition against parthenium weed, while not causing so much intra-crop competition that it results in reduced crop yields. For example, in Ethiopia manipulation of crop varieties, sowing densities and planting dates was used to improve competition against parthenium weed in sorghum (Sorghum bicolor (L.) Moench) crops (Tamado and Milberg, 2004; Besufekad et al., 2005).

8.2.4 Crop rotation

Crop rotation involves alternating different crops on the same land and is an important strategy for weed management (Singh et al., 2003). Weeds tend to thrive and build up in density in paddocks that are continually grown with a crop that has similar growth requirements to them. By implementing a crop rotation process, the germination and growth cycles of weeds can be disrupted through the different cultural practices (such as tillage, planting dates and competition, etc.) associated with each crop (Singh et al., 2003). For parthenium weed management, some options could include rotating between summer and winter grown crops, and/or between monocot and dicot crops. In areas where cool season temperatures are less favourable for germination and growth of parthenium weed, introducing a cool

season crop may help break the cycle. However, in tropical environments parthenium weed can proliferate all year round and be a significant problem for both warm and cool season crops, as is the case in central Queensland (Australia) (Osten et al., 2007). Because parthenium weed is a dicot species, planting a monocot crop would broaden the range of selective herbicides that could be used to control it.

A more severe rotation option could be the introduction of a pasture phase before then returning back to cropping. In India, it has also been suggested that instead of traditional crops, a rotation phase using marigold (Tagetes spp.) during the rainy season can effectively reduce parthenium weed in cultivated areas (Kaur et al., 2014; Kumari, 2014).

8.2.5 Cover crops

Cover (or smother) crops can be used for weed suppression in two main ways. First, they can be grown between periods of regular cropping and can negatively impact on the weeds present by competing with them (e.g. for light, moisture and nutrients), stimulating microbial activity, changing physical factors of the soil (e.g. pH, temperature) or through the release of allelochemicals (Singh et al., 2003). Second, they can be planted between rows of crops to minimize the need for inter-row control of weeds. If a legume species is used, it can have the added advantage of improving soil fertility. These positive effects of cover crops have seen them become a useful part of integrated weed management strategies (Singh et al., 2003). In eastern Ethiopia, Tamado and Milberg (2004) investigated the use of a legume cover crop (cowpea; Vigna unguiculata L.) in sorghum crops for control of parthenium weed. They found that growing cowpea in the inter-row suppressed parthenium weed (particularly if used in combination with hoeing) but sorghum grain and stalk yields were reduced significantly. It was concluded that the reduced crop yields could have been associated with insufficient soil moisture being available to support the combined demand of both the sorghum and the cowpea (Tamado and Milberg, 2004). This reduced crop production is less likely in higher rainfall environments.

8.3 Physical Management

8.3.1 Manual removal

Manual removal is generally only feasible for treating small and isolated areas of parthenium weed, such as in residential zones/ areas, roadsides or small agricultural fields. It is not considered economical or practical for large areas (e.g. grazing areas, national parks and protected forests), particularly in countries where the cost of labour is high (Dhileepan, 2009; Adkins and Shabbir, 2014).

Pulling out plants, including the root system, when the soil is sufficiently moist or hoeing (Fig. 8.2) to remove a section of the root system are the most effective manual control options (Tamado and Milberg, 2004;

Gnanavel, 2013). Treatments that only cut off the above ground part but leave the root system can result in the plant reshooting (Adkins and Shabbir, 2014). Where possible, control should be undertaken before plants produce seed and start replenishing the seed bank. If plants have already flowered, they should be destroyed after being removed, with burning being an effective option (Kohli et al., 1997; Gnanavel, 2013). Also, to be effective in the longer term, it will need to be repeated regularly or followed up with some other techniques to control seedling regrowth that generally follows the disturbance created by removing the parent plants. In cropping situations, manual control is generally most effective in the young crop. As the crop matures, it is better able to suppress any new parthenium weed plants by actively competing for available resources, especially soil moisture and light.

Several studies investigating integrated weed management strategies in cropping situations, either specifically for parthenium weed or where it is a component of a weed infestation, have frequently found repeated

Fig. 8.2. Manual hoeing of parthenium weed in a field at Kikopey, Kenya. (Sarah Hilliar.)

manual hoeing to be one of the most effective control options, followed by hoeing in combination with other control techniques (Tamado and Milberg, 2004; Pratap *et al.*, 2013). In some instances, hoeing is not only effective for weed control but also seems to provide an advantage to the crop, with yields frequently higher if this technique is used (Tamado and Milberg, 2004; Pratap *et al.*, 2013). Tamado and Milberg (2004) suggested that this could be due to improved soil aeration and water infiltration.

Given the health risks associated with parthenium weed (Allan *et al.*, Chapter 6, this volume), direct contact should be avoided by wearing adequate protective clothing, particularly a light face mask to reduce risk of an allergic respiratory reaction, long sleeves and gloves (Goodall *et al.*, 2010). Using a hoe instead of pulling out plants by hand will also help avoid direct contact with plants and reduce the risk of contracting dermatitis (Goodall *et al.*, 2010). Pulling plants out when they are in flower poses the greatest risk as this may aid in the dispersal of pollen grains, resulting in further allergic reactions (Gnanavel, 2013; Kumari, 2014). In some countries, such as India, it is becoming harder to get labourers to undertake manual removal of parthenium weed as they are becoming increasingly aware of, and concerned about, the health risk of this plant (Batish *et al.*, 2012; Gnanavel, 2013).

8.3.2 Mechanical removal

Mechanical treatments can be effective at killing a range of weed species but they often create favourable seed beds and promote large-scale seedling regrowth, which may exacerbate the problem if follow-up control is not undertaken. This applies to parthenium weed, where methods such as grading, slashing and ploughing of large infestations can provide some relief over the short term, but are not effective in long-term management as they are known to stimulate germination (Haseler, 1976). Mowing or slashing can also result in the rapid regrowth of plants from the lateral parts of the shoot that are close to the ground (Bhowmik and Sarkar, 2005).

Mechanical control (such as ploughing and mulching) would be most appropriate for controlling parthenium weed and other weeds while preparing land for planting with a crop or improved pasture, particularly if it can be done at the vegetative stage before flowering occurs (Masum *et al.*, 2013; Kaur *et al.*, 2014; Kumari, 2014). However, even then some level of follow-up control will probably be needed to treat seedling regrowth until the crop or pasture can establish sufficiently to actively compete with the parthenium weed. In some crops that are planted in wide enough rows, such as sugarcane (*Saccharum officinarum* L.) or sorghum, inter-row cultivation may be possible while the crop is young and the machinery can drive over the top of it without causing damage.

8.3.3 Fire and heat

The response of weeds to fire can vary markedly, from those that are deleteriously affected, to others that incur nil or minimal damage, right up to ones that proliferate afterwards. In susceptible species, fire may have an impact on one, several or all life stages (i.e. seedlings and juvenile plants, mature plants, the seed bank) with responses likely to vary considerably depending on the particular fire regime imposed (Campbell, 2009). Intensity, frequency and season of application are three major factors that can influence how effective fire will be on susceptible species.

Current knowledge on the effect of fire on parthenium weed is restricted to a combination of some anecdotal evidence, case studies (Butler and Fairfax, 2003) and the findings from a trial undertaken in northern Australia (Vogler *et al.*, 2006). In areas where parthenium weed is dense, there is unlikely to be sufficient fuel to carry a fire (Haseler, 1976). In areas that have been burnt, greater parthenium weed incidence has been observed and measured (Butler and Fairfax, 2003) in burnt areas as compared with nearby unburnt areas. Based on such findings, burning of parthenium-infested areas has tended to be discouraged.

To gain a better understanding of the effects of fire on parthenium weed in pasture situations, Vogler *et al.* (2006) undertook a comprehensive study in northern Australia. One of the experiments included a series of two and three annual fires implemented in either autumn or spring, respectively. An untreated control plus a mechanical mowing and plant removal treatment were also included for comparison, with the latter incorporated to test whether plant responses following fires were directly associated with burning or defoliation. Another experiment exposed seed banks collected near the study site to combinations of smoke and heat to test their impacts on parthenium weed seed germination and that of other species present at the site. The results showed that fire did not affect the size of the parthenium weed seed bank, nor did it stimulate germination through the effect of smoke or heat. A single fire (particularly in spring) tended to increase the germination and subsequent frequency and density of parthenium weed, though this occurred indirectly through the removal of the above ground pasture cover, with a similar response occurring following mowing. Parthenium weed populations declined after subsequent fires applied to the same area (Vogler *et al.*, 2006).

The findings of Vogler *et al.* (2006) confirmed that fire is not an effective control technique for parthenium weed in northern Australia. This is similar to other one-off control options that can exacerbate the problem in some instances. The findings also indicate that the effect of fire on parthenium weed populations in grass pastures is similar to that of biomass removal by mowing or heavy grazing. Therefore, maintaining strong pasture competitiveness is potentially the most important factor influencing parthenium weed populations in grass pastures. Nevertheless, further research on fire is recommended to cover a broader range of situations, as there is some contrasting information in the literature. For example, Shabbir (2007) observed that in some wheat (*Triticum aestivum* L.) and rice (*Oryza sativa*

L.) fields, patches of stubble that were burnt had minimal parthenium weed seedling emergence, compared with patches that avoided being burnt (see Shabbir *et al.*, Chapter 12, this volume). The quantity of litter/fuel and the location of seeds (e.g. on the surface or buried) are two factors that could have a large influence on how parthenium weed responds to fires.

For small patches, the use of flame-thrower-style equipment may be an effective option, particularly as an alternative for land managers who do not want to use chemicals, such as organic farmers. The efficacy of a hand-held burner (Artarus Ranger®) was tested on three woody weeds in northern Australia – rubber vine (*Cryptostegia grandiflora* R.Br.), bellyache bush (*Jatropha gossypiifolia* L.) and parkinsonia (*Parkinsonia aculeata* L.) – with high mortality obtained on bellyache bush and parkinsonia if the base of plants was exposed to the flame for 10 s or more. Rubber vine required 60 s exposure to achieve high kill rates, with the variation between species attributed to differences in the bark (e.g. thickness and moisture condition) (Vitelli and Madigan, 2004). In the case of parthenium weed, directly heating the soil using a flame may also kill some seeds in the soil seed bank, particularly those located on or close to the soil surface.

The use of radiation generated through microwave technology is another technique that has been tested on weeds in recent years, and while still in the developmental stage it could have application for control of weeds such as parthenium weed in the future. A laboratory-based experiment demonstrated that radiation could penetrate into the soil and deactivate weed seeds of several species, including parthenium weed (Bebawi *et al.*, 2007).

8.4 Chemical Control

8.4.1 Synthetic herbicides

Research to identify effective herbicides for control of parthenium weed in a range of situations has been underway for over 40 years and is constantly evolving, with some original herbicides no longer available and newer ones having been developed (Table 8.2). There are generally several options available to effectively control parthenium weed in different situations, including crops, pastures, roadsides, waste lands, industrial areas, urban areas and roadsides (Fig. 8.3a,b). Permissible herbicides may vary between and within countries and they are often controlled through a formal registration process. It is therefore important that those undertaking spraying of parthenium weed check to ensure that they are compliant.

Chemicals can be expensive to buy and apply over large areas, which has tended to restrict their use mainly to eradication programmes, control in urban areas and along roadsides and for minimizing the impacts of parthenium weed in high-value crops. They are much less frequently used in grazing areas, public and uncultivated areas and forests. Some people are also hesitant to use them due to either perceived or actual risks that they pose to beneficial and native plants, the environment and human health.

Best results are generally achieved at the pre-emergent stage of the weed, or when plants are small and prior to reproductive maturity and replenishing the soil seed bank (Navie et al., 1996, 1998a; Gnanavel, 2013; Kumari, 2014). However, the long-lived nature of parthenium weed seeds means that even if a good kill of initial plants is achieved, parthenium weed will generally reappear from the soil seed-bank reserves (Navie et al., 1996, 1998a). Residual herbicides (e.g. atrazine) will provide some reprieve for a period, but new seedlings will emerge once the residual effects expire, necessitating follow-up control. Maintaining competition is important for control of parthenium weed, so spraying with a pre- or post-emergent selective herbicide that will not kill other species is recommended (Gnanavel, 2013).

Some of the most frequently tested and used chemicals on parthenium weed have included formulations of 2,4-D, atrazine,

glyphosate, glufosinate, metribuzon and metsulfuron-methyl. Several formulations of 2,4-D herbicides provide post-emergent control of a wide range of broadleaf weeds in both crop and non-crop situations. They have been widely tested for post-emergence control of parthenium weed with highly variable results, ranging from low to high mortality, depending on the formulation and rates applied, the size of plants at the time of application and the application method used (Armstrong and Orr, 1984; Mishra and Bhan, 1995; Singh et al., 2004; Reddy et al., 2007; Goodall et al., 2010; Fernandez et al., 2015). For example, Singh et al. (2004) reported poor results (30–60% mortality) using 1–3 kg ae [acid equivalent]/ha of 2,4-D ethyl ester on large plants. In some instances, the poor results reported may be due to large-scale seedling regrowth that can occur after control of initial plants (Armstrong and Orr, 1984; Tamado and Milberg, 2004). To overcome this, the inclusion of more persistent herbicides such as atrazine (Armstrong and Orr, 1984) is often recommended and in some instances chemical companies sell pre-mixed products, such as 2,4-D + picloram.

Atrazine is a selective herbicide that can be used both before and after the emergence of several crops or trees to control grasses and broadleaf weeds. In Australia, it is registered for the control of parthenium weed in crop (e.g. maize (Zea mays L.), sorghum) and non-crop situations at rates ranging between 1.53 and 3.00 kg ai [active ingredient]/ha, with a maximum of 3.00 and 4.50–7.20 kg ai/ha (depending on soil type) able to be applied annually in crop and forest situations, respectively. Atrazine can give pre-emergent control of parthenium weed for several months (Muniyappa et al., 1980; Armstrong and Orr, 1984; Brooks et al., 2004), as well as some post-emergent control, although mortality of emerged plants can be variable and decreases markedly as plant size increases (Singh et al., 2004; Khan et al., 2012a). It is sometimes recommended to be mixed with a more effective post-emergent herbicide if larger plants are

Table 8.2. Some herbicides and the associated rates that have been successfully tested or recommended in different situations for control of at least one growth stage of parthenium weed. Some of these herbicides may be no longer available or recommended for control of parthenium weed in certain countries.

Herbicide	Rate (g ae or ai/ha)[a]	Growth stage	Situation	Country	Reference
2,4-D	806	S	N	U	Reddy *et al.* (2007)
2,4-D amine	200/100 l	S, M	P, R, Cl	A	DAF (2016)
2,4-D amine	2240	L	N	U	Fernandez *et al.* (2015)
2,4-D ethyl ester	1500–2000	M	N	I	Mishra and Bhan (1995)
2,4-D + picloram	900 + 225	M	N, P, R, Cl	A	DAF (2016)
Ametryn + atrazine	625–2500[b]	M, L	PT	P	Javaid (2007)
Aminocyclopyrachlor + chlorsulfuron	83 + 33 to 166 + 66	L	N	U	Fernandez *et al.* (2015)
Aminopyralid	70–123	L	N	U	Fernandez *et al.* (2015)
Aminopyralid + metsulfuron methyl	3.75 + 3/100 l	S, M, L	P, Cl, R	A	DAF (2016)
Atrazine	1800–3000	P, S, M	F, C, R, Cl	A	DAF (2016)
	250–1000[b]	M, L	PT	P	Javaid (2007)
Atrazine + 2,4-D	2000–4000 + 1000–2000	P, S, M, L	P	A	Armstrong and Orr (1984)
Atrazine + dicamba	2000–4000 + 200–400	P, S, M, L	P	A	Armstrong and Orr (1984)
Atrazine + s-metolachlor	840 + 660	P	Maize	P	Khan *et al.* (2014)
Bentazon	1500	–	Soybean	I	Mishra and Bhan (1996)
Bromoxynil	560	S	N	U	Reddy *et al.* (2007)
Bromoxynil + MCPA	250–1000[b]	M, L	PT	P	Javaid (2007)
Butachlor	500–2000[b]	M, L	PT	P	Javaid (2007)
Chlorimuron-ethyl	10	P	N	U	Reddy *et al.* (2007)
	20 to 40	M	N	I	Mishra and Bhan (1995)
Clomazone	1120	P	N	U	Reddy *et al.* (2007)
Clopyralid	210	L	N	U	Fernandez *et al.* (2015)
Dicamba	140–560	S, M	P	A	DAF (2016)
Diuron	1790	P	N	U	Reddy *et al.* (2007)
Fluometuron	2240	P	N	U	Reddy *et al.* (2007)
Flumioxazin	90	S, L	N	U	Reddy *et al.* (2007)
	70–110	P	Sorghum	U	Grichar (2006)
Glyphosate	840	S, L	N	U	Reddy *et al.* (2007)
	1000–1500	M	N	I	Mishra and Bhan (1995)
	4000	S, L	N	P	Khan *et al.* (2012)
	375–500[b]	M	PT, N	P	Shabbir (2014)
	1000–4000[b]	M	PT, N	P	Javaid (2007)
	2700–5400	L	N	I	Singh *et al.* (2004)
	2500	M	WL	I	Kathiresan (2008)
Glufosinate-ammonium	410	S, L	N	U	Reddy *et al.* (2007)
	2500–3000[b]	M	WL	I	Kathiresan (2008)
Halosulfuron	70	S	N	U	Reddy *et al.* (2007)
Hexazinone	875	S, M, L	A	A	DAF (2016)
	560–1120	L	N	U	Fernandez *et al.* (2015)
Isoproturon	2000[b]	M	PT, N	P	Shabbir (2014)
Metribuzin	700	P	N	U	Reddy *et al.* (2007)
	2000	S	N	P	Khan *et al.* (2012)

continued

Table 8.2. *continued.*

Herbicide	Rate (g ae or ai/ha)[a]	Growth stage	Situation	Country	Reference
Metsulfuron-methyl	3.0–4.2	S, M	P, R, CI	A	DAF (2016)
	3.0–4.2	S, M	wheat, rye barley, triticarle	A	Dupont (2016)
	3.5–4.5	M	N	I	Mishra and Bhan (1995)
	18–45	M	N	SA	Goodall *et al.* (2010)
Norflurazon	2240	P	N	U	Reddy *et al.* (2007)
Oxyfluorfen	150	P	onion	I	Dalavai *et al.* (2008)
Picloram	270–360	M	N	SA	Goodall *et al.* (2010)
Quinclorac	560	P	N	U	Reddy *et al.* (2007)
Saflufenacil	5.7–27	M, L	N	U	Odero (2012)
Saflufenacil + dimethenamid	40 + 350– 90 + 790	L	N	U	Fernandez *et al.* (2015)
S-metolachlor	1920	P	maize	P	Khan *et al.* (2014)
Trifloxysulfuron	8	S, L	N	U	Reddy *et al.* (2007)

Situations: A, around agricultural buildings; P, pastures; C, crops; F, fallows; N, non-crops; R, rights of way; CI, commercial and industrial; PT, pot trial; WL, waste land.
Growth stage: P, pre-emergence; S, small; M, medium; L, large.
Country: A, Australia; I, India; P, Pakistan; SA, South Africa; U, USA.
[a] ae, acid equivalent; ai, active ingredient. [b]Could not be confirmed whether the rate supplied was the quantity of product or active ingredient.

Fig. 8.3. Manual hoeing in fallow land in Pakistan (a); spot spraying of parthenium weed in a rangeland environment in Australia (b). (Zahid Atta Cheema and Pukallus Kelli.)

present. Armstrong and Orr (1984) reported much better results if 2,4-D or dicamba was mixed with atrazine than if it was applied alone.

Glyphosate is a systemic herbicide used for control of both broadleaf weeds and grasses in cropped and non-cropped situations. Given that it is a non-selective herbicide (i.e. kills both broadleaf weeds and grasses), in cropped situations it is mainly used to control weeds when the land is being rested (i.e. fallow) and not planted with a crop. The more recent development of some glyphosate-resistant crops has enabled farmers in some countries to use it to control weeds while a crop is growing. Generally, high efficacy of glyphosate is reported following its use on small to large parthenium weed plants (Mishra and Bhan, 1995; Singh *et al.*, 2004; Javaid, 2007; Reddy *et al.*, 2007;

Khan et al., 2012a; Shabbir, 2014). However, its non-selective nature results in bare patches that remove competition and create a highly favourable environment for germination and growth of new parthenium weed plants, necessitating ongoing follow-up control.

Glufosinate-ammonium is another non-selective herbicide that will control a range of broadleaf weeds and grasses. Effective control of parthenium weed has been reported at rates ranging from 0.70 to 1.52 kg ai/ha (Crane et al., 2006; Reddy et al., 2007; Kathiresan, 2008), but there have also been instances where results have been poor (Singh et al., 2004; Barbier et al., 2012). It seems to work best on small plants and to obtain better efficacy on larger plants it has sometimes been mixed with other herbicides, such as flumioxazin (Barbier et al., 2012). Like glyphosate, it does not have any residual activity and creates bare patches that provide a favourable environment for seedling regrowth of parthenium weed (Crane et al., 2006).

Metribuzon is a selective herbicide that can give pre- and post-emergence control of many broadleaf weeds and some grasses in a range of crops and pastures. It has provided good pre-emergence control of parthenium weed when applied at a rate of 0.70 kg ai/ha (Reddy et al., 2007). When applied as a post-emergence herbicide, high mortality can be achieved on small plants up to around the rosette stage, but mortality decreases markedly thereafter (Singh et al., 2004; Khan et al., 2012a) .

Metsulfuron-methyl is another selective herbicide that can control a range of broadleaf weeds. In Australia, it is registered for control of parthenium weed in several crops and in a range of non-crop situations such as pastures, rights of ways and commercial and industrial areas at rates ranging between 3.0 and 4.2 g ai/ha. Like some of the other herbicides mentioned above, it appears to work best on small to medium-sized plants (Mishra and Bhan, 1995) with larger mature plants more tolerant (Singh et al., 2004). In more recent times, higher rates of metsulfuron-methyl have been found to be highly effective at providing

some residual control of seedling regrowth (Brooks et al., 2004) and in Queensland (Australia) it is being used as an alternative to atrazine in some situations for control along roadsides, often in combination with a relatively fast-acting post-emergence herbicide like dicamba. In Queensland, the Australian Pesticides and Veterinary Medicines Authority has also approved a minor use permit (PER10367; current until 30 June 2018) for aerial applications at a rate of 4.2 g ai/ha in non-crop areas that are inaccessible to ground-based application methods.

Some other herbicides that have been reported to provide effective pre-emergence control of parthenium weed in certain situations include clomazone, chlorimuron-ethyl, diuron, fluometuron, norflurazon, quinclorac (Reddy et al., 2007), flumioxazin (Grichar, 2006), oxyfluorfen (Dalavai et al., 2008) and S-metolachlor (Khan et al., 2014). For post-emergence control, isoproturon (Shabbir, 2014), saflufenacil (Odero, 2012), bromoxynil, halosulfuron, trifloxysulfuron (Reddy et al., 2007), picloram (Goodall et al., 2010), aminopyralid, clopyralid, hexazinone (Fernandez et al., 2015), bentazon (Mishra and Bhan, 1996), chlorimuron-ethyl and dicamba (DAF, 2016) are some other herbicides that have been reported to provide high mortality of parthenium weed at some growth stages (Table 8.2).

In open wasteland, non-cropped areas and along railway track and roadsides, the spraying of a solution of common salt (sodium chloride) at a 15–20% concentration has been found effective (Gnanavel, 2013). However, if this practice was to be repeated on a regular basis in the same area, monitoring should be undertaken to ensure there are no deleterious impacts to the surrounding environment.

As for many weeds, herbicide resistance has been identified as an issue for management of parthenium weed. To date, an ongoing international survey of herbicide-resistant weeds has identified 250 resistant weed species that have been found in 86 crops in 66 countries. A total of 160 herbicides are affected within 23 of the 26 known herbicide sites of action (Heap, 2016). Glyphosate, which belongs to the group of herbicides known as EPSP synthase inhibitors (G/9) is particularly susceptible to development of resistance and a number of cases involving parthenium weed have been reported dating back to 2004 (Heap, 2016). The first case was reported in a fruit orchard in Colombia that had been intensively treated with glyphosate applications for more than 15 years (Vila-Aiub et al., 2008; Heap, 2016). More recently, a high level of glyphosate resistance in parthenium weed was discovered in the Everglades Agricultural Area of South Florida, USA (Odero, 2012; Fernandez et al., 2015), then in the Dominican Republic (Jimenez et al., 2014) and more recently it has been reported in perennial crops in Cuba (Bracamonte et al., 2016).

Another confirmed case of herbicide resistance was reported in Brazil in 2004 (Gazziero et al., 2006; Heap, 2016). Parthenium weed growing in a soybean crop was found to be resistant to several herbicides belonging to the group B/2, known as ALS inhibitors (inhibition of acetolactate synthase). They included chlorimuron-ethyl, cloransulam-methyl, foramsulfuron, imazethapyr and iodosulfuron-methyl-sodium (Gazziero et al., 2006; Heap, 2016). Metsulfuron-methyl also belongs to this herbicide group and is another chemical that could lose its effectiveness on parthenium weed if not managed properly. There are currently 73 reported cases of resistance to metsulfuron-methyl, involving 37 weed species (Heap, 2016). To reduce the risk of herbicide resistance developing, it is recommended that land managers rotate between chemicals with different modes of action (Vitelli and Pitt, 2006; Osten et al., 2007).

8.4.2 Natural products

While there are a range of synthetic herbicides available to control parthenium weed, concerns about their effects on the environment and human health have led to significant research into the development of natural alternatives (Singh et al., 2003; Javaid, 2010; Tanveer et al., 2015).

Extracts, residues and essential oils of many allelopathic plant species (herbs, grasses and trees) have been found to reduce the germination and/or growth of many weeds species, including parthenium weed (Singh et al., 2003; Javaid, 2010). These include Imperata cylindrica (L.) P. Beauv., blady grass (Anjum et al., 2005); Ficus bengalensis L., Indian banyan and Melia azedarach L., Chinaberry, white cedar (Javaid et al., 2006); O. sativa L., rice (Javaid et al., 2008); Senna tora L., sickle senna, Alstonia scholaris (L.) R. Br., devil tree, blackboard tree, shaitan and Azadirachta indica Juss., neem (Javaid, 2010); Mangifera indica L., mango (Javaid et al., 2010); Hyptis suaveolens L., hyptis, pignut (Kapoor, 2011); Tagetes erectus L., marigold (Shafique et al., 2011); Eucalyptus citriodora Hook., lemon-scented eucalypt (Singh et al., 2005; Javaid, 2010); Ricinus communis L., castor oil plant and Ziziphus mauritiana L., desi ber, Chinee apple (Safdar et al., 2013); Achyranthes aspera L., prickly chaff flower, puthkanda and Alternanthera philoxeroides Gris., alligator weed (Safdar et al., 2016). In addition, the metabolites of many fungal species have also been found to have herbicidal effects on the germination and growth of parthenium weed (Javaid, 2010).

Commercialization or the use of these natural products on a large scale does not appear to have occurred yet. Singh et al. (2003) suggest a number of reasons, including complexities in the structure of the allelochemicals, cost effectiveness, mammalian toxicology, poor results in field trials, rapid degradation and intellectual property rights issues. There are instances where plant material of some of the plant species has been spread over areas and over time the material breaks down and allelochemicals leach out and suppress weeds such as parthenium weed. Generally the effects are short-lived (e.g. 4–6 weeks) for most crop residues (Singh et al., 2003).

8.5 Integrated Weed Management

As mentioned previously, the ecology of parthenium weed, particularly its persistent soil seed bank, the multiple dispersal mechanisms and the ability to reach reproductive maturity quickly, allows it to persist in areas following initial control activities. To attain the desired level of management requires several follow-up treatments and to do this most effectively, consideration should be given to all available options and how they can be integrated. However, before starting a management programme, several factors need to be considered, such as the overall objective, the scale of the problem, the location and habitat of infestations and the resources available. The objective of weed management could include completely eradicating parthenium weed, containing it or undertaking a level of control that will minimize its impacts. The cost and level of resourcing tends to increase markedly as the objective moves towards eradication, with the techniques used and frequency of control activities needing to kill all plants before they have a chance to set seeds and replenish the soil seed bank. Control activities also need to be continued for long enough to treat any seedling regrowth that will inevitably appear, particularly if the infestation is old and a large persistent seed bank has established. If minimizing impacts is the objective, less severe techniques could be incorporated as it is not as critical that all plants are killed each time treatments are imposed.

Irrespective of the objective, preventative and early intervention strategies should be implemented to minimize the introduction and spread of parthenium weed into new areas and may require a combination of good hygiene practices, regular surveillance of susceptible or high-risk areas, awareness creation and early control if plants are found.

In most instances, the introduction of biological control agents should be considered if they have been approved for release in respective countries. As previously discussed (see Dhileepan et al., Chapter 7, this volume), there are a number of effective agents that can adversely affect the growth, development and reproduction of parthenium weed. These agents can help slow down the spread of parthenium weed and also reduce its vigour.

The suppressive ability of plants growing in association with parthenium weed should also be used advantageously, particularly in areas where biological control agents are adversely affecting the vigour of parthenium weed (see Dhileepan et al., Chapter 7, this volume). Minimizing disturbance and maintaining a dense ground cover of desirable species will create a more resilient environment and greatly reduce the opportunity for parthenium weed to proliferate (Shabbir et al., 2013). In pastures, this will require the use of grazing strategies that enable the grasses and legumes present to remain healthy and competitive. If growing improved pastures, several highly competitive species against parthenium weed have been identified (as discussed in Section 8.2.3) and they would greatly increase the resilience of these pastures if introduced into suitable environments. If resources permit, control activities should also be implemented to treat any outbreaks of parthenium weed that may occur or to reduce the initial level of parthenium weed if it is too dense. For example, some landholders in Queensland (Australia) have used broad-scale applications of herbicides containing metsulfuron-methyl to reduce the initial populations of parthenium weed in pastures and thereafter a combination of light grazing and biological control has maintained these pastures in a healthy and productive state (Chamberlain and Gittens, 2004). Parthenium weed is still present as a minor component of the vegetation but it is not affecting production from these pastures. As mentioned previously, in cropping situations competition against parthenium weed can also be incorporated, through selection of the most competitive crops and varieties, and manipulation of plant densities.

The use of physical, mechanical and chemical techniques in combination with cultural practices can all be integrated in cropping situations, with the most applicable combinations dependent on the resources available and the scale of the farming operation. In countries where the cost of labour is relatively low, manual control has been successfully integrated with initial ploughing and cultural practices such as growing of a green manure crop or mulching. In countries with higher labour costs there is often an increased reliance on the use of machinery and pre- and post-emergent herbicides.

It is important to note that in cropping situations there are often numerous weed species growing in association with each other and if parthenium weed is present its control will often occur as part of the most appropriate techniques used to reduce the impact of the weeds generally. This may mean that the treatment imposed might vary slightly to that used if parthenium weed was the only weed present. Management of parthenium weed in neighbouring areas should also be given consideration to reduce the risk of further re-invasion following implementation of control activities.

Individual landowners or managers dealing solely with parthenium weed on their land may be able to achieve a satisfactory level of control for their situation, but the likelihood of success of broad-scale management will be increased through a critical mass of people working together (see Shabbir et al., Chapter 9, this volume). This could include a group of farmers within a specified area or a whole community trying to manage parthenium weed in a region. To be successful, those involved need to be motivated, educated about the biology and ecology of the weed, and committed to long-term management (Batish et al., 2004).

Acknowledgements

We are grateful to Kelsey Hosking, Rose Easton and Dannielle Brazier for their assistance with compilation of reference material and to Joe Scanlan, Steve Adkins, Dhileepan Kunjithapatham and Asad Shabbir for reviewing drafts of the manuscript. Paul Gray also kindly supplied photos of control activities.

References

Adkins, S. and Shabbir, A. (2014) Biology, ecology and management of the invasive parthenium weed (Parthenium hysterophorus L.). Pest Management Science 70, 1023–1029.

Anjum, T., Bajwa, R. and Javaid, A. (2005) Biological control of parthenium 1: Effect of Imperata cyclindrica on distribution, germination and seedling growth of Parthenium hysterophorus L. International Journal of Agriculture and Biology 7, 448–450.

Armstrong, T.R. and Orr, L.C. (1984) Parthenium hysterophorus control with knockdown and residual herbicides. In: Madin, R.W. (ed.) Proceedings of the 7th Australian Weeds Conference: Volume 2. The Weed Society of Western Australia, Perth, Australia, pp. 8–13.

Auld, B.A., Hosking, J. and McFadyen, R.E. (1983) Analysis of the spread of tiger pear and parthenium weed in Australia. Australian Weeds 2, 56–60.

Australian Weeds Committee (2012) Parthenium (Parthenium hysterophorus L.) Strategic Plan 2012–17. Weeds of National Significance, Australian Government Department of Agriculture, Fisheries and Forestry, Canberra, Australia.

Barbier, M., Crane, J., Castillo, J. and Yuncong, L. (2012) Effectiveness of flumioxazin alone and in combination with glufosinate-ammonium for control of parthenium (Parthenium hysterophorus) under grove conditions in Homestead, Florida. Proceedings of the Florida State Horticultural Society 125, 1–5.

Batish, D.R., Singh, H.P., Kohli, R.K., Johar, V. and Yadav, S. (2004) Management of invasive exotic weeds requires community participation. Weed Technology 18, 1445–1448.

Batish, D.R., Kohli, R.K., Singh, H.P. and Kaur, G. (2012) Biology, ecology and spread of the invasive weed Parthenium hysterophorus in India. In: Bhatt, J.R., Singh, J.S., Singh, S.P., Tripathi, R.S. and Kohli, R.K. (eds) Invasive Alien Plants: An Ecological Appraisal for the Indian Subcontinent. CPI Group (UK) Ltd, Croydon, UK, pp. 10–18.

Bebawi, F.F., Cooper, A.P., Brodie, G.I., Madigan, B.A., Vitelli, J.S., Worsley, K.J. and Davis, K.M. (2007) Effect of microwave radiation on seed mortality of rubber vine (Cryptostegia grandiflora R.Br.), parthenium (Parthenium hysterophorus L.) and bellyache bush (Jatropha gossypiifolia L.). Plant Protection Quarterly 22, 136–142.

Besufekad, T., Das, T.K., Mahadevappa, M., Taye, T. and Tamado, T. (2005) Parthenium distribution, biology, hazards and control measures in Ethiopia. Pest Management Journal of Ethiopia 9, 1–15.

Bhowmik, P.C. and Sarkar, D. (2005) Parthenium hysterophorus: Its world status and potential management. In: Prasad, T.V.R. (ed.) *Proceedings of the 2nd International Conference on Parthenium Management, Volume 2*. University of Agricultural Sciences, Bangalore, India, pp. 1–5.

Blackmore, P.J. and Johnson, S.B. (2010) Continuing successful eradication of parthenium weed (*Parthenium hysterophorus*) from New South Wales, Australia. In: Zydenbos, S.M. (ed.) *Proceedings of the Seventeenth Australasian Weeds Conference*. Christchurch, New Zealand, pp. 382–385.

Bracamonte, E., Fernández-Moreno, P.T., Barro, F. and De Prado, R. (2016) Glyphosate-resistant *Parthenium hysterophorus* in the Caribbean Islands: Non target site resistance and target site resistance in relation to resistance levels. *Frontiers in Plant Science* 7, 1845.

Brooks, S.J., Vitelli, J.S. and Rainbow, A.G. (2004) Developing best practice roadside *Parthenium hysterophorus* L. control. In: Sindel, B.M. and Johnson, S.B. (eds) *Proceedings of the 14th Australian Weeds Conference*. Wagga Wagga, New South Wales, Australia, pp. 195–198.

Butler, D.W. and Fairfax, R.J. (2003) Buffel grass and fire in a Gidgee and Brigalow woodland: A case study from central Queensland. *Ecological Management and Restoration* 4, 120–125.

Butler, J.E. (1984) Longevity of *Parthenium hysterophorus* L. seed in the soil. *Australian Weeds* 3, 6.

Campbell, S.D. (2009) Experiences associated with the use of fire in weed management. In: Baker, T., Campbell, S., Grice, T., Kinnear, S. and Sandral, C. (eds) *Proceedings of the 10th Queensland Weed Symposium*. Yeppoon, Queensland, Australia, pp. 87–90.

Chamberlain, J. and Gittens, A. (2004) *Parthenium Weed Management: Challenges, Opportunities and Strategies*. Department of Natural Resources, Mines and Energy, Brisbane, Australia.

Crane, J.H., Stubblefield, R. and Meister, C.W. (2006) Herbicide efficacy to control parthenium (*Parthenium hysterophorus*) under grove conditions in Homestead, Florida. *Proceedings of the Florida State Horticultural Society* 119, 9–12.

Dalavai, B.L., Kandasamy, O.S., Hanumanthappa, M. and Arasumallaiah, L. (2008) Effect of herbicides and their application techniques on weed flora in onion at Coimbatore. *Environment and Ecology* 26, 2157–2160.

Department of Agriculture and Food (2016) *Parthenium hysterophorus* L. Available at: https://www.agric.wa.gov.au/organisms/106333 (accessed 6 December 2016).

Department of Agriculture, Fisheries and Forestry (DAFF) (2013) Vehicle and machinery inspection procedure, the State of Queensland, Department of Agriculture, Fisheries and Forestry, Brisbane. Available at: http//:www.daf.qld.gov.au (accessed 29 October 2016).

Department of Agriculture, Fisheries and Forestry (DAFF) (2014) Vehicle and machinery checklists, the State of Queensland, Department of Agriculture, Fisheries and Forestry, Brisbane. Available at: http//:www.daf.qld.gov.au (accessed 29 October 2016).

Department of Agriculture and Fisheries (DAF) (2016) Parthenium weed (*Parthenium hysterophorus*), the State of Queensland, Department of Agriculture and Fisheries, Brisbane. Available at: http//:www.daf.qld.gov.au (accessed 15 August 2016).

Department of Primary Industries, Parks, Water and Environment (2016) Parthenium Weed. Available at: http://dpipwe.tas.gov.au/invasive-species/weeds/weeds-index/weeds-index-declared-weeds/parthenium-weed (accessed 6 December 2016).

Dhileepan, K. (2009) Managing parthenium weed across diverse landscapes: prospects and limitations. In: Inderjit, R (ed.) *Management of Invasive Weeds*. Springer, Dordrecht, The Netherlands, pp. 227–259.

Dupont (2016) Ally herbicide label. Available at: www.dupont.com.au (accessed 18 August 2016).

Environment, Planning and Sustainable Development Directorate (2015) ACT Weeds Strategy. Available at: http://www.environment.act.gov.au/cpr/conservationstrategies/act_weeds_strategy (accessed 6 December 2016).

Fernandez, J.V., Odero, D.C., MacDonald, G.E., Ferrell, J. and Gettys, L.A. (2015) Confirmation, characterization, and management of glyphosate-resistant ragweed parthenium (*Parthenium hysterophorus* L.) in the Everglades agricultural area of South Florida. *Weed Technology* 29, 233–242.

Gazziero, D.L.P., Brighenti, A.M. and Voll, E. (2006) Ragweed parthenium (*Parthenium hysterophorus*) cross-resistance to acetolactate synthase inhibiting herbicides. *Planta Daninha* 24, 157–162.

Gnanavel, I. (2013) *Parthenium hysterophorus* L.: a major threat to natural and agro eco-systems in India. *Science International* 1, 186–193.

Goodall, J., Braack, M., de Klerk, J. and Keen, C. (2010) Study on the early effects of several weed-control methods on *Parthenium hysterophorus* L. *African Journal of Range and Forage Science* 27, 95–99.

Grichar, W.J. (2006) Weed control and grain sorghum tolerance to flumioxazin. *Crop Protection* 25, 174–177.

Haseler, W.H. (1976) *Parthenium hysterophorus* L. in Australia. *PANS* 22, 515–517.

Heap, I. (2016) *The International Survey of Herbicide-resistant Weeds.* Available at: www.weedscience.com (accessed 18 August 2016).

Holman, D.J. and Dale, I.J. (1981) Parthenium weed threatens Bowen Shire. *Queensland Agricultural Journal* 107, 57–60.

Javaid, A. (2007) Efficacy of some common herbicides against parthenium weed. *Pakistan Journal of Weed Science Research* 13, 93–98.

Javaid, A. (2010) Herbicidal potential of allelopathic plants and fungi against *Parthenium hysterophorus*; a review. *Allelopathy Journal* 25, 331–344.

Javaid, A., Bajwa, R., Shafique, S. and Shafique, S. (2006) Chemical, phytochemical and biological control of *Parthenium hysterophorus* in Pakistan. In: Preston, C., Watts, J.H. and Crossman, N.D. (eds) *Proceedings of the 15th Australian Weeds Conference: Managing Weeds in a Changing Climate.* Adelaide, South Australia, Australia, pp. 876–879.

Javaid, A., Shafique, S., Shafique, S. and Riaz, T. (2008) Effects of rice extracts and residue incorporation on *Parthenium hysterophorus* management. *Allelopathy Journal* 22, 353–362.

Javaid, A., Shafique, S., Kanwal, Q. and Shafique, S. (2010) Herbicidal activity of flavonoids of mango leaves against *Parthenium hysterophorus* L. *Natural Product Research* 24, 1865–1875.

Jimenez, F., Fernandez, P., Rosario, J., Gonzalez-Torralva, F. and De Prado, R. (2014) First case of glyphosate resistance in the Dominican Republic. *AGRO-554 in 248th ACS National Meeting and Exposition.* American Chemical Society, San Francisco, California.

Kapoor, R.T. (2011) Bioherbicidal potential of leaf-residue of *Hyptis suaveolens* on the growth and physiological parameters of *Parthenium hysterophorus* L. *Current Research Journal of Biological Sciences* 3, 341–350.

Kathiresan, R.M. (2008) Ecology and control of *Parthenium hysterophorus* invasion in Veeranum. *Indian Journal of Weed Science* 40, 78–80.

Kaur, M., Aggarwal, N.K., Kumar, V. and Dhiman, R. (2014) Effects and management of *Parthenium hysterophorus*: A weed of global significance. *International Scholarly Research Notices* 2014, 1–12.

Kenya Law (2010) *The Kenya Gazette*, Gazette Notice No. 4423. Available at: http://kenyalaw.org/kenya_gazette/gazette/notice/17263 (accessed 6 December 2016).

Khan, H., Marwat, K.B., Hassan, G. and Khan, M.A. (2012a) Chemical control of *Parthenium hysterophorus* L. at different growth stages in non-cropped area. *Pakistan Journal of Botany* 44, 1721–1726.

Khan, H., Marwat, K.B., Khan, M.A. and Hashim, S. (2014) Herbicidal control of parthenium weed in maize. *Pakistan Journal of Botany* 46, 497–504.

Khan, I. (2012) Spread of weed seeds and its prevention. PhD thesis. University of Queensland, Brisbane, Australia.

Khan, I., O'Donnell, C., Navie, S., George, D., Nguyen, T. and Adkins, S. (2013) Weed seed spread by vehicles. In: O'Bryan, M., Vitelli, J. and Thornby, D. (eds) *Proceedings of the 12th Queensland Weed Symposium.* Hervey Bay, Queensland, Australia, pp. 94–97.

Khan, N., O'Donnell, C., George, D. and Adkins, S.W. (2012b) Suppressive ability of selected fodder plants on the growth of *Parthenium hysterophorus.* *Weed Research* 53, 61–68.

Kohli, R.K., Batish, D.R. and Singh, H.P. (1997) Management of *Parthenium hysterophorus* L. through an integrated approach. In: Mahadevappa, M. and Patil, V.C. (eds) *First International Conference on Parthenium Management, Volume 2.* University of Agricultural Sciences, Dharwad, India, pp. 60–62.

Kohli, R.K., Batish, D.R., Singh, H.P. and Dogra, K.S. (2006) Status, invasiveness and environmental threats of three tropical American invasive weeds (*Parthenium hysterophorus* L., *Ageratum conyzoides* L., *Lantana camara* L.) in India. *Biological Invasions* 8, 1501–1510.

Kumari, M. (2014) *Parthenium hysterophorus* L.: a noxious and rapidly spreading weed of India. *Journal of Chemical, Biological and Physical Sciences* 4, 1620–1628.

Mahadevappa, M. (2009) *Parthenium: Insight into its Menace and Management.* Stadium Press (India) Pvt Ltd, New Delhi.

Masum, S.M., Hasanuzzaman, M. and Ali, M.H. (2013) Threats of *Parthenium hysterophorus* on agro-ecosystems and its management: A review. *International Journal of Agriculture and Crop Sciences* 6, 684–697.

Mishra, J.S. and Bhan, V.M. (1995) Efficacy of sulfonylurea herbicides against *Parthenium hysterophorus* L. *Indian Journal of Weed Science Research* 27, 45–48.

Mishra, J.S. and Bhan, V.M. (1996) Chemical control of carrot grass (*Parthenium hysterophorus*) and associated weeds in soybean (*Glycine max*). *Indian Journal of Agricultural Sciences* 66, 518–521.

Muniyappa, T.V., Prasad, T.V.R. and Krishnamurthy, K. (1980) Comparative efficacy and economics of mechanical and chemical methods of control of *Parthenium hysterophorus* Linn. *Indian Journal of Weed Science* 12, 137–144.

Navie, S.C., McFadyen, R.E., Panetta, F.D. and Adkins, S.W. (1996) The biology of Australian weeds. 27. *Parthenium hysterophorus* L. *Plant Protection Quarterly* 11, 76–88.

Navie, S.C., McFadyen, R.E., Panetta, F.D. and Adkins, S.W. (1998a) *Parthenium hysterophorus* L. In: Panetta, F.D., Groves, R.H. and Shepherd, R.C.H. (eds) *The Biology of Australian Weeds Volume 2*. R.G. and F.J. Richardson, Meredith, Australia, pp. 157–176.

Navie, S.C., Panetta, F.D., McFadyen, R.E. and Adkins, S.W. (1998b) Behaviour of buried and surface-sown seeds of *Parthenium hysterophorus*. *Weed Research* 38, 335–341.

Nguyen, T.L. (2011) The invasive potential of parthenium weed (*Parthenium hysterophorus* L.) in Australia. PhD thesis. University of Queensland, Brisbane, Australia.

NSW Department of Primary Industries (2014) Parthenium weed (*Parthenium hysterophorus*). Available at: http://weeds.dpi.nsw.gov.au/Weeds/Details/165#declarations (accessed 6 December 2016).

Odero, D.C. (2012) Response of ragweed parthenium (*Parthenium hysterophorus*) to saflufenacil and glyphosate. *Weed Technology* 26, 443–448.

O'Donnell, C. and Adkins, S.W. (2005) Management of parthenium weed through competitive displacement with beneficial plants. *Weed Biology and Management* 5, 77–79.

Osten, V.A., Walker, S.R., Storrie, A., Widderick, M., Moylan, P., Robinson, G.R. and Galea, K. (2007) Survey of weed flora and management relative to cropping practices in the north-eastern grain region of Australia. *Australian Journal of Experimental Agriculture* 47, 57–70.

Parsons, W.T. and Cuthbertson, E.G. (2001) *Noxious Weeds of Australia*, 2nd edn. CSIRO Publishing, Collingwood, Australia.

Pratap, T., Singh, R., Pal, R., Yadaw, S. and Singh, V. (2013) Integrated weed management studies in sugarcane ratoon. *Indian Journal of Weed Science* 45, 257–259.

Primary Industries and Regions SA (2015) Weeds in South Australia. Available at: http://www.pir.sa.gov.au/biosecurity/weeds_and_pest_animals/weeds_in_sa (accessed 6 December 2016).

Reddy, K.N., Bryson, C.T. and Burke, I.C. (2007) Ragweed parthenium (*Parthenium hysterophorus*) control with pre-emergence and post-emergence herbicides. *Weed Technology* 21, 982–986.

Republic of South Africa Government Gazette (2014) Government Notice 599, Alien and Invasive Species List, 2014. Available at: https://www.environment.gov.za/sites/default/files/gazettednotices/nemba10of2004_alienandinvasive_speciesrelist.pdf (accessed 6 December 2016).

Safdar, M.E., Tanveer, A., Khaliq, A., Naeem, M.S. and Ahmad, S. (2013) Tree species as a potential source of bio-herbicides for controlling *Parthenium hysterophorus* L. *Natural Products Research* 27, 2154–2156.

Safdar, M.E., Tanveer, A., Khaliq, A., Ali, H.H. and Burgos, N.R. (2016) Exploring herbicidal potential of aqueous extracts of some herbaceous plants against parthenium weed. *Planta Daninha* 34, 109–116.

Shabbir, A. (2007) Burning crop residues reduces the parthenium weed emergence: A study. *Parthenium News* 2, 2.

Shabbir, A. (2014) Chemical control of *Parthenium hysterophorus* L. *Pakistan Journal of Weed Science Research* 20, 1–10.

Shabbir, A., Dhileepan, K., O'Donnell, C. and Adkins, S.W. (2013) Complementing biological control with plant suppression: Implications for improved management of parthenium weed (*Parthenium hysterophorus* L.). *Biological Control* 64, 270–275.

Shafique, S., Bajwa, R. and Shafique, S. (2011) *Tagetes erectus* L. – a potential resolution for management of *Parthenium hysterophorus* L. *Pakistan Journal of Botany* 43, 885–894.

Singh, H.P., Batish, D.R. and Kohli, R.K. (2003) Allelopathic interactions and allelochemicals: New possibilities for sustainable weed management. *Critical Reviews in Plant Sciences* 22, 239–311.

Singh, H.P., Batish, D.R., Setia, N. and Kohli, R.K. (2005) Herbicidal activity of volatile oils from *Eucalyptus citriodora* against *Parthenium hysterophorus*. *Annals of Applied Biology* 146, 89–94.

Singh, S., Yadav, A., Balyan, R.S., Malik, R.K. and Singh, M. (2004) Control of ragweed parthenium (*Parthenium hysterophorus*) and associated weeds. *Weed Technology* 18, 658–664.

Tamado, T. and Milberg, P. (2004) Control of parthenium (*Parthenium hysterophorus*) in grain sorghum (*Sorghum bicolor*) in the smallholder farming system in eastern Ethiopia. *Weed Technology* 18, 100–105.

Tanveer, A., Khaliq, A., Ali, H.H., Mahajan, G. and Chauhan, B.S. (2015) Interference and management of parthenium: The world's most important invasive weed. *Crop Protection* 68, 49–59.

Vila-Aiub, M.M., Vidal, R.A., Balbi, M.C., Gundel, P.E., Trucco, F. and Ghersa, C.M. (2008)

Glyphosate-resistant weeds of South America cropping systems: An overview. *Pest Management Science* 64, 366–371.

Vitelli, J.S. and Madigan, B.A. (2004) Evaluation of a hand-held burner for the control of woody weeds by flaming. *Australian Journal of Experimental Agriculture* 44, 75–81.

Vitelli, J.S. and Pitt, J.L. (2006) Assessment of current weed control methods relevant to the management of the biodiversity of Australian rangelands. *The Rangeland Journal* 28, 37–46.

Vogler, W., Navie, S., Adkins, S. and Setter, C. (2006) *Use of Fire to Control Parthenium Weed.* A report for the Rural Industries Research and Development Corporation, Kingston, Australia.

9 Coordination of Management

Asad Shabbir,[1]* Sushilkumar,[2] Ian A. W. Macdonald[3] and Colette Terblanche[4]

[1]University of the Punjab, Lahore, Pakistan; current affiliation: Plant Breeding Institute, the University of Sydney, Narrabri, New South Wales, Australia; [2]ICAR, Directorate of Weed Research Adhartal, Jabalpur, India; [3]International Environmental Consultant, KwaZulu-Natal, South Africa; [4]Colterra Environmental Consultants, KwaZulu-Natal, South Africa

9.1 Introduction

Throughout the world, except in a very few extreme environments, the challenge posed by invasive alien organisms far exceeds the capacity to manage them. Unfortunately, all the indications are that the problem is increasing and is likely to continue to do so for the foreseeable future, given current trends. Under such a scenario it is crucially important that the effectiveness of management is maximized. In this chapter, the important role that coordination can play in managing one of the world's most invasive and harmful alien plant species, parthenium weed (*Parthenium hysterophorus* L.), is addressed.

Parthenium weed negatively impacts agricultural production, human health and native community biodiversity (Wise *et al.*, 2007) and is extremely difficult to control. Only a coordinated, strategic approach to its management will increase the chances of achieving sustainable and effective control (Terblanche *et al.*, 2016). In most cases coordination can only be achieved by governments, with the necessary legal framework, and needs to be organized at a national scale. This national level is essential because human-aided dispersal of parthenium weed, mainly by vehicles, occurs over long distances. In fact, so serious is the problem posed by parthenium weed globally that coordination of management is also required at the international level.

For coordination to be effective, it should be based on a strategy (or plan) with objectives that are measurable and that have clear time frames allowing progress to be assessed. The responsibilities of stakeholders need to be clearly identified and agreed to by all stakeholders during the formulation of the national or international strategy. It is vital that regular monitoring of outcomes should be established from the start, so that appropriate adjustments can be made to the strategy and lessons can be derived for the implementation of similar strategies elsewhere (Terblanche *et al.*, 2016).

Once a strategy has been agreed and adopted as 'the' strategy for all stakeholders, a person/organization/department needs to be appointed to coordinate all activities according to the strategy. This is necessary to ensure targets are met and that stakeholders play their allotted roles in the management strategy. Without such a formally appointed coordinator or coordinating agency, the strategy will generally not be effectively implemented.

The Australian Weeds Strategy (NRMMC, 2007) lists seven key principles,

* asad.shabbir@sydney.edu.au

one of which states that weed management requires coordination among all levels of government in partnerships with industry, land and water management, and the community, regardless of their tenure. The coordinated approach for effective parthenium weed management was emphasized by Austin (2005) in Australia.

The objective of the 'South African Strategy for the Management of Parthenium Weed' is to ensure collaboration, coordination and the prioritization of resources to achieve eight broad goals. One of these goals is to ensure national cross-sectoral coordination and collaboration between different domains of government and stakeholders within South Africa, and with countries bordering South Africa. In fact, coordination for parthenium weed management is not for individual activities but is a process to apply to integrate different activities at different levels and between different stakeholder groups.

Effective coordination between active groups in different countries and between regional groups within countries has been enabled by the International Parthenium Weed Network, which was founded in Australia (Adkins and Shabbir, 2014) and currently enables the exchange of information on parthenium weed management between the Network's more than 350 members in 30 countries.

9.2 Benefits of Coordination

One of the key lessons that has emerged from recent experiences with the management of invasive alien plant species is the crucial importance of setting priorities for such management programmes if they are to be effective. First, it has to be agreed which alien species should be prioritized for control and which areas should be selected for what kind of management. In the case of parthenium weed, the management of the species is so difficult that it is almost certainly impossible for any invaded country to simply decide to eradicate the species throughout its entire range in that country. What has evolved as the most effective strategy for parthenium weed management is to agree on different parthenium weed management objectives for different areas within the country or region, such as active surveillance in some areas, local eradication of new infestations in others, containment of infestations in patchily invaded areas, and asset protection in localized high-value sites (Fig. 9.1). Unless there is effective coordination between all the relevant stakeholders, such a system of spatially differentiated management objectives would simply not be possible.

In the very important prevention of the spread of parthenium weed into previously uninvaded areas, coordination once again plays a crucial role. That is, the active surveillance programmes that are generally involved in detecting new invasions into such areas are expensive to implement and require coordination between agencies to ensure: (i) such surveillance is not unnecessarily duplicated; and (ii) the agencies responsible for eradicating any such new invasions are in close and immediate communication with the surveillance programme operators. Authorities responsible for awareness-creation programmes should also be coordinating their activities so as to focus their efforts in high-risk uninvaded areas so that land holders and members of the public can identify and report new invasions in these areas to the appropriate management authorities.

In the 'asset protection' zones in such a spatially differentiated management programme, it is essential that all the relevant stakeholders act in a coordinated manner, first to agree on what are the high-value sites requiring parthenium weed management and second, on the integrated approach to the management of each high-value site.

Another lesson that is emerging from current experience in managing invasive alien plant species is the importance of managing 'invasion pathways'. In the case of parthenium weed, one of the key 'pathways' for its long-distance dispersal is vehicular traffic along transportation corridors (Haseler, 1976). Accordingly, it is extremely important in attempting to limit the invasion of currently uninvaded areas that

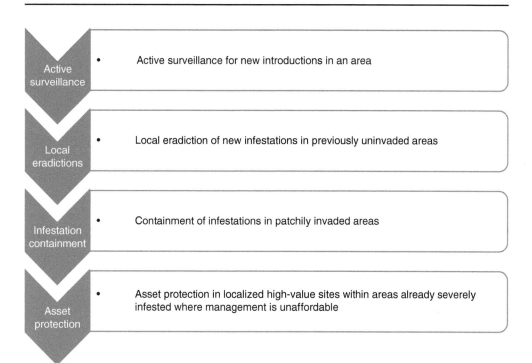

Fig. 9.1. A hierarchy of different strategies for parthenium weed management. (A. Shabbir.)

parthenium weed control is prioritized in areas alongside these transportation corridors, especially when they lead from heavily infested into uninfested areas. Without coordination, it would be impossible to have all the authorities responsible for the management of the different sections of these long invasion 'pathways' working in concert to minimize parthenium weed seed production and hence the dispersal of the species along these corridors.

The coordination will encourage team spirit among different stakeholders of different regions and countries by giving proper direction to make optimum utilization of resources, which in turn will lead to higher efficiency. Coordination of parthenium weed management may further be improved by proper documentation of different activities, capacity building, and use of local languages to develop literature to educate people and create awareness programmes, which subsequently will lead to improved management.

In the case of parthenium weed, the only real possibility for sustainable management in the long term is biological control aimed at reducing the invasiveness and impacts of this species. It is in the field of effective biological control that coordination plays several fundamental roles. International coordination between cooperating countries massively reduces the costs to countries which are able to take advantage of biological control and its expertise already developed in another country. At the national level, coordination between all the agencies that have an interest in managing parthenium weed must include a step to ensure that each agency provides the appropriate level of support (finance, expertise etc.) to guarantee the rapid development of an effective national biological control programme. At the local level, coordination between stakeholders is essential to ensure that areas that are being utilized for field testing and release of new biological control agents are not inadvertently subjected to mechanical or chemical control programmes before the released agents have established themselves, have had their success evaluated and have begun spreading to new sites.

9.3 Coordination as an Essential Component of Management

Effective coordination will be required for the long-term implementation of a parthenium weed management programme. It is essential that such management strategies should be continued at least for 5–10 years to exhaust the available seed bank in the soil. It is not expected that 1 or 2 years of effort would reduce the population of parthenium weed to a tolerable level. Further, adequate quarantine measures need to be adopted to check the immigration and emigration of the weed within and outside a region. Therefore, combined efforts by all the government departments, such as agriculture, forestry, horticulture, water resources, transport and communications and health, are required (Sushilkumar, 2005; Sushilkumar and Varshney, 2007; Adkins and Shabbir, 2014).

9.3.1 The Australian experience

Parthenium weed is a major invasive weed in Queensland, with the potential for it to spread to all medium-to-low rainfall rangelands, the low rainfall channel country and summer cropping areas. In Australia, parthenium weed has major impacts on pasture and to a lesser extent cropping industries, with estimated losses of AUS$109 million per year for Queensland pastoralists (Australian Weeds Committee, 2012). Parthenium weed also poses a risk to most grazing and cropping areas in eastern and northern Australia. In addition to livestock production, the weed is also causing health concerns in regional communities (Goldsworthy, 2005).

Australia has developed a 'National Weed Awareness Action Plan'. This plan focuses on improving awareness as a prerequisite to achieve acceptable long-term management of weeds. Increased awareness depends on participation by landowners, land managers, industry, the wider community and local, state and national government. Australia has also developed action plans against a series of 'Weeds of National Significance' (WONS). The WONS are a nationally agreed priority plant species for management. Species are selected based on their high rankings for invasiveness, potential to spread and impact on socioeconomic and environmental assets. There are currently 32 WONS, which include parthenium weed.

The First National Strategic Plan on parthenium weed was made in 2001 and a National Parthenium Weed Action Group was initiated. A coordinator was appointed in 2002 to look after the implantation of the plan. In 2009, the programme was reviewed and various achievements were identified and some barriers to performance recognized. Among several, one of the key challenges was the coordination of the detection and eradication steps. The review further emphasized a need for ongoing national coordination, with particular focus on the establishment of incentive- and compliance-based detection and eradication in risk areas and maintaining the awareness of best management practices in Queensland.

In Australia, the Parthenium Weed Action Plan was revised for 2012–2017 to continue to overcome the parthenium weed menace through focused research and an improved coordination and awareness programme (Australian Weed Committee, 2012). The strategy documents are needed for continued commitment from the various stakeholders and for the provision of a clear framework for coordinated management of parthenium weed throughout Australia. This strategy was set up with three main goals and associated objectives: (i) to prevent new infestations from establishing; (ii) to strategically manage existing infestations; and (iii) to increase landholder capability and willingness to manage parthenium weed. The Australian Weeds Committee will be responsible for monitoring and evaluating this plan, which is subject to review after five years.

Preventing establishment of new infestations

The main objectives of this goal are to reduce the movement and spread of parthenium weed into new areas. This necessitates the establishment of effective vehicle washdown and inspection facilities and the enhancement of landholder and land manager awareness of the relevant legislation and of preventing parthenium weed spread by the high-risk vectors. It is important to detect and record infestations through a surveying and mapping programme and to eradicate new infestations based on 'management zones'. These management zones are identified based on their location with respect to the core infestation zone, and the containment, prevention and eradication plans. Finally, the use of compliance as a management tool will encourage all stakeholders to comply with legislative requirements and take action when new infestations occur.

Strategically managing existing infestations

This goal requires the containment of parthenium weed by strategically managing it in different zones. The identification of a zonal management plan is key to containing the spread of parthenium weed to new locations. The goal also describes the importance of awareness and the implementation of the best-practice parthenium weed management plan. Key actions to achieve this goal include developing and implementing a targeted national communication programme that includes TV and radio advertising and including pest management in the education curricula. Finally, it is very important to investigate the impacts of parthenium weed in a changing climate. This requires research, especially in the areas of weed biology and ecology, and assessment of management strategies (biological control, suppressive plants etc.) under changing climate. However, as yet, funding has not been found to do this.

Increasing landholder capability and willingness to manage parthenium weed

This goal demands the stakeholders' increased commitment to parthenium weed management. This can be achieved through raising awareness of the problems of parthenium weed as significant threats that need to be managed effectively. This also emphasizes facilitating the implementation of the strategic plan by reporting on progress annually to Australian Weeds Committee and stakeholders. It is also very important to monitor the implementation of the strategy by establishing an evaluation methodology.

9.3.2 An Indian experience

After noticeable populations of parthenium weed were observed in Pune (Maharashtra) in 1956 (Rao, 1956), the weed has spread rapidly throughout India. On the basis of published information, the estimated area infested with parthenium weed is now about 35 million ha (Sushilkumar and Varsheny, 2010). Awareness and community participation are a central part of the Indian strategy to tackle invasive weeds such as parthenium weed (Batish *et al.*, 2004; Chinnusamy *et al.*, 2010; Sushilkumar, 2014).

The organization of a large-scale national 'Parthenium Weed Awareness Week' has been an example to help motivate people in other countries with parthenium weed to undertake action in its management. In 2004, the Directorate of Weed Research (DWR; previously known as the National Research Centre for Weed Science) in Jabalpur organized the first 'Parthenium Weed Awareness Day' (Gogoi *et al.*, 2005). This activity involved 24 State Agricultural Universities (SAU) and the Institutes in the All India Coordinated Research Project on Weed Control (AICRP-WC). However, it was discovered that a day was not sufficient to run such an activity and from 2005 onwards the activity became the Parthenium Weed Awareness Week. In this new form it included all of the country's AICRP-WC and SAUs, around 550 Krishi Vigyan Kendra (KVK) Centres, in addition to another 100 Institutes under the Indian Council of Agricultural Research (ICAR), many NGOs, a number of environmental agencies, as well as many schools, colleges and farmers (Fig. 9.2).

Fig. 9.2. Parthenium Weed Awareness posters and leaflets in Hindi and English describing identification and strategies to manage parthenium weed. (Directorate of Weed Science Research, India.)

Imparting education for parthenium weed management

The stakeholder's knowledge of different management strategies is central to achieve the goal of management through coordination. Different categories of stakeholders, such as farmers, urban public, scientists, teachers, administrative groups, school children, college students, should be involved in parthenium weed management in India. These stakeholders may be educated by demonstrating different types of control techniques, such as manual, mechanical, chemical, cultural and biological methods (Fig. 9.3).

Selection of partners

India is a country of rich cultural heritage with a diversity of religions and languages. Therefore, it is a significant challenge to reach all of the potential partners by any one approach. Parthenium weed is now abundant along roadsides and railway tracks, and on community lands, but it has also invaded cropping land. Therefore, in the selection of partners for the awareness and management strategy, it has been necessary to involve farmers and all the major institutes that advise them. These include ICAR, which takes care of most agricultural research through its more than 100 centres, the

Fig. 9.3. Different activities during parthenium weed awareness weeks in India: (a) a rally of school children; (b) seminar; (c) weed uprooting; (d) distribution of biological control agents. (Sushilkumar.)

SAUs, which are responsible for agricultural research and teaching, and the 600 KVKs, which are an integral part of extending agriculture ideas directly to farmers. In addition, representation from the state agricultural departments, a number of NGOs, schools and colleges are all necessary so the parthenium weed awareness and management strategy can be communicated to the general public.

Preparation of literature for effective awareness

In the national parthenium weed awareness and management strategy, it has been necessary to develop a number of educational literature items to allow for the participation of the major stakeholders in the programme. This has included leaflets created using simple language, both in Hindi and English, a short video showing the problems parthenium weed can cause, and parthenium weed awareness week posters advertising upcoming activities. Posters on 'Wanted' and 'Reward' themes have also been developed to attract the attention of people during the awareness week, as well as newspaper articles on rewards for destroying parthenium weed plants.

Distribution of material with appeal to stakeholders

The hard copies of most materials (pamphlets, banners etc.) were dispatched to pre-identified stakeholders with a request for them to be used during the week's activities. Simultaneously, soft copies of these materials, including the video, were also sent to the stakeholders. The video, developed in both Hindi and English, was uploaded to YouTube for further dissemination. The stakeholders were also requested to organize awareness rallies, set up photo exhibitions, arrange demonstrations, organize workshops, undertake the physical removal of plants and release biological control agents. The release of the biological control insect (*Zygogramma bicolorata* Pallister) was a very popular activity; therefore, in future years, consignments of the beetle will be sent early

to the different stakeholders so they can multiply before the week's activities take place.

Initially, the response to the national Parthenium Weed Awareness Week was poor, but it has now gained significant momentum and now all stakeholders are undertaking some kind of awareness activity during the rainy season, when parthenium weed grows abundantly. Information received from all over India indicates that millions of people are now becoming involved in rallies, demonstrations, photo exhibitions, quiz contests, essay writing and painting competitions, and programmes broadcast on radio and television. The activities now carried out by the different stakeholders are being documented and circulated to motivate them to do even more activities in future (Gogoi *et al.*, 2005; Varshney and Sushilkumar, 2006, 2007, 2008, 2009, 2010; Sushilkumar and Ranganatha, 2011; Sushilkumar and Sharma, 2012).

9.4 International Networking and Collaboration

In a special session during the First International Conference on Parthenium Weed Management in 1997, participants unanimously agreed on the formation of an International Parthenium Network (IPAN). It was agreed to locate the headquarters of IPAN in India. The recommendations of the Conference were finalized in this special session and approved. The recommendations included the creation of a public awareness campaign through extension and community activities. It also drew the attention of scientists, encouraging them to look for alternative uses of parthenium weed and not to rely on the preliminary results and anecdotal data sets. It was also felt that the creation of a 'network' that involved other countries was necessary. Thus, it was recommended that a steering committee should be formed and that memoranda of understanding should be drawn up between various countries to strengthen the research and management activities on parthenium weed.

Interdisciplinary and inter-institutional approaches for the management of parthenium weed were also suggested and it was proposed to hold a second conference on parthenium weed in the near future. Further, it was suggested to secure financial support from ICAR to help develop the network to tackle the problem on a global basis. Some time later, with little activity taking place, the International Parthenium Weed Network (IPaWN) was formed at the University of Queensland, Australia.

9.4.1 Societies and action groups on parthenium weed

International Parthenium Weed Network

In 2009, and with research involvement in three continents, the Tropical and Subtropical Weed Research Unit at the University of Queensland (Australia) started to develop information packages on the weed and to send them out to more than 20 countries that had, or were under threat from having, the weed. After an overwhelming positive response to this circulation, the value of setting up an international network became clear. The IPaWN came into existence in 2009 under the management of Professor Steve Adkins and Dr Asad Shabbir. IPaWN is an international network of expert volunteers devoted to creating awareness about the parthenium weed threat, and to sharing information on how to reduce its adverse impacts upon agro-ecosystems, the environment and human health. IPaWN's mission is to coordinate and disseminate information regarding the global invasion of parthenium weed, its diverse impacts on agro-ecosystems, the environment and human health, and its management. There are now more than 350 members from 30 different countries. It was decided that meetings of IPaWN would be held at the same time as major international conferences such as those of the International Weed Science Society, the Asian Pacific Weed Science Society and the International Parthenium Weed Management Conferences (http://apwss.org/apwss-ipawn.htm).

International Organization for Biological Control working group on parthenium weed

The aim of this working group is to promote the use of sustainable, environmentally safe, economically feasible, and socially acceptable control methods, including biological control, of *P. hysterophorus* L. in the invasive plant's introduced range (Africa, Asia and Australia). Drs Lorraine Strathie, Wondi Mersie and Rangaswamy Muniappan are the most recent conveners of this working group.

International Parthenium Research News Group

International Parthenium Research News Group is a group initiated in India to create awareness on parthenium weed. This Yahoo group was formed to discuss aspects of parthenium weed management. Mr Pankaj Oudhia is the convener of this group (http://www.pankajoudhia.com/iprng/IPRNG-parthenium_a%26w.htm)

Parthenium Action Group

The Parthenium Action Group (PAG) is a community group that was formed in the Central Highlands of Queensland, Australia in 1994. This group was funded by the National Heritage Trust of Australia to promote the best parthenium weed management practice in the Fitzroy Basin grazing lands. Initially they employed two project officers to promote the message of best management practices for parthenium weed. PAG compiled information on the advances in parthenium weed management, on its biological control and management with herbicides. Such information has been provided to many researchers and landholders to further help improve management approaches.

EPPO Expert Working Group

The European and Mediterranean Plant Protection Organization (EPPO) included parthenium weed in its pest alert list in 2012. EPPO established a working group on parthenium weed in 2012 under the coordination of Ms Sarah Brunel and performed a pest risk analysis for the weed in Europe (https://gd.eppo.int/taxon/PTNHY/documents).

published regularly (Fig. 9.4). The past issues of this newsletter may be downloaded from the APWSS website (http://www.apwss.org/apwss-ipawn.htm).

9.4.2 Publications

International Parthenium Weed Newsletter

After successfully organizing a national symposium on parthenium weed in 2004, the Institute of Plant Pathology, University of the Punjab started to publish a newsletter on parthenium weed entitled 'Parthenium News'. In Australia from about 2000, a Parthenium Weed Newsletter was also being circulated nationally. With the inception of IPaWN in 2009, Parthenium News was merged with the Parthenium Weed Newsletter and International Parthenium News was formed. The first issue was published in January 2010 and since then it has been

9.4.3 Conferences and symposia

Scientific conferences and symposia help participants to develop new ideas, disseminate their own ideas and develop more focused research programmes. Parthenium weed was emerging as such a problematic weed in India, Australia and many other countries that it was felt that an international conference exclusively on this problematic plant was justified. The First International Conference on Parthenium Management was held at the University of Agricultural Sciences, Dharwad, India, 6–8 October 1997. The conference was organized into five technical sessions, wherein 19 oral papers and 54 poster papers covering

Fig. 9.4. International newsletter (left) and identification kit for parthenium weed (right) to raise awareness. (International Parthenium Weed Network, Australia.)

important themes were presented. The Second International Conference on Parthenium Management was held at University Agricultural Sciences, Bangalore, India, 5–7 December 2005. The theme of this second international conference was 'Integrated Management of Parthenium to Sustainable Conservation of Bio-diversity in our Ecosystem'. The Third International Conference on Parthenium Management was held at the Indian Agricultural Research Institute, New Delhi, 8–10 December 2010. The Third International Conference (ICP 2010) focused on the theme 'Integrated Management of Parthenium Linked Hazards to Plant, Human and Animal Health for Sustainable Biodiversity' (Fig. 9.5).

The First International Organisation for Biological Control (IOBC) International Workshop on Biological Control and Management of *Parthenium hysterophorus* was held in Nairobi, Kenya, 1–5 November 2010. The Second International Workshop of the IOBC Parthenium Working Group was held in Addis Ababa and Adama, Ethiopia, 13–17 July 2014.

A seminar entitled 'Parthenium, a Positive Danger' was held on 4 September 1976 at University of Agricultural Science, Bengaluru. A national symposium on the awareness of parthenium weed with a theme 'Parthenium Weed: Let's Fight the Menace' was held at the University of the Punjab in Lahore, Pakistan, 7–8 August 2004. Three international symposia on parthenium weed have been organized by IPaWN during International Weed Science Conferences. The first symposium was held in Cairns, Australia in 2011, the second in Hangzhou, China in 2012 and the third at Bandung, Indonesia in 2013.

9.5 Conclusions

It can be concluded that to achieve the goal of sustainable and effective parthenium

Fig. 9.5. Local and international meetings and field workshops on parthenium weed. (a: I. Hossain; b: I. Macdonald; c: Gul Hassan; d: A. Witt.)

weed management, coordination must take place between the major stakeholders of each country and for countries being invaded by this extremely invasive and harmful alien plant to coordinate at the international level. In each of the affected countries, there should be a National Parthenium Weed Working Group that coordinates a National Parthenium Management Plan and liaises with all other relevant international groups.

Acknowledgements

The authors are grateful to Steve Adkins for providing information on coordination of management of parthenium weed in Australia and photos of international workshops and symposia on parthenium weed.

References

Adkins, S.W. and Shabbir, A. (2014) Biology, ecology and management of the invasive Parthenium weed (*Parthenium hysterophorus* L.). *Pest Management Science* 70, 1023–1029.

Austin, P.J. (2005) Strategy for Parthenium management: Australian experience of the National Parthenium Weed Management Group in a coordinated approach. *Proceedings of the Second International Conference on Parthenium Management*. Bangalore, India.

Australian Weeds Committee (2012) *Parthenium (Parthenium hysterophorus L.) Strategic Plan 2012–17*. Weeds of National Significance, Australian Government Department of Agriculture, Fisheries and Forestry, Canberra.

Batish, D.R., Singh, H.P., Kohli, R.K., Vandana, J. and Yadav, S. (2004) Management of invasive exotic weeds requires community participation. *Weed Technology* 18, 1445–1448.

Chinnusamy, C., Murali Arthanari, P. and Nithya, C. (2010) Awareness creation activities on the ill effects and management of *Parthenium* in Tamil Nadu. In: *Proceedings of the 3rd International Conference on Parthenium: Indian Agricultural Research Institute*, New Delhi, 8–10 December, pp. 197–198.

Goldsworthy, D. (2005) Technical report: Parthenium weed and health. University of Central Queensland, Rockhampton, Australia.

Gogoi, A.K., Mishra, J.S. and Dubey, R.P. (2005) *A Report on Parthenium Awareness Week*. Directorate of Weed Research, Jabalpur, India.

Haseler, W.H. (1976) *Parthenium hysterophorus* L. in Australia. *PANS* 22, 515–517.

NRMMC (National Resource Management Ministerial Council) (2007) Australian Weeds Strategy: A national strategy for weed management in Australia. Australian Government Department of the Environment and Water Resources, Canberra. Available at: https://evergladesrestoration.gov/content/ies/meetings/091713/Australian_Weeds_Strategy.pdf (Accessed 17 May 2018).

Rao, R.S. (1956) Parthenium, a new record for India. *Journal Bombay Natural History Society* 54, 218–220.

Sushilkumar (2005) *Biological Control of Parthenium Through Zygogramma bicolorata*. National Research Centre for Weed Science, Jabalpur, India.

Sushilkumar (2014) Spread, menace and management of Parthenium. *Indian Journal of Weed Science* 46, 205–219.

Sushilkumar and Ranganatha, A.R.G. (2011) *Community Approach for Parthenium Management: A Report of Parthenium Awareness Week 2011*. Directorate of Weed Science Research, Jabalpur, India.

Sushilkumar and Sharma, A.R. (2012) *Creating Awareness on Parthenium Management: A Report on Parthenium Awareness Week:* Directorate of Weed Science Research, Jabalpur, India.

Sushilkumar and Varshney, J.G. (2007) *Gajar ghas ka jaivik niyantrana: Vartman sthathi and sambhavnaie* (in Hindi) [Biological control of Parthenium: present and future]. National Research Centre for Weed Science, Jabalpur, India.

Sushilkumar and Varshney, J.G. (2010) Parthenium infestation and its estimated cost management in India. *Indian Journal of Weed Science* 42(1&2), 73–77.

Terblanche, C., Nänni, I., Kaplan, H., Strathie, L.W., McConnachie, A.J., Goodall, J. and Van Wilgen, B.W. (2016) An approach to the development of a national strategy for controlling invasive alien plant species: The case of *Parthenium hysterophorus* in South Africa. *Bothalia* 46, 1–11.

Varshney, J.G. and Sushilkumar (2006) *A Report on Parthenium Awareness Week*. National Research Centre for Weed Science, Jabalpur, India.

Varshney, J.G. and Sushilkumar (2007) *People Participatory Approach to Manage Parthenium: A Report on Parthenium Awareness Week*.

National Research Centre for Weed Science, Jabalpur, India.

Varshney, J.G. and Sushilkumar (2008) *Community Approach for Parthenium Management: A Report on Awareness Week 2008*. National Research Centre for Weed Science, Jabalpur, India.

Varshney, J.G. and Sushilkumar (2009) *Motivation Approach for Parthenium Management: A Report on Awareness Week 2009*. National Research Centre for Weed Science, Jabalpur, India.

Varshney, J.G. and Sushilkumar (2010) *Marching Ahead for Parthenium Management: A Report on Awareness Week 2009*. Directorate of Weed Science Research, Jabalpur, India.

Wise, R.M., van Wilgen, B.W., Hill, M.P., Schulthess, F., Tweddle, D., Chabi-Olay, A. and Zimmermann, H.G. (2007) The economic impact and appropriate management of selected invasive alien species on the African continent. Report Number: CSIR/NRE/RBSD/ER/2007/0044/C. Council for Scientific and Industrial Research, Stellenbosch, South Africa.

10 Parthenium Weed: Uses and Abuses

Nimal Chandrasena[1]* and Adusumilli Narayana Rao[2]

[1]GHD Water Sciences, Parramatta, New South Wales, Australia;
[2]ICRISAT, Hyderabad, India

10.1 Introduction

Parthenium weed (*Parthenium hysterophorus* L.), a plant of the Asteraceae family, has long been recognized as a weed of global significance (Aneja *et al.*, 1991; Towers and Subba Rao, 1992; Evans, 1997; Pandey *et al.*, 2003). It is an annual herb, native to the area around the Gulf of Mexico, including the Caribbean islands and central South America. After introductions and spread in other regions, parthenium weed now has a pantropical distribution. It normally grows fast, producing an adult plant, about 1.5 m in height, which produces flowers early, and sets a large number of seeds in its lifetime (Adkins and Shabbir, 2014). The weed can also grow under wide ecological conditions – from sea level up to 3000 m (K. Dhileepan, Australia, 2017, personal communication). Parthenium weed is now present in 91 countries around the globe, of which only 44 appear to be possibly in its native range. It is regarded as one of the worst weeds in several parts of Africa, Asia and Australia where it has been introduced (Evans, 1997). In the case of India, parthenium weed was first recorded in 1956 (Rao, 1956) and may have entered the country as a contaminant of wheat imported from the USA. Since then, it has become a major weed within a short period, spreading to over 35 million ha in 60 years (Sushilkumar and Varsheny, 2010).

The weed occupies wastelands and disturbed habitat, including roadsides and railway tracks, and grows well in native grasslands, open scrub vegetation, floodplains, cultivated fields and grazed pastures, often forming pure stands (Dale, 1981; Evans, 1997; EPPO, 2014).

The negative impacts of parthenium weed are well documented, with the most profound effects being on livestock farming, productivity of grain crops and human health. Knox *et al.* (2011) estimated that parthenium weed in India caused yield declines of 50–55% in agricultural crops (>5–10 million rupees per annum) and a 90–92% reduction in forage production (1–2 million rupees per annum). Unlike with most other weeds, parthenium weed poses a serious added social dimension, the adverse health problems it causes for humans and animals. These are in addition to significant economic losses, biodiversity losses and habitat destruction it can cause. Parthenium weed causes severe dermatitis, allergy and toxicity in humans (Towers and Subba Rao, 1992). Most domesticated animals also dislike the weed. If eaten, however, the meat is tainted, causing economic losses. Other chapters in this book cover this subject (e.g. see Allan *et al.*, Chapter 6, this volume).

Despite the well-documented negative impacts of parthenium weed, there is a large volume of published research, over three

* nimal.chandrasena@gmail.com

decades, that indicates both actual and potential uses of the weed and opportunities for further exploitation. There are a number of reviews of beneficial uses, which have largely originated from India (Pandey, 2009; Patel, 2011; Kushwaha and Maurya, 2012; Saini et al., 2014). The current review adds to the above, providing an update and a critical appraisal, focusing on the actual uses of the plant in different countries, and its demonstrable potential uses, based on published results. We also explore the question: should utilization be considered an effective management tool in countries like India, to manage existing infestations and prevent its further spread?

10.2 Chemical Constituents

In assessing the potentially useful or harmful aspects of a weed, such as parthenium weed, it is important to understand its dominant, bioactive chemicals. Over several decades, various researchers have described the chemical constituents of parthenium weed, establishing an impressive array of compounds (Table 10.1). Early research by Herz and Watanabe (1959), Dominguez and Sierra (1970), Picman et al. (1979, 1980, 1982) and Kanchan and Jayachandra (1979, 1980a, 1980b) indicated that various parts of parthenium weed contain parthenin a sesquiterpene lactone of pseudoguanolide

Table 10.1. The major secondary plant products of parthenium weed.

Chemical group	Chemical	Plant part	References
Sesquiterpene lactones (Terpenoids)	Parthenin	Stems, leaves, pollen	Herz and Watanabe (1959), Dominguez and Sierra (1970), Kanchan and Jayachandra (1980a, 1980b, 1980c), Belz (2008), Belz et al. (2007), Reinhardt et al. (2004, 2006), Ramesh et al. (2003), Chen et al. (2011)
	Coronopilin	Stem, flowers, trichomes	Picman et al. (1980), Ramesh et al. (2003), Das et al. (1999)
	Pseudoguananolides	Stem, leaves	de la Fuente et al. (2000)
	Hysterin	Stem	Wickham et al. (1980)
	Acetylated pseudoguananolides	Flower	Das et al. (2007)
	Charminarone	Whole plant	Venkataiah et al. (2003)
	Hysterones A–D	Flower	Ramesh et al. (2003), Das et al. (1999)
Minor sesquiterpenes	Ambrosonalides; 2β-hydroxycoronopilin 1,3 hydroyparthenin; tetraneurin A	Flowers	Sethi et al. (1987), Das et al. (1999), Ramesh et al. (2003)
Phenolics	Caffeic acid; p-coumaric acid; ferulic acid; vanillic acid; anicic acid; fumaric acid	Roots, leaves	Kumar et al. (2013a, 2013b), Kumar and Pruthi (2015)
Flavonoids	Quercelagetin; 3,7-dimethylether; 6-hydroxyl kaempferol; p-hydroxy benzoin sitosterol	Aerial parts	Shen et al. (1976), Yadava and Khan (2013)
Volatile oils	Germacrene-D; trans-β-myrcene; camphor; camphene; p-cymene; borneol; bornyl acetate; β-pinene; euginol etc.	Aerial parts	Kumamoto et al. (1985), de Miranda et al. (2014)

nature. More than 45 sesquiterpene lactones have subsequently been recorded from the plant (Wickham *et al.*, 1980; Patil and Hegde, 1988; Towers and Subba Rao, 1992). It is now established that all parts of the plant contain biologically active sesquiterpene lactones (Picman and Towers, 1982; Picman and Picman, 1984; Chhabra *et al.*, 1999; de la Fuente *et al.*, 2000; Ramos *et al.*, 2002; Belz, 2008). Reinhardt *et al.* (2006) showed that parthenin is synthesized during the entire life of parthenium weed, reaching maximum levels during flowering and seed formation stages. It is sequestered in capitate-sessile trichomes on leaves, stems and the achene complex (Reinhardt *et al.*, 2004). Saxena *et al.* (1991) had earlier demonstrated that parthenin is readily transformed by chemical or photochemical reactions into other derivatives, some of which have stronger bioactivity than parthenin itself.

Parthenium weed also releases a range of water-soluble phenolic acids (Table 10.1), from living roots, leaves and seeds, as well as from dead or decaying residues (Kanchan and Jayachandra, 1979, 1980a, 1980b, 1981; Batish *et al.*, 2002a, 2002b). Analysing leaf extracts, Kumamoto *et al.* (1985) recorded the occurrence of many well-known essential oils in the leaves, yielding 0.033% oil. Subsequently, de Miranda *et al.* (2014) identified 27 essential oils in parthenium weed. Other phytotoxic compounds found in parthenium weed include various flavonoids (Shen *et al.*, 1976; Yadava and Khan, 2013).

Rodriguez *et al.* (1975, 1976) had earlier reported that sesquiterpene lactones in plants, such as parthenium weed, exhibit a wide spectrum of biological activities, which include cytotoxic, antitumour, allergic, antimicrobial, phytotoxic, antifeedant and insecticidal properties. The production of such an array of bioactive chemicals as secondary metabolites (Table 10.1) is not unique to parthenium weed. However, combinations of these chemicals and their concentrations in various parts of the plant indicate that they may be of ecological significance, possibly as part of the defences against herbivory. The same chemicals may also be involved in the invasion success of parthenium weed through allelopathic interactions with neighbours. The fact that parthenium weed's phytochemicals elicit strong effects upon other organisms is the reason why numerous studies have attempted to determine if they are of any beneficial use in human health, crop protection, insect control or in other areas.

10.3 Uses of Parthenium Weed

From the large volume of published literature available, the discussions below focus on some of the most significant findings of actual uses and potential uses demonstrated in *in vitro* and *in vivo* experiments, as well as in various field studies. It is important to highlight that some of the experimental results we have reviewed only indicate potential uses and applications and are seriously constrained by the lack of comparisons against benchmarks. The suitability of the species being used in any given area is a critical aspect of utilizing any species with 'colonizing' attributes for beneficial purposes. Instead of any uniqueness of the species, our assessment reveals that the primary motivation for promoting utilization, particularly in India, is the integration of uses into a broader, national weed management effort. However, despite the extensive research, most authors acknowledge that the actual practical uses of the weed's demonstrated beneficial uses would require considerably more research to establish the cost effectiveness of developing useful products and/or applications. In dealing with a species like parthenium weed, it would also be important to consider not just economics, but also associated ecological and environmental considerations, which are quite significant (EPPO, 2014).

10.3.1 Medicinal uses and medical applications

Reports indicate that the word parthenium is derived from the Latin word parthenice, which suggests medicinal uses (Bailey, 1960). According to Lindley (1838) in *Flora Medica*: 'The whole plant is bitter and strong-scented, reckoned tonic, stimulating and anti-hysteric. It was once a popular remedy in ague. Its odour is said to be peculiarly disagree to bees and that insects may be easily kept at a distance by carrying a handful of the flower heads'.

There is ethnobotanical evidence that parthenium weed is used as a folk remedy in the Caribbean and Central American countries (Cuba, Guyana, Trinidad, Jamaica and Mexico), the USA and sub-Saharan Africa, where it is applied externally to cure skin disorders or taken internally, often as a decoction in the treatment of a variety of ailments (Dominguez and Sierra, 1970). The *Dictionary of Economic Plants in India* (Singh *et al.*, 1996) records the use of parthenium weed as a tonic, febrifuge, an emmenagogue, and for the treatment of inflammation, eczema, skin rashes, herpes, colds, heart problems, amoebiasis, gynaecological ailments, muscular rheumatism, neuralgia and dysentery. An ethnobotanical study in Mauritius and Rodrigues (Gurib-Fakim *et al.*, 1993) reported that tea made from parthenium weed is used as a tonic, febrifuge, analgesic and emmenagogue. A similar survey in Venezuela recorded the use of a decoction from dried roots as an antimalarial drug (Caraballo *et al.*, 2004).

Table 10.2 provides a summary of recent laboratory-based evidence of potential medical uses, major effects reported and the sources. The studies indicate that apart from the traditional medicinal uses, parthenium weed extracts may also have other potential applications. Although the 'causes and effects' are not quite proven, the bioactivity recorded by these studies may justify the uses of parthenium weed in traditional medicine. These studies have non-specifically attributed the properties to the occurrence of flavonoids, terpenoids, alkaloids and phenolic compounds in the weed's extracts (Kumar *et al.*, 2014; Kumar and Pruthi, 2015). However, despite the evidence from largely *in vitro* and *in vivo* studies indicating the anticancer, antioxidant and antibacterial potential in parthenium weed extracts,

developing these properties towards modern medicines is still a long way off.

10.3.2 Non-medicinal uses: potential uses as a pesticide

Secondary metabolites of plants are part of their chemical defences against natural enemies, such as fungi, bacteria and insects; these compounds are often antimicrobial. Given the array of phytochemicals found as secondary metabolites in parthenium weed, it has featured strongly in the search for alternative and 'eco-friendly' pesticides. Datta and Saxena (2001) demonstrated that pesticidal bioactivity was evident at relatively low concentrations of parthenin and its derivatives, as pure compounds – in the range above 25 mg/l up to 1000 mg/l. Table 10.3 provides a summary of recent studies and their findings, which demonstrate the pesticidal potential of parthenin and its derivatives. However, it should be noted that despite the large amount of research and empirical demonstrations, over nearly two decades, commercial production of a 'botanical pesticide' based on parthenin or its derivatives, as an alternative to synthetic chemicals, is yet to occur.

10.3.3 Herbicidal potential of parthenium weed allelochemicals

Parthenin is released from the plant by being washed from ruptured trichomes or from decomposing tissues and may contribute to parthenium weed's interference with surrounding neighbours. However, after its release into the soil environment, the persistence and phytotoxicity to neighbours of parthenin or any other phytochemical would be significantly modified by physical, chemical and biological soil properties. Therefore, whether allelochemicals in parthenium weed can be utilized in various applications depends on their fate and persistence in soil, and soil concentrations (Belz *et al.*, 2007). Despite the promising allelopathic potential demonstrable in laboratory experiments

Table 10.2. Potential medicinal uses of parthenium weed extracts.

Medicinal use	Plant part/extract	Major outcome reported	References
Diabetes mellitus treatment	Aqueous extracts of dried leaves and flowers, hypoglycaemic activity tested in diabetic rats (blood glucose 280–310 mg/dl)	A dose of 100 mg/kg body weight reduced blood glucose in test animals to below 240 mg/dl at 2 h; this compared favourably with the reduction achieved by a standard diabetic drug in control animals, at 2 h	Patel *et al.* (2008)
Antimalarial activity	Ethanolic leaf extract tested against the malarial parasite	Antimalarial activity demonstrated (*in vitro*), but not linked to any specific phyto-constituent	Valdés *et al.* (2010)
Antimalarial activity	Parthenin, extracted from whole plant tested against malarial parasite	Significant antimalarial activity (*in vitro*) against a multidrug resistant strain of *Plasmodium falciparum*	Hooper *et al.* (1990)
Amoebicidal activity	Parthenin, extracted from the whole plant tested *in vitro* against *Entamoeba histolytica*	Parthenin was amoebicidal and as effective as the standard drug used in treating amoebiasis	Sharma and Bhutani (1988)
Anticancer activity	Parthenin tested on mice (*Mus musculus* L.) injected with cancer cells	Sublethal parthenin doses either cured mice or increased their survival time	Mew *et al.* (1982)
Cytotoxicity and antitumour potential	Extracts of dried aerial parts; tested *in vivo* and *in vitro* using bacterial cell lines, lymphocytes and mice	Weed extracts and parthenin showed no mutagenicity in the Ames Salmonella/ microsomal assay, but demonstrated potent cytotoxicity	Ramos *et al.* (2002)
Cytotoxic and anticancer properties	Methanol extracts of dried flowers tested on human cell lines	Extracts showed significant cytotoxicity against T lymphocytes and T-cell leukaemia, HL-60 (leukaemia) and Hela (human cervical carcinoma) cell lines	Das *et al.* (2007)
Anticancer properties	Ethanolic extracts of dried leaves tested *in vivo* (rat kidney cells) and in *in vitro* models	Potent cytotoxicity against MCF-7 and THP-1 human cancer cell lines at 100 µg/ ml; concentration-dependent inhibition of HL-60 cancer cell lines; moderately anti-HIV activity	Kumar *et al.* (2013a, 2013b)
Anti-inflammatory activity	Extracts of parthenium weed	Ferulic acid (FA) extracted from the weed inhibited the enzyme cyclooxygenase-2 (COX-2) by molecular docking; this may lead to developing anti-inflammatory drugs	Kumar and Pruthi (2015)
Antibacterial activity	Ethanolic extracts of parthenium weed, compared with extracts from 20 other species	Parthenium weed extracts had the highest antibacterial activity; antibacterial activity was strongest against gram-positive, pathogenic bacteria	Nair and Chanda (2006)
Antibacterial activity	Ethanolic extracts of parthenium weed	Extracts showed greater antibacterial properties against some bacteria compared with standard antibiotics (Azithromycin and Cepaxim)	Fazal *et al.* (2011)
Antibacterial activity	Solvent extracts of leaves and other parts	Leaf extracts were significantly higher in antibacterial activity against several common pathogenic bacteria compared with stem, flowers or root extracts	Kumar *et al.* (2014)
Antibacterial, antifungal activity	Petroleum-ether extracts of dried aerial parts	Strong antibacterial and antifungal activity against several common, pathogenic bacteria and fungi	Madan *et al.* (2011)

Table 10.3. Potential non-medicinal uses of parthenium weed extracts.

Potential use	Plant part/extract	Major outcome reported	References
Antifungal activity	Aqueous leaf extracts against rice blast fungus *Pyricularia grisea* Sacc. using *in vitro* studies	Strong inhibition of mycelial growth of *P. grisea* on rice (*Oryza sativa* L.) seedlings by a 10% aqueous extract; no adverse effects on rice	Pedroso *et al.* (2012)
Dengue fever vector control	Leaf extracts with several solvents tested against *Aedes aegypti* (L.)	High concentration extracts (1000 ppm) were selectively effective against female mosquitoes; potential of developing as an oviposition deterrent and ovicidal agent	Kumar *et al.* (2011, 2012)
Malarial vector control	Leaf extracts tested against *Anopheles stephensi* Liston	Strong larvicidal potential against the fourth instar larvae of *A. stephensi*	Ahmad *et al.* (2011)
Nematicidal activity	Parthenin and derivatives tested against root-knot nematode (*Meloidogyne incognita* Fab.)	Lethal concentration (LC50) at 72 h was 512 mg/l; an acid-converted derivative was five times more nematicidal (LC50 104 mg/l at 72 h)	Datta and Saxena (2001)
Insecticidal activity	Parthenin and derivatives tested against the cowpea beetle (*Callosobruchus maculatus* Fab.)	The pyrazoline derivative was 17 times more insecticidal than parthenin, and was as toxic as azadirachtin from neem (*Azadirachta indica* A. Juss.)	Datta and Saxena (2001)
Insecticidal activity	Whole plant extracts tested on mustard aphid (*Lipaphis erysimi* Kaltenbach)	More effective than other plant extracts in reducing the mustard aphid populations on mustard (*Brassica juncea* L.)	Bhattacharyya *et al.* (2007)

and some field situations, short half-lives and low field levels of major phytotoxins would mean that allelochemicals may not always be involved in interference mechanisms between parthenium weed and its neighbours.

The early research of Kanchan and Jayachandra (1979, 1980a, 1980b, 1980c), Kohli *et al.* (1996), Pandey *et al.* (1996a, 1996b) and others recorded strong allelopathic effects of parthenium weed on a range of crops and weeds, although species varied considerably in their sensitivity to weed extracts or exudates. The effects were largely attributed to the bioactivity to parthenin in leaf washings and root exudates, while acknowledging the possible inhibitory role of other terpenoids, phenolic acids, such as *p*-hydrobenzoic acid, in the extracts. Several laboratory studies in India (Batish *et al.*, 2002a, 2002b, 2007) have documented parthenin phytotoxicity towards a range of weeds and crops. Based on such results, several research groups in India have promoted parthenium weed as a 'botanical

herbicide', for both pre-emergent and post-emergent activity, without much selectivity. However, the selectivity of parthenin, or just parthenium weed extracts, against crops remains largely unknown. Consequently, despite an early suggestion by Datta and Saxena (2001) that parthenin and its derivatives may be developed as commercial herbicides, this is yet to occur.

A summary of major studies that have recorded phytotoxic effects on other weeds is given in Table 10.4. It is clear from these studies that parthenium weed extracts are toxic to some species. However, most studies indicate relatively low persistence of parthenin in aquatic environments. For instance, Pandey (1994a, 1994b, 2009) showed that the phytotoxicity of the extracts was gradually lost in water within about 30 days under outdoor conditions. The possibility of incorporating parthenium weed biomass for weed management in the field has not been widely tested. At least in one study, Marwat *et al.* (2008) tested the herbicidal potential of parthenium weed in

Table 10.4. Selected studies showing the herbicidal potential of parthenium weed extracts.

Potential use	Plant part/extract	Major outcome reported	References
Botanical herbicide	Parthenin and derivatives tested against sickle pod (*Senna tora* L.)	Parthenin reduced seed germination by 50% at 364 mg/l, a propenyl derivative and a cyclopropyl derivative were much more effective (50% inhibition at 136 mg/l and 284 mg/l, respectively)	Datta and Saxena (2001)
Water hyacinth (*Eichhornia crassipes* (Mart.) Solms.) control	Dried leaf powder; aqueous extract	Low concentrations (0.25% w/v) reduced growth; higher concentrations (0.5% w/v) killed water hyacinth in 2–4 weeks; death was due to leakage of solutes from roots, loss of dehydrogenase activity in roots and chlorophyll in the leaves	Pandey *et al.* (1993a, 1993b)
Salvinia (*Salvinia molesta* Mitchell)	Dried leaf powder; aqueous extract	Higher concentrations (0.75% w/v) killed Salvinia within 5–15 days	Pandey (1994a, 1994b)
Submerged aquatic weeds	Dried leaf powder; aqueous extract	Submerged aquatics were sensitive to parthenin at 25 ppm; however, effects were short-lived, as parthenin degraded	Pandey (1996b)
Laboratory studies and field applications	Dried leaf powder, soaked in water for 24 h, at various concentrations (10 to 250 g/l)	Differential response of weed species to parthenium weed extracts; pre-emergent applications were more effective in reducing the abundance of some weeds in field plots than post-emergence sprays	Marwat *et al.* (2008)

field applications, and despite its low persistence in soil, still suggested that the extracts could be developed as a bio-herbicide in Pakistan.

Parthenin is rapidly degraded in the environment to various metabolites that have little or no phytotoxicity (Belz *et al.*, 2009). Parthenin concentrations declined to less than 50% of the initial levels within 60 h in soil, and the degradation was accelerated by soil pre-conditioning with parthenin, higher clay content and higher temperatures. While parthenin is likely to contribute to allelopathic effects, high parthenium weed densities are required to have high levels of parthenin in soil and favourable soil conditions for parthenin or its derivatives to persist in bioactive forms (Belz *et al.*, 2009). Overall, the short-lived nature of parthenin in soil and loss of its phytotoxicity in the aquatic environment is a major limitation. This raises considerable doubts whether the bioactivity demonstrable in parthenium weed extracts can lead to a commercially viable 'eco-friendly' bio-herbicide.

10.3.4 Use in phyto-remediation of heavy-metal-contaminated soils

The interest in using parthenium weed for phyto-extraction of heavy metals and other pollutants has been growing in the past few years, stimulated by the observations that parthenium weed has the capacity to grow well even in contaminated sites. Bapat and Jaspal (2016) have recently summarized much of the available research. The important findings of some studies are summarized in Table 10.5, as examples. Collectively, these studies confirm parthenium weed's ability to tolerate high levels of soil contamination and conditions, which are relatively unfavourable to the growth of most plants. They also demonstrate the capacity of parthenium weed to take up, accumulate and sequester heavy metals in its tissues. However, we find that most of the published studies have not benchmarked parthenium weed against other known hyper-accumulators. This makes it difficult to draw firm conclusions as to the comparative

Table 10.5. Examples of removal of heavy metals from contaminated soil by parthenium weed.

Pollutant	Experimental set-up	Major outcome reported	References
Lead (Pb)	Parthenium weed grown in soil spiked with lead nitrate	Parthenium weed extracted Pb from soil in significant amounts. Foliar applications of gibberellic acid GA3 aided the uptake and accumulation of Pb in stems	Hadi and Bano (2009)
Zinc (Zn)	Parthenium weed grown in soil spiked with zinc sulphate	Parthenium weed 'hyper-accumulated' Zn with a bio-concentration factor (BCF) >1.0 and a translocation factor >1.0. Addition of Ethylenediaminetetraacetic acid (EDTA) at 0.1 g/kg of soil increased the Zn uptake	Sanghamitra et al. (2012)
Cadmium (Cd)	Parthenium weed grown in soil spiked with Cd	Parthenium weed 'hyper-accumulated' Cd with a high BCF (1.85); addition of EDTA to the soil (40 mg/kg of soil) and foliar sprays of GA3 increased the uptake and translocation of Cd	Ali and Hadi (2015)
Heavy metals: Fe, Zn, Cu, Pb, Ni and Cd	Parthenium weed grown in fly-ash mixed soil	Parthenium weed accumulated considerable amounts of the heavy metals in different parts of the plant. Heavy-metal accumulation by 90 days was in the order Fe > Zn > Cu > Pb > Cd > Ni. Translocation of Pb, Ni and Cd was much higher than Fe, Zn and Cu	Ahmad and Al-Othman (2014)

advantages of using parthenium weed for remediating contaminated soil. Therefore, whether the phyto-extractive capacity of parthenium weed can be put into actual practice in ecological restoration – to reduce heavy-metal pollution in contaminated soils, on any scale, is still largely untested in field situations.

10.3.5 Use of parthenium weed carbon as bio-adsorbent in pollution removal

Many agricultural and wood wastes, such as sugar cane pith, sawdust, coconut husks, wheat shells, corncobs and similar materials are considered useful as bio-adsorbents of organic pollutants from waste effluents of industrial processes. These materials are of considerable value in developing countries, because activated carbon (AC) – the most widely used adsorbent – is quite expensive and needs to be regenerated. On the other hand, in countries like India, parthenium weed biomass is freely available in large quantities, throughout the year (Shrivastava, 2010), and is potentially useful after conversion into a bio-adsorbent

form. In addition, the adsorbent can be safely discarded after use and does not need costly regeneration. There is a large volume of literature available, mostly from India (summarized in Table 10.6), that demonstrates adsorbent properties of parthenium weed in laboratory-based experiments. These studies characterize the nature of the adsorbent material that it can generate, and provide details of varied adsorption processes. Despite the demonstration of adsorbent properties in laboratories, we find this research largely academic and there is no evidence of practical uses up to the time of this review.

Collectively, the studies show that parthenium weed biomass is a low-cost biomaterial that can be used as an alternative to costly adsorbent for dye removal in wastewater, or for extracting heavy-metal pollutants from effluents. Adsorption efficiency is highly dependent on initial dye or heavy-metal concentration; the lower the initial pollutant concentration, the higher the adsorption. Particle size of parthenium activated carbon (typically in the range of 0.3–1.0 mm) is also a significant factor; the smaller the particle size, the higher the surface area and sorption. Other influential

Table 10.6. Summary of potential uses of parthenium weed in removing industrial dyes and heavy-metal pollutants from aqueous solution.

Industrial pollutant	Adsorbent	Major outcome reported	References
Methylene Blue	Sulfuric acid-treated carbon (SWC); phosphoric acid-treated carbon (PWC)	Compared with standard activated carbon (AC); both SWC and PWC effectively removed the dye; order of adsorption capacity: AC > PWC > SWC	Lata *et al.* (2007)
Rhodamine-B	SWC, PWC and formaldehyde-treated carbon	All three were quite effective in removing the dye	Lata *et al.* (2008a, 2008c)
Safranine	Dried and crushed biomass (particle size 60–250 μm)	Maximum adsorption 89.3 mg/g from a dye concentration 400 mg/l in wastewater	Shrivastava (2010)
p-Cresol	Sulfuric acid-treated parthenium activated carbon (PAC) particle size < 0.5 mm	Adsorption of p-cresol from wastewater by PAC was as good as commercial-grade AC	Singh *et al.* (2008a)
Cadmium (II)	Dried powder made from the whole plant	Removal of 99% of Cd (II) from wastewater; an endothermic process; maximal at pH 4.0	Ajmal *et al.* (2006)
Nickel (II)	Sulfuric acid-activated weed ash	Effectively removed Ni; maximal removal 17.2 mg/g of ash at pH 5.0	Lata *et al.* (2008b)
Nickel (II)	Weed ash from whole plant	Effectively adsorbed and removed Ni; maximal removal at pH 11.0	Singh *et al.* (2009)
Chromium (VI)	Weed ash from whole plant	Effectively adsorbed Cr; maximal removal 64% at pH 2.0	Singh *et al.* (2008b)

factors are adsorbent dose, pH and contact time. Ajmal *et al.* (2006) also suggested that the inexpensive material could be used to sequester Cd ions in contaminated soil and potentially reduce Cd uptake by agricultural crops grown in polluted areas. The claim from the research groups that have conducted highly detailed studies (summarized above) is that parthenium weed biomass could become a useful adsorbent carbon. Sivaraj *et al.* (2010) have described the preparation of activated carbon from parthenium weed biomass using various physical (thermal) and chemical methods, along with characteristics of the material. Their work demonstrated $ZnCl_2$-impregnated parthenium weed carbon as the most efficient adsorbent, due to its porous nature and higher adsorption area. However, we find that comparative research against less controversial species demonstrating a unique capacity of the parthenium weed biomass to be of exceptional value as bio-adsorbent is yet to be presented.

10.3.6 Use of parthenium weed biochar as soil amendment

Kumar *et al.* (2013c) showed that parthenium weed biomass could be converted to biochar by burning at different temperatures (200–500°C) for varying periods. With increased temperature, biochar yield decreased, but its stability was highest at 300–350°C and charring for 30–45 min. Incorporation of this biochar up to 20 g/kg of soil increased the soil microbial biomass and several important soil enzymes. The charring also removed allelochemicals of parthenium weed, which was demonstrated using a maize (*Zea mays* L.) seedling assay. It is reasonable to expect that parthenium weed residues could be converted to usable carbon residues via pyrolysis. However, any future adoption of this as a wide-scale practical use would depends on a number of factors, not least the energy cost of pyrolysis.

10.3.7 Parthenium weed uses in biofuel production

Utilization of biomass of various common weeds, including parthenium weed, for biofuel production are promoted as methods that would help rural communities in obtaining energy from cheap feedstock. A large volume of research, mainly from India, confirms that this is a valid premise with practical application. Opinion is also strong that large-scale utilization of parthenium weed for biofuel generation will also pave the way for its eradication. Proponents point out that after generating biogas, the digested parthenium weed could also be used as organic manure, as the nutrients (N-P-K) are largely conserved.

In some early research, Gunaseelan (1987) used dried parthenium weed biomass (10%) and cattle manure slurry as feedstock and produced methane by anaerobic digestion. When parthenium weed biomass alone was anaerobically digested, it produced a maximum of 35 l methane/kg of biomass at a total solids (TS) loading of 5% (Gunaseelan and Lakshmanaperumalsamy, 1990). The low methane yield was due to the high lignin content of parthenium weed (Gunaseelan, 1994). Sodium hydroxide treatment of the dried biomass for 24 h significantly enhanced the digestibility, cellulose reduction and methane production. Pre-treatment doubled the methane yield (214 ml of gas/g of solids, at 10-day retention time and 40°C) compared with untreated biomass (Gunaseelan, 1994), proving the feasibility of biogas production by anaerobic fermentation of parthenium weed at minimal cost.

Recent research has focused on using parthenium weed biomass for bioethanol production. Ghosh et al. (2013) converted the lignocellulosic biomass of parthenium weed (cellulose 28%; hemicellulose 21% and lignin 14%) into a fermentable sugar mixture by a pre-treatment with dilute sulfuric acid at 150–210°C, which was then efficiently fermented by yeasts to produce bioethanol. Pandiyan et al. (2014) subsequently showed that pre-treatment with 1% NaOH enhanced lignin recovery, increasing the yields of reducing sugars. Rana et al. (2013) demonstrated that basidiomycetes fungi could be used to de-lignify parthenium weed biomass, generating large amounts of reducing sugars.

Singh et al. (2014) reported that the large amount of cellulose in the parthenium weed biomass makes it a suitable substrate for bioethanol production. A hot acid hydrolysis was the most effective pre-treatment, yielding 398 mg/g of total fermentable sugars from the raw biomass. While acknowledging the high cost of commercial cellulose enzymes and slow kinetics, Singh et al. (2014, 2015) obtained a final bioethanol yield of c.203 mg/g through an optimized enzymatic process, aided by an ultrasound sonication process. When the hot acid hydrolysis and ultra-sonication-aided digestion with two enzymes was combined with yeast fermentation of the sugars, an even higher bioethanol yield of 260 mg/g raw biomass was obtained (Bharadwaja et al., 2015). Further optimization of the combined process and fermentation by different strains of yeasts achieved a maximum fermentable sugar yield of 615 mg/g of biomass, producing 240–270 mg of bioethanol/g biomass (Tavva et al., 2016).

Comparisons with published literature on other lignocellulosic biomass, including some weeds, reveal that parthenium weed biomass compares favourably as biofuel feedstock with other conventional biomass from agricultural and non-agricultural residues. The consensus of these studies is that parthenium weed biomass is highly suitable for bioethanol generation, as a cheap feedstock, where it is plentifully available. As the processes are cost effective, utilization of parthenium weed biomass for biogas (methane) or bioethanol production are valid applications for the large amounts of weed biomass removed during control efforts in countries such as India.

10.3.8 Utilizing parthenium weed as compost, vermi-compost or green manure

Parthenium weed biomass has long been considered as a useful source of compost or green manure to improve soil health and crop yields (Raju and Gangwar, 2004; Biradar and Patil, 2001). Although the compost contains abundant macro- and micro-nutrients, and is much richer than farmyard manure, there has been concern that high levels of parthenin, phenolics and other allelochemicals in parthenium weed may adversely affect seed germination and seedling development of sensitive crop species. Therefore, effective composting is essential to break down the constituent phytochemicals (DWSR, 2010). At the same time, effective composting is needed to kill parthenium weed seeds, preventing further spread of the weed through compost.

Over the past two decades, several research groups in India have examined the process of composting parthenium weed, often with cow dung slurry or mixture (Fig. 10.1). The research has also attempted to quantify the benefits of incorporating the parthenium weed compost either alone or in combination with inorganic fertilizer or other organic manures. The results of numerous studies indicate variable nutrient compositions in the compost, but generally positive effects in improving the growth of crop species, provided the compost is well prepared and has undergone mineralization for more than about 60 days. In one study (Channappagoudar *et al.*, 2007) the nutrient contents from pre-flowering parthenium weed biomass were much higher than those made from older, post-flowering plants.

In some early research, Biradar and Patil (2001) showed that parthenium weed biomass, mixed with cow dung, provided a good substrate for the growth of the earthworm *Eudrilus eugeniae* Kinberg and for vermi-composting. The nutrient composition of

Fig. 10.1. Compost prepared from parthenium weed in India. (S. Adkins.)

vermi-compost, produced by *E. eugeniae*, in a 1:1 mixture of weed biomass to cow dung, was higher than that of its individual substrates (Sharma *et al.*, 2008). Although this vermi-compost, incorporated at 10 t/ha, increased the yield of wheat (*Triticum aestivum* L.), a comparison showed that its performance was inferior to vermi-compost from lantana (*Lantana camara* L.), farmyard manure and a reduced inorganic fertilizer (Sharma *et al.*, 2008).

Sivakumar *et al.* (2009) evaluated the efficiency of vermi-composting of parthenium weed by the earthworm *Eisenia fetida* (Savigny), in mixture with neem and cow dung. While varying levels of neem leaves had no effect on the earthworms, parthenium weed significantly reduced their growth and number of castings above a rate of 75 g in 500 g of cow dung. Recently, Yadav and Garg (2011) confirmed this adverse effect and found a 25% parthenium weed biomass mixed with 75% cow dung was the optimal feed material for *E. fetida*. In another study, Rajiv *et al.* (2013b) found that the biomass gain, cocoon production and antioxidant enzyme production of *E. eugeniae* were adversely affected by a high concentration of parthenium weed without cow dung.

The utilization of parthenium weed as green manure has also received attention. Research in India by Saravanane *et al.* (2012) using a mixture of both pre-flowering and flowering parthenium weed biomass, incorporated at 5.0 t/ha into soil, significantly increased the productivity of rice compared with the control (no added fertilizer). Incorporation of weed biomass, combined with 75% of the recommended NPK fertilizer dose for rice, produced grain and straw yields similar to that of application of 100% of the recommended fertilizer dose. In another study, addition of parthenium weed green leaf manure to the first crop of a sequential cropping system, potato (*Solanum tuberosum* L.), markedly improved the grain yield of the succeeding finger millet (*Eleusine coracana* (L.) Gaertn.) in Karnataka, India (Saravanane *et al.*, 2011).

In promoting parthenium weed biomass as compost, vermi-compost or green manure, the consensus in India is that this ecologically friendly use will also help better manage the spread of the weed. Utilization of parthenium weed biomass in these ways, without letting it go to waste, are valid practical applications where the weed is abundantly available. Anecdotal evidence is that the practices are widely used in different states and regions in India. However, there is little evidence of such uses from other regions of the world.

10.4 Other Potential Uses

10.4.1 Parthenium weed biomass as a source of cellulose

Renewable and non-conventional raw materials, such as fibres from weeds, grasses, bamboos, and agricultural and forest wastes, are gaining interest as alternative sources of cellulose. Naithani *et al.* (2008) showed that parthenium weed biomass is a rich source of lignocellulose that can be extracted cost effectively. The cellulose could be readily converted into derivatives, such as ethers (carboxy-methyl cellulose, CMC), hydroxy-methyl cellulose (HMC), or cross-linked with formaldehyde and other chemicals to produce much stronger cellulose for textile and paper products (Varshney and Naithani, 2011). However, parthenium weed is not a species unique as a source of cellulose and comparative assessments with other less controversial sources are needed to justify such a use.

10.4.2 Nanoparticles from parthenium weed and their uses

Nanotechnology is the field of science that includes synthesis and utilization of various nanoparticles, which are objects ranging in size from 1 to 100 nm. Nanoparticles have the potential to revolutionize various fields of endeavour, due to their exceptional stability, high resistance to oxidation, high thermal conductivity, and other properties. Various plant extracts can be used to biologically reduce metallic ions, such as silver, gold and zinc, into their corresponding nanoparticles, with potential applications in medicine and other fields (Sanvicens and Marco, 2008). Exploring this potential as a 'green synthesis', Ranjani and Sakthivel (2013) used parthenium weed leaf extracts to reduce silver ions and form silver nanoparticles. Mondal *et al.* (2014) also synthesized silver nanoparticles using parthenium weed root extracts, and suggested that soluble carbohydrates in the aqueous extracts were involved in the reduction of silver ions to highly stable, spherical nanoparticles, which showed considerable larvicidal activity against the filarial vector mosquito (*Culex quinquefasciatus* Say). In other work, Rajiv *et al.* (2013a) synthesized highly stable zinc oxide nanoparticles using parthenium weed leaf extracts, and showed these to possess significantly higher antifungal activity against plant pathogens *Aspergillus flavus* Link and *Aspergillus niger* van Tieghem than a standard antifungal drug (amphotericin B). Nanotechnology is an emerging field of endeavour. However, the risks of applying nanotechnology to cosmetics and human and veterinary medicines are still under review. The research to date has not indicated evidence of any special attributes of parthenium weed extracts that will justify their use as reducing agents over other non-controversial species. Therefore, the use of parthenium weed extracts for the production of nanoparticles remains to be further explored and justified.

10.5 Potential Abuses

Despite the various actual and potential uses of parthenium weed, promoted by several research groups, our view is that its utilization presents significant challenges outside its native range, particularly in regions where the weed's expansion has been spectacularly rapid. Our review finds that most medicinal uses of parthenium weed in different countries fall within the realm of traditional medicine, which is currently undergoing a global revival. However, as with other modern medicines, large-scale use of a potentially toxic species like parthenium weed requires more rigorous clinical testing of toxicity and safety of dosages, validation of cause and effect, and regulation, so that its improper use can be prevented. Despite records of potential medicinal applications, there is very little information on actual uses and clinical results. The well-known allergic sensitivity of humans to parthenium weed indicates that its medicinal use would be a contraindication to at least part of the population, introducing a serious public health risk for some individuals.

Other potential abuses or harm could occur if the biomass of parthenium weed growing in metal contaminated sites is harvested and used as fodder for animals. Heavy metals are poorly secreted and not well metabolized within animal systems, so there is a potential risk of unacceptable levels of heavy metals entering the food chain via livestock (Ahmad *et al.*, 2013). Similarly, parthenium weed obtained from heavy-metal-contaminated sites could also pose risks to its traditional medicinal uses (Rehman *et al.*, 2013).

When used as a soil amendment, partially burnt parthenium biomass may not be good enough to improve soil conditions and could be detrimental to sensitive crop species. If the compost-making process is substandard, parthenium weed seeds will not be fully killed (DWSR, 2010) and the use of poorly prepared compost in agricultural fields would greatly increase the risk of spread of the weed across all landscapes. In promoting the use of parthenium biomass to produce biofuel, Ghosh *et al.* (2013) cautioned that parthenin and other allelochemicals hinder the growth of microorganisms, resulting in reduced fermentation and bioethanol yield. They also warned that the collection of parthenium weed causes negative health effects on workers if they are sensitive to the allergens.

Recent research from Africa has demonstrated an additional potential problem that can arise from sustaining populations of parthenium weed. In the malaria-endemic regions of East Africa, the malaria-transmitting mosquito vector *Anopheles gambiae* Giles feeds on nectar from various plants and honeydew to obtain sugars (Nyasembe *et al.*, 2012, 2015). Controlled feeding assays showed that the fitness, energy reserves and survival of female *A. gambiae* mosquitoes increased substantially when fed on parthenium weed nectar compared with sugar from two other weeds, castor oil (*Ricinus communis* L.) and cobbler's pegs (*Bidens pilosa* L.). The females tolerated the toxins parthenin produced by parthenium weed and 1-phenylhepta-1,3,5-triyne produced by cobblers pegs, but not ricinine produced by castor oil. The authors suggest that if parthenium weed suppresses other weed species that are less suitable host plants for the malaria disease vector, this could lead to a higher transmission of the disease in endemic areas (Nyasembe *et al.*, 2015). Other potential abuses of parthenium weed is in the creation of floral bouquets (Fig. 10.2) and in packaging of goods (Fig. 10.3).

10.6 Conclusions

Parthenium weed is a unique colonizing species, possessing an array of strongly bioactive chemical compounds that can pose considerable problems to sensitive plant and animal species. The phytochemicals are part of the robust and adaptive defence system of the plant, which allow it to compete with other organisms, survive and reproduce in new environments and spread further. Despite the evidence of relatively low persistence in the environment, under certain circumstances the biologically active secondary metabolites extruded from the plant may be implicated in discouraging or actively displacing other species occupying similar ecological niches. The same compounds are implicated in causing undesirable health impacts on sensitive humans and in the medicinal values of the plant. However, whether or not the same array of phytochemicals in parthenium weed can be extracted and exploited more broadly for a variety of practical uses remains a question.

The intensive attention given to parthenium weed in India is because it has greatly increased its invaded territory in the subcontinent, and continues to spread widely. The second major reason is the management of parthenium weed over the past three decades has been broadly ineffective, despite major efforts (DWSR, 2010). However, it is noteworthy that the existing reviews on parthenium weed (Patel, 2011; Kushwaha and Maurya, 2012; Saini *et al.*, 2014; Bapat and Jaspal, 2016) have refrained from commenting specifically on the issue of deliberate cultivation, or sustaining its populations for utilization. The majority of published studies, collated in the above reviews, only demonstrate bioactivity in parthenium weed extracts in laboratory or greenhouse studies, which is not unique to this species.

Fig. 10.2. Parthenium weed collected for its use in floral bouquets in Pakistan. (A. Shabbir.)

Fig. 10.3.   A pile of parthenium ready for use as packaging material in Ethiopia. (A. Witt.)

Many of the studies stop short of extending potential applications to actual proving of practical uses. A large number also suffer from inadequate benchmarking of parthenium weed against less controversial species that can also be harnessed for the specific uses being promoted.

However, potential uses may expand in the near future based on the emerging evidence in various treatments, such as diabetes and malaria. Understanding the scientific basis behind the traditional usage may also improve with more ethno-pharmacological studies. Further exploration of the antioxidant, cytotoxic and anticancer potential is also likely. Nevertheless, the role of different bioactive compounds must be separated, and the 'causes and effects' better understood to make the traditional practices of using parthenium weed a modern reality. Thus far, there is no evidence of explorations of parthenium weed by the pharmaceutical industry for the development of any

medicines, despite the recent spike in pharmacological research.

Given the short persistence of parthenin (Belz *et al.*, 2007), the commercial production of an 'eco-friendly', 'botanical' bioherbicide from parthenium weed with wide application in agriculture appears unlikely. Nevertheless, with additional research there is a possibility of developing the fungicidal, insecticidal and nematicidal bioactivity of some parthenium weed phytochemicals as 'eco-friendly' natural products with targeted, commercial applications.

Among other uses, parthenium weed biomass can yield nutritionally rich compost or green manure, which can be used as a partial substitute for inorganic fertilizers, provided care is taken to properly compost the material or use plants prior to flowering. The lignocellulose-rich biomass of the weed can also be exploited for biofuel and biogas production, or for a low-cost substrate for the production of cellulose and for the pulp and paper industry. The potential of parthenium weed for phyto-extraction of heavy metals and pollutants, biochar preparation and soil amendment also suggests other ways of using the plant. However, in nanotechnology, the evidence of a comparative advantage of parthenium weed extracts over the reducing power of extracts of less controversial species is yet to be presented.

The utilization potential and abuse potential of parthenium weed pose dilemmas that should be resolved. Reporting from the Botanical Survey of India, Singh and Garg (2014) suggested that parthenium weed has been 'a victim of ignorance and misconceptions' (p. 1260) and that the emerging vista of applications 'may endorse the species as a fast growing medicinal herb with immense multifarious relevance, out numbering the harmful properties' (2014, p. 1261). In our view, the above suggestion is yet to be proven. The premise that utilization should be considered as a new strategy for controlling this unmanageable weed (Saini et al., 2014) has to be balanced with the established evidence of negative effects of the species on the welfare of societies, environment and agriculture.

Given the strongly negative, known social impacts and the potential to cause even larger-scale economic damage in cropping and non-agricultural situations, we agree with Marwat et al. (2008, p. 1940), who pointed out from Pakistan 'all efforts should be made to restrict its further spread and eliminate it in a planned way by declaring it as a noxious weed under the Seed Act'. In our view, the focus on parthenium weed should be on containment and reduction of its abundance, possibly leading to its eradication, instead of commercial exploitation. This would be particularly true for all non-native areas and countries to which the weed has spread. Outside the field of traditional medicine, the economic potential of parthenium weed and practical uses remain largely exploratory, descriptive and speculative. While some of the laboratory findings, such as the antidiabetic and anticancer activity of the extracts, are compelling, we find most other research unconvincing with regard to future practical value for uses, until further,

systematic evaluations are carried out. Any parthenium weed uses promoted can only be acceptable on the grounds that they will not contribute to further spread of the weed, and perhaps might be of short-term value to its overall management in a given region or area. Finally, our assessment is that the risks of spread and the overall negative economic, social and environmental impacts of parthenium weed may far outweigh its potential benefits, particularly in regions susceptible to new invasions. Therefore, a precautionary approach is suggested in promoting utilization of parthenium weed, not least because failure to do so would undermine the current efforts to contain its spread across many countries in Asia, South-east Asia, Africa and Australia.

Acknowledgements

The authors wish to thank the editors – Drs Steve Adkins, Asad Shabbir and K. Dhileepan for the invitation to write this chapter and for their editorial assistance. They also provided several articles on parthenium weed, which we were able to review and include in the chapter. Additional information was kindly provided by Sushilkumar (ICAR-Directorate of Weed Research, Jabalpur, Madhya Pradesh, India), R.K. Ghosh (Bidhan Chandra Krishi Viswavidyalaya, Kalyani, West Bengal, India) and several other colleagues who helped improve our review. Their assistance is also gratefully acknowledged.

References

Adkins, S. and Shabbir, A. (2014) Biology, ecology and management of the invasive parthenium weed (*Parthenium hysterophorus* L.). *Pest Management Science* 70, 1023–1029.

Ahmad, A. and Al-Othman, A.A.S. (2014) Remediation rates and translocation of heavy metals from contaminated soil through *Parthenium hysterophorus*. *Chemistry and Ecology* 30, 317–327.

Ahmad, N., Fazal, H., Abbasi, B.H. and Iqbal, M. (2011) *In vitro* larvicidal potential against

Anopheles stephensi and antioxidative enzyme activities of *Ginkgo biloba, Stevia rebaudiana* and *Parthenium hysterophorus. Asian Pacific Journal of Tropical Medicine* 4, 169–175.

Ahmad, K., Shaheen, M., Khan, Z.I. and Bashir, H. (2013) Heavy metals contamination of soil and fodder: a possible risk to livestock. *Science Technology and Development* 32, 140–148.

Ajmal, M., Rao, R.A.K., Ahmad, R. and Khan, M.A. (2006) Adsorption studies on *Parthenium hysterophorus*: removal and recovery of Cd(II) from wastewater. *Journal of Hazardous Materials B* 135, 242–248.

Ali, N. and Hadi, F. (2015) Phytoremediation of cadmium improved with the high production of endogenous phenolics and free proline contents in *Parthenium hysterophorus* plant treated exogenously with plant growth regulator and chelating agent. *Environmental Science & Pollution Research* 22, 13305–13318.

Aneja, K.R., Dhawan, S.R. and Sharma, A.B. (1991) Deadly weed *Parthenium hysterophorus* L. and its distribution. *Indian Journal of Weed Science* 23(3–4), 14–18.

Bailey, L.H. (1960) *Manual of Cultivated Plants,* Macmillan, New York.

Bapat, S.A. and Jaspal, D.K. (2016) *Parthenium hysterophorus*: novel adsorbent for the removal of heavy metals and dyes. *Global Journal of Environmental Science and Management* 2, 135–144.

Batish, D.R., Singh, H.P., Saxena, D.B. and Kohil, R.K. (2002a) Weed suppressing ability of parthenin: A sesquiterpene lactone from *Parthenium hysterophorus. New Zealand Plant Protection* 55, 218–221.

Batish, D.R., Singh, H.P., Kholi, R.K., Saxena, D.B. and Kaur, S. (2002b) Allelopathic effects of parthenin against two weedy species, *Avena fatua* and *Bidens pilosa. Environmental and Experimental Botany* 47, 149–155.

Batish, D.R., Singh, H.P., Kohli, R.K., Kaur, S., Saxena, D.B. and Yadava, S. (2007) Assessment of phytotoxicity of parthenin. *Zeitschrift für Naturforschung* 62C, 367–372.

Belz, R.G. (2008) Stimulation versus inhibition – bioactivity of parthenin, a phytochemical from *Parthenium hysterophorus* L. *Dose Response* 6, 80–96.

Belz, R.G., Reinhard, C.F., Foxcroft, L.C. and Hurle, K. (2007) Residue allelopathy in *Parthenium hysterophorus* L. Does parthenin play a leading role? *Crop Protection* 26, 237–245.

Belz, R.G., van der Laan, M., Reinhardt, C.F. and Hurle, K. (2009) Soil degradation of parthenin – does it contradict the role of allelopathy in the invasive weed *Parthenium*

hysterophorus? Journal of Chemical Ecology 35, 1137–1150.

Bharadwaja, S.T.P., Singh, S. and Moholkar, V.S. (2015) Design and optimization of a sono-hybrid process for bioethanol production from *Parthenium hysterophorus. Journal of Taiwan Institute of Chemical Engineers* 51, 71–78.

Bhattacharyya, A., Adhikary, S., Roy, S. and Goswami, A. (2007) Insecticidal efficacy of *Parthenium hysterophorus* on *Lipaphis erysimi* – a field study. *Journal of Ecotoxicology & Environmental Monitoring* 17, 113–118.

Biradar, A.P. and Patil, M.B. (2001) Studies on utilization of prominent weeds for vermi-culturing. *Indian Journal of Weed Science* 33(3–4), 229–230.

Caraballo, A., Caraballo, B. and Rodriguez-Acosta, A. (2004) Preliminary assessment of medicinal plants used as anti-malarials in the southeastern Venezuelan Amazon. *Revista da Sociedade Brasileira de Medicina Tropical* 37, 186–188.

Channappagoudar, B.B., Biradar, N.R., Patil, J.B. and Gasimani, C.A.A. (2007) Utilization of weed biomass as an organic source in sorghum. *Karnataka Journal of Agricultural Science* 20, 245–248.

Chen, Y., Wang, J., Wu, X., Sun, J. and Yang. N. (2011) Allelopathic effects of *Parthenium hysterophorus* L. volatiles and its chemical components. *Allelopathy Journal* 27, 217–223.

Chhabra, B.R., Kohli, J.C. and Dhillon, R.S. (1999) Three ambrosanolides from *Parthenium hysterophorus. Phytochemistry* 52, 1331–1334.

Dale, I.J. (1981) Parthenium weed in the Americas. *Australian Weeds* 1, 8–14.

Das, B., Venkataiah, B. and Kashinatham, A. (1999) Chemical and biochemical modifications of parthenin. *Tetrahedron* 55, 6585–6594.

Das, B., Reddy, V.S., Krishnaiah, M., Sharma, A.V.S., Kumar, R., *et al.* (2007) Acetylated pseudoguananolides from *Parthenium hysterophorus* and their cytotoxic activity. *Phytochemistry* 68, 2029–2034.

Datta, S. and Saxena, D.B. (2001) Pesticidal properties of parthenin (from *Parthenium hysterophorus* L.) and related compounds. *Pest Management Science* 57, 95–101.

de la Fuente, J.R., Uriburu, M.L., Burton, G. and Sosa, V.E. (2000) Sesquiterpene lactone variability in *Parthenium hysterophorus* L. *Phytochemistry* 55, 769–772.

de Miranda, C.A.S.F., das G. Cardoso, M., de Carvalho, M.L.M., Figueiredo, A.C.S., Nelson, D.L., *et al.* (2014) Chemical composition and allelopathic activity of *Parthenium hysterophorus* and *Ambrosia polystachya* weeds essential oils. *American Journal of Plant Sciences* 5, 1248–1257.

Dominguez, X.A. and Sierra, A. (1970) Isolation of a new diterpene alcohol and parthenin from *Parthenium hysterophorus*. *Planta Medica* 18, 275–277.

DWSR (2010) Compost making from *Parthenium*. Technical Extension Bulletin. Directorate of Weed Science Research, Jabalpur, India.

EPPO (2014) Pest risk analysis for *Parthenium hysterophorus*. European and Mediterranean Plant Protection Organization, Paris. Available at: https://www.eppo.int/QUARANTINE/Pest_ Risk_Analysis/PRAdocs_plants/15-21049_ PRA_record_Parthenium_hysterophorus.pdf (accessed 10 February 2016).

Evans, H.C. (1997) *Parthenium hysterophorus*: A review of its weed status and the possibilities for biological control. *Biocontrol News and Information* 18, 389–398.

Fazal, H., Ahmad, N., Ullah, I., Inayat, H., Khan, L., *et al.* (2011) Antibacterial potential in *Parthenium hysterophorus*, *Stevia rebaudiana* and *Ginkgo biloba*. *Pakistan Journal of Botany* 43, 1307–1313.

Ghosh, S., Haldar, S., Ganguly, A. and Chatterjee, P.K. (2013) Review on *Parthenium hysterophorus* as a potential energy source. *Renewable and Sustainable Energy Review* 20, 420–429.

Gunaseelan, V.N. (1987) Parthenium as an additive with cattle manure in biogas production. *Biological Wastes* 21, 195–202.

Gunaseelan, V.N. (1994) Methane production from *Parthenium hysterophorus* L., a terrestrial weed, in semi-continuous fermenters. *Biomass and Bioenergy* 6, 391–398.

Gunaseelan, V.N. and Lakshmanaperumalsamy, P. (1990) Biogas production potential of parthenium. *Biological Wastes* 33, 311–314.

Gurib-Fakim, A., Swerab, M.D., Gueho, J. and Dullo, E. (1993) Medical ethnobotany of some weeds in Mauritius and Rodrigues. *Journal of Ethnopharmacology* 39, 175–185.

Hadi, F. and Bano, A. (2009) Utilization of *Parthenium hysterophorus* for the remediation of lead-contaminated soil. *Weed Biology and Management* 9, 307–314.

Herz, W. and Watanabe, H. (1959) Parthenin, a new guaianolide. *Journal of American Chemical Society* 81(22), 6088–6089.

Hooper, M., Kirby, G.C., Kulkarni, M.M., Kulkarni, S.N., Nagasampagi, B.A., O'Neill, M.J., *et al.* (1990) Antimalarial activity of parthenin and its derivatives. *European Journal of Medicinal Chemistry* 25, 717–723.

Kanchan, S.D. and Jayachandra, J. (1979) Allelopathic effects of *Parthenium hysterophorus* L. I.

Exudation of inhibitors through roots. *Plant and Soil* 53, 27–35.

Kanchan, S.D. and Jayachandra, J. (1980a) Allelopathic effects of *Parthenium hysterophorus* L. Part II. Leaching of inhibitors from aerial vegetative parts. *Plant and Soil* 55, 61–66.

Kanchan, S.D. and Jayachandra, J. (1980b) Allelopathic effects of *Parthenium hysterophorus* L. Part IV. Identification of inhibitors. *Plant and Soil* 55, 67–75.

Kanchan, S. and Jayachandra, J. (1980c) Pollen allelopathy – a new phenomenon. *New Phytologist* 84, 739–746.

Kanchan, S. and Jayachandra, J. (1981) Effects of *Parthenium hysterophorus* on nitrogen-fixing and nitrifying bacteria. *Canadian Journal of Botany* 59, 199–202.

Knox, J., Jaggi, D. and Paul, M.S. (2011) Population dynamics of *Parthenium hysterophorus* (Asteraceae) and its biological suppression through *Cassia occidentalis* (Caesalpiniaceae). *Turkish Journal of Botany* 35, 111–119.

Kohli, R.K., Rani, D., Singh, H.P. and Kumar, S. (1996) Response of crop seeds towards the leaf leachates of *Parthenium hysterophorus* L. *Indian Journal of Weed Science* 28, 104–106.

Kumamoto, J., Scora, R.W. and Clerx, W.A. (1985) Composition of leaf oils in the genus *Parthenium* L., Compositae. *Journal of Agricultural and Food Chemistry* 33, 650–652.

Kumar, N. and Pruthi, V. (2015) Structural elucidation and molecular docking of ferulic acid from *Parthenium hysterophorus* possessing COX-2 inhibition activity. *Biotech* 5, 541–551.

Kumar, S., Singh, A.P., Nair, G., Batra, S., Seth, A., *et al.* (2011) Impact of *Parthenium hysterophorus* leaf extracts on the fecundity, fertility and behavioural response of *Aedes aegypti* L. *Parasitology Research* 108, 853–859.

Kumar, S., Nair, G., Singh, A.P., Batra, S., Wahab, N. and Warikoo, R. (2012) Evaluation of the larvicidal efficiency of stem, roots and leaves of the weed, *Parthenium hysterophorus* (Family: Asteraceae) against *Aedes aegypti* L. *Asian Pacific Journal of Tropical Disease* 2, 395–400.

Kumar, S., Chashoo, G., Saxena, A.K. and Pandey, A.K. (2013a) *Parthenium hysterophorus*: A probable source of anticancer, antioxidant and anti-HIV agents. *BioMed Research International*, Article ID 810734 (http://dx.doi.org/10.1155/2013/810734).

Kumar, S., Mishra, A. and Pandey, A.K. (2013b) Antioxidant mediated protective effect of *Parthenium hysterophorus* against oxidative damage using *in vitro* models. *BMC Complementary and Alternative Medicine* 13, 120.

Kumar, S., Masto, R.E., Ram, L.C., Pinaki, S., Joshy, G. and Selvi, V.A. (2013c) Biochar preparation from *Parthenium hysterophorus* and its potential use in soil application. *Ecological Engineering* 55, 67–72.

Kumar, S., Pandey, S. and Pandey, A.K. (2014) *In vitro* antibacteria, antioxidant and cytotoxic activities of *Parthenium hysterophorus* and characterization of extracts by LC-MS analysis. *BioMed Research International*, Article ID 495194 (http://dx.doi.org/10.1155/2014/495154).

Kushwaha, V.B. and Maurya, S. (2012) Biological utilities of *Parthenium hysterophorus*. *Journal of Applied and Natural Science* 4, 137–143.

Lata, H., Garg, V.K. and Gupta, R.K. (2007) Removal of a basic dye from aqueous solution by adsorption using *Parthenium hysterophorus*: An agricultural waste. *Dyes and Pigments* 74, 653–658.

Lata, H., Garg, V.K. and Gupta, R.K. (2008a) Adsorptive removal of basic dye by chemically activated parthenium biomass: Equilibrium and kinetic modeling. *Desalination* 219(1/3), 250–261.

Lata, H., Garg, V.K. and Gupta, R.K. (2008b) Sequestration of nickel from aqueous solution onto activated carbon prepared from *Parthenium hysterophorus* L. *Journal of Hazardous Materials* 157, 503–509.

Lata, H., Mor, S., Garg, V.K. and Gupta, R.K. (2008c) Removal of a dye from simulated wastewater by adsorption using treated parthenium biomass. *Journal of Hazardous Materials* 153, 213–220.

Lindley, J. (1838) *Flora Medica* (Indian Reprint, 1985). Ajay Book Service, New Delhi.

Madan, H., Gogia, S. and Sharma, S. (2011) Antimicrobial and spermicidal activities of *Parthenium hysterophorus* Linn. and *Alstonia scholaris* Linn. *Indian Journal of Natural Products and Resources* 2, 458–463.

Marwat, K.B., Khan, M.A., Nawaz, A. and Anees, A.A. (2008) *Parthenium hysterophorus* L. a potential source of bioherbicide. *Pakistan Journal of Botany* 40, 1933–1942.

Mew, D., Balza, F., Towers, G.H.N. and Levy, I.G. (1982) Anti-tumour effects of the sesquiterpene lactone parthenin. *Planta Medica* 45, 23–27.

Mondal, N.K., Chowdhury, A., Dey, U., Mukhopadhya, P., Chatterjee, S., *et al.* (2014) Green synthesis of silver nanoparticles and its application for mosquito control. *Asian Pacific Journal of Tropical Disease* 4 (Suppl. 1), S204–S210.

Nair, R. and Chanda, S. (2006) Activity of some medicinal plants against certain pathogenic bacterial strains. *Indian Journal of Pharmacology* 38, 142–144.

Naithani, S., Chhetri, R.B., Pande, P.K. and Naithani, G. (2008) Evaluation of *Parthenium* for pulp and paper making. *Indian Journal of Weed Science* 40(3–4) Supplementary, 188–191.

Nyasembe, V.O., Teal, P.E.A., Mukabana, W.R., Tumlinson, J.H. and Torto, B. (2012) Behavioural response of the malaria vector *Anopheles gambiae* to host plant volatiles and synthetic blends. *Parasites & Vectors* 5, 234 (http://dx.doi.org/10.1186/1756-3305-5-234).

Nyasembe, V.O., Cheseto, X., Kaplan, F., Foster, W.A., Teal, P.E.A., Tumlinson, J.H. *et al.* (2015) The invasive American weed *Parthenium hysterophorus* can negatively impact malaria control in Africa. *PLoS One* 10(9), e0137836 (http://dx.doi.org/10.1371/journal.pone.0137836).

Pandey, D.K. (1994a) Inhibition of salvinia (*Salvinia molesta* Mitchell) by parthenium (*Parthenium hysterophorus* L.). I. Effect of leaf residue and allelochemicals. *Journal of Chemical Ecology* 20, 3111–3122.

Pandey, D.K. (1994b) Inhibition of salvinia (*Salvinia molesta* Mitchell) by parthenium (*Parthenium hysterophorus* L.). II. Relative effect of flower, leaf, stem and root residue on salvinia and paddy. *Journal of Chemical Ecology* 20, 3123–3131.

Pandey, D.K. (1996a) Phytotoxicity of sesquiterpene lactone parthenin on aquatic weeds. *Journal of Chemical Ecology* 22, 151–160.

Pandey, D.K. (1996b) Relative toxicity of allelochemicals to aquatic weeds. *Allelopathy Journal* 3, 240–246.

Pandey, D.K. (2009) Allelochemicals in parthenium in response to biological activity and the environment. *Indian Journal of Weed Science* 41, 111–123.

Pandey, D.K., Kauraw, L.P. and Bhan, V.M. (1993a) Inhibitory effect of parthenium (*Parthenium hysterophorus* L.) residue on growth of water hyacinth (*Eichhornia crassipes* (Mart.) Solms.). I. Effect of leaf residue. *Journal of Chemical Ecology* 19, 2651–2662.

Pandey, D.K., Kauraw, L.P. and Bhan, V.M. (1993b) Inhibitory effect of parthenium (*Parthenium hysterophorus* L.) residue on growth of water hyacinth (*Eichhornia crassipes* Mart Solms.). II. Relative effect of flower, leaf, stem and root residue. *Journal of Chemical Ecology* 19, 2663–2670.

Pandey, D.K., Pani, L.M.S. and Joshi, S.C. (2003) Growth, reproduction and photosynthesis of ragweed parthenium (*Parthenium hysterophorus* L.). *Weed Science* 51, 191–201.

Pandiyan, K., Tiwari, R., Rana, S., Arora, A., Singh, S., *et al.* (2014) Comparative efficiency of different pre-treatment methods on enzymatic digestibility of *Parthenium* sp. *World Journal of Microbiology & Biotechnology* 30, 55–64.

Patel, S. (2011) Harmful and beneficial aspects of *Parthenium hysterophorus*: An update. *3 Biotech* 1, 1–9.

Patel, V.S., Chitra, V.P., Prasanna, L. and Krishnaraju, V. (2008) Hypoglycemic effect of aqueous extract of *Parthenium hysterophorus* L. in normal and alloxan induced diabetic rats. *Indian Journal of Pharmacology* 40, 183–185.

Patil, T.M. and Hegde, B.A. (1988) Isolation and purification of a sesquiterpene lactone from the leaves of *Parthenium hysterophorus* L. – its allelopathic and cytotoxic effects. *Current Science* 57, 1178–1181.

Pedroso, A.T.R., Arrebato, M.A.R., Ravieso, R.M.C., Gonzalez, D.R., Triana, A.C. and Baños, S.B. (2012) *In vitro* and *in vivo* antifungal activity of the aqueous extract of *Parthenium hysterophorus* L. against *Pyricularia grisea* Sacc. [in Spanish]. *Revista Científica UDO Agrícola* 12, 839–844.

Picman, J. and Picman, A.K. (1984) Autotoxicity in *Parthenium hysterophorus* and its possible role in control of germination. *Biochemical Systematics and Ecology* 12, 287–292.

Picman, A.K. and Towers, G.H.N. (1982) Sesquiterpene lactones in various populations of *Parthenium hysterophorus*. *Biochemical Systematics and Ecology* 10, 145–153.

Picman, A.K., Rodriguez, E. and Towers, G.H.N. (1979) Formation of adducts of parthenin and related sesquiterpene lactones with cysteine and glutathione. *Chemical and Biological Interactions* 28, 83–89.

Picman, A.K., Towers, G.H.N. and Subba Rao, P.V. (1980) Coronopilin – another major sesquiterpene lactone in *Parthenium hysterophorus*. *Phytochemistry* 19, 2206–2207.

Picman, A.K., Balza, F. and Towers, G.H.N. (1982) Occurrence of hysterin and di-hydroisoparthenin in *Parthenium hysterophorus*. *Phytochemistry* 21, 1801–1802.

Rajiv, P., Sivaraj, R. and Rajendran, V. (2013a) Biofabrication of zinc oxide nanoparticles using leaf extract of *Parthenium hysterophorus* L. and its size-dependent antifungal activity against plant fungal pathogens. *Spectrochimica Acta Part A: Molecular and Biomolecular Spectroscopy* 112, 384–387.

Rajiv, P., Rajeshwari, S., Hiranmai Yadav, R. and Rajendran, V. (2013b) Vermiremediation: Detoxification of parthenin toxin from *Parthenium* weed. *Journal of Hazardous Materials* 262, 489–495.

Raju, R.A. and Gangwar, B. (2004) Utilization of potassium-rich green-leaf manures for rice (*Oryza sativa*) nursery and their effect on crop productivity. *Indian Journal of Agronomy* 49, 244–247.

Ramesh, C., Ravindranath, N., Das, B., Prabhakar, A., Bharatam, J., *et al.* (2003) Pseudoguaino-lides from the flowers of *Parthenium hysterophorus*. *Phytochemistry* 64, 841–844.

Ramos, A., Rivero, R., Visozo, A., Piloto, J. and Garcia, A. (2002) Parthenin, a sesquiterpene lactone of *Parthenium hysterophorus* L. is a high toxicity clastogen. *Mutation Research/Genetic Toxicology and Environmental Mutagenesis* 514(1/2), 19–27.

Rana, S., Tiwari, R., Arora, A., Singh, S., Kaushik, R., Saxena, A.K. *et al.* (2013) Prospecting *Parthenium* sp. pretreated with *Trametes hirsuta*, as a potential bioethanol feedstock. *Biocatalysis and Agricultural Biotechnology* 2, 152–158.

Ranjani, S. and Sakthivel, P. (2013) Biosynthesis and characterization of silver nanoparticles using *Parthenium hysterophorous* leaves. *Asian Journal of Chemistry* 25(Suppl.), S340–S342.

Rao, R.S. (1956). Parthenium: A new record for India. *Journal of Bombay Natural History Society* 54, 218–220.

Rehman, A., Iqbal, T., Ayaz, S. and Rehman, H.U. (2013) Investigations of heavy metals in different medicinal plants. *Journal of Applied Pharmaceutical Science* 3, 72–74.

Reinhardt, C., Kraus, S., Walker, F., Foxcroft, L., Robbertse, P. and Hurle, K. (2004) The allelochemical parthenin is sequestered at high level in capitate-sessile trichomes on leaf surfaces of *Parthenium hysterophorus*. *Journal of Plant Diseases and Protection* 19, 253–261.

Reinhardt, C., Van der Laan, M., Belz, R.G., Hurle, K. and Foxcroft, L. (2006) Production dynamics of the allelochemical parthenin in leaves of *Parthenium hysterophorus* L. *Journal of Plant Diseases and Protection* Special Issue XX, 427–433.

Rodriguez, E., Dillon, M.O., Mabry, T.J., Mitchell, J.C. and Towers, G.H.N. (1975) Dermatologically active sesquiterpene lactones in trichomes of *Parthenium hysterophorus* L. (Compositae). *Experientia* 32, 236–238.

Rodriguez, E., Towers, G.H.N. and Mitchell, J.C. (1976) Biological activities of sesquiterpene lactones. *Phytochemistry* 15, 1573–1580.

Saini, A., Aggarwal, N.K., Sharma, A., Kaur, M. and Yadav, A. (2014) Utility potential of *Parthenium hysterophorus* for its strategic management. *Advances in Agriculture* Vol 2014 Article ID 381859 (http://dx.doi.org/10.1155/2014/381859).

Sanghamitra, K., Prasada Rao, P.V.V. and Naidu, G.R.K. (2012) Uptake of Zn (II) by an invasive weed species *Parthenium hysterophorus* L. *Applied Ecology and Environmental Research* 10, 267–290.

Sanvicens, N. and Marco, M. (2008) Multifunctional nanoparticles – properties and prospects for their use in human medicine. *Trends in Biotechnology* 26, 425–433.

Saravanane, P., Nanjappa, H.V., Ramachandrappa, B.K. and Soumya, T.M. (2011) Effect of residual fertility of preceding potato crop on yield and nutrient uptake of finger millet. *Karnataka Journal of Agricultural Science* 24, 234–236.

Saravanane, P., Poonguzhalan, R. and Chellamuthu, V. (2012) Parthenium (*Parthenium hysterophorus* L.) distribution and its bio-resource potential for rice production in Puducherry, India. *Pakistan Journal of Weed Science Research* 18, 551–555.

Saxena, D.B., Dureja, P., Kumar, B., Daizy, R. and Kohli, R.K. (1991) Modification of parthenin. *Indian Journal of Chemistry* 30B, 849–852.

Sethi, V.K., Koul, S.K., Taneja, S.C. and Dhar, K.L. (1987) Minor sesquiterpenes of flowers of *Parthenium hysterophorus*. *Phytochemistry* 26, 3359–3361.

Sharma, G.L. and Bhutani, K.K. (1988) Plant based anti-amoebic drugs. Part II. Amoebicidal activity of parthenin isolated from *Parthenium hysterophorus*. *Planta Medica* 54, 20–22.

Sharma, V., Pandher, J.K. and Kamla, K. (2008) Bio management of lantana (*Lantana camara* L.) and congress grass (*Parthenium hysterophorus* L.) through vermicomposting and its response on soil fertility. *Indian Journal of Agricultural Research* 42, 283–287.

Shen, M.C., Rodriguez, E., Kerr, K.T.J. and Marby, K. (1976) Flavonoids of four species of Parthenium (Compositae). *Phytochemistry* 15, 1045–1047.

Shrivastava, V.S. (2010) The biosorption of safranine onto *Parthenium hysterophorus* L: Equilibrium and kinetics investigation. *Desalination and Water Treatment* 22(1–3), 146–155.

Singh, R.K. and Garg, A. (2014) *Parthenium hysterophorus* L. – neither noxious nor obnoxious weed. *Indian Forester* 140, 1260–1262.

Singh, R.K., Kumar, S., Kumar, S. and Kumar, A. (2008a) Development of parthenium-based activated carbon and its utilization for adsorptive removal of p-cresol from aqueous solution. *Journal of Hazardous Materials* 155, 523–535.

Singh, R.S., Singh, V.K., Mishra, A.K., Tiwari, P.N., Singh, U.N. and Sharma, Y.C. (2008b) *Parthenium hysterophorus*: A novel adsorbent to remove Cr (VI) from aqueous solutions.

Journal of Applied Sciences in Environmental Sanitation 3, 177–189.

Singh, R.S., Singh, V.K., Tiwari, P.N., Singh, J.K. and Sharma, Y.C. (2009) Biosorption studies of nickel on *Parthenium hysterophorus* ash. *Environmental Technology* 30, 355–364.

Singh, S., Khanna, S., Moholkar, V. and Goval, A. (2014) Screening and optimization of pretreatments for *Parthenium hysterophorus* as feedstock for alcoholic biofuels. *Applied Energy* 129, 195–206.

Singh, S., Agarwal, M., Bhatt, A. Goyal, A. and Moholkar, V.S. (2015) Ultrasound enhanced enzymatic hydrolysis of *Parthenium hysterophorus*: a mechanistic investigation. *Bioresource Technology* 192, 636–645.

Singh, U., Wadhwani, A.M. and Johri, B.M. (1996) *Dictionary of Economic Plants in India*. Indian Council of Agricultural Research, New Delhi.

Sivakumar, S., Kasthuri, H., Prabha, D., Senthilkumar, P., Subbhuraam, C.V. and Song, C. (2009) Efficiency of composting parthenium plant and neem leaves in the presence and absence of an oliogochaete, *Eisenia fetida*. *Iran Journal of Environmental Health, Science and Engineering* 6, 201–208.

Sivaraj, R., Venckatesh, R. and Gunalan, G.S. (2010) Preparation and characterization of activated carbons from *Parthenium* biomass by physical and chemical activation techniques. *E-Journal of Chemistry* 7, 1314–1319.

Sushilkumar and Varshney, J.G. (2010) Parthenium infestation and its estimated cost management in India. *Indian Journal of Weed Science* 42(1&2), 73–77.

Tavva, S.S.M.D., Deshpande, A., Durbha, S.R., Palakollu, V.A.R., Goparaju, A.U., *et al.* (2016) Bioethanol production through separate hydrolysis and fermentation of *Parthenium hysterophorus* biomass. *Renewable Energy* 86, 1317–1323.

Towers, G.H.N. and Subba Rao, P.V. (1992) Impact of the pantropical weed *Parthenium hysterophorus* L. on human affairs. In: Richardson, R.G. (ed.), *Proceedings of the 1st International Weed Control Congress*, Melbourne, Australia. Weed Science Society of Victoria, Melbourne, Australia, pp. 135–138.

Valdés, A.F.C., Martínez, J.M., Lizama, R.S., Gaitén, Y.G., Rodríguez, D.A. and Payrol, J.A. (2010) *In vitro* antimalarial activity and cytotoxicity of some selected Cuban medicinal plants. *Revista do Instituto de Medicina Tropical de Sao Paulo* 52, 197–201.

Varshney, V.K. and Naithani, S. (2011) Chemical functionalization of cellulose derived from non-conventional sources. In: Kalia, S., Kaith, B.S.

and Kaur, I. (eds), *Cellulose Fibers: Bio- and Nano-Polymer Composites*. Springer, Berlin, pp. 43–60.

Venkataiah, B., Ramesh, C., Ravindranath, N. and Das, B. (2003) Charminarone, a seco-seudoguaianolide from *Parthenium hysterophorus*. *Phytochemistry* 63, 383–386.

Wickham, K., Rodriguez, E. and Arditti, J. (1980) Comparative phytochemistry of *Parthenium hysterophorus* L. (Compositae tissue cultures). *Botanical Gazette* 141, 435–439.

Yadav, A. and Garg, V.K. (2011) Vermicomposting – an effective tool for the management of invasive weed *Parthenium hysterophorus*. *Bioresource Technology* 102, 5891–5895.

Yadava, R.N. and Khan, S. (2013) Isolation and characterisation of a new allelochemical from *Parthenium hysterophorus* L. *International Journal of Pharmaceutical Sciences and Research* 4, 311–315.

11

History and Management – Australia and Pacific

Rachel McFadyen,[1]* Kunjithapatham Dhileepan[2] and Michael Day[2]

[1]PO Box 88, Mt Ommaney, Queensland, Australia; [2]Biosecurity Queensland, Department of Agriculture and Fisheries, Brisbane, Queensland, Australia

11.1 Introduction

Parthenium weed (*Parthenium hysterophorus* L.) is now recognized as a major invasive weed worldwide. Yet in the 1950s, when it was first discovered in Australia, it was an almost unknown plant. International research on parthenium weed did not start until the 1970s, after reports of increasing health problems caused by the dense infestations in central India (Chandras and Vartak, 1970). Australian policy makers in 1973–1975 were therefore working in an information vacuum when trying to manage this new weed, which was rapidly spreading south from the northern cattle zone. Their response was to establish one of the largest long-term and well-funded weed management programmes ever seen against a single weed, and the outcome has been startlingly successful, with parthenium weed ceasing to be one of the top ten weeds of both cropping and grazing lands in the affected zone. This chapter outlines the history and background of parthenium in Australia and the management tools used to produce this success.

11.2 History

11.2.1 Initial introduction

Parthenium weed was first recorded in the scientific literature as coming from Toogooloowah, north-west of Brisbane in 1955 (Fig. 11.1; Table 11.1; Everist, 1976; Haseler, 1976). However, an earlier specimen, collected from Toogooloowah in 1950, is held in the Queensland Herbarium. This infestation, probably from seed brought in from the USA on machinery during the Second World War, did not spread significantly. A second infestation in 1960 resulted from contaminated pasture seed imported from Texas, USA, into the pastoral property 'Elgin Downs', north of Clermont in central Queensland (Armstrong, 1978). Subsequent research has demonstrated that these infestations represent two different biotypes, with the second Clermont biotype being the highly invasive one (Navie *et al.*, 1996b).

At the time, the pastoral country of north and central Queensland was largely undeveloped, with poor roads and a sparse population. Rainfall averages 500–700 mm/

* rachel.mcfadyen@live.com.au

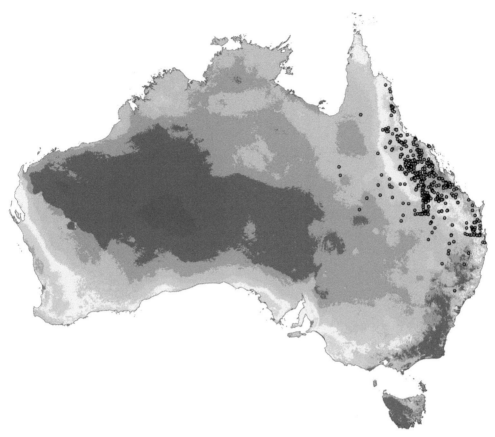

Fig. 11.1. Prediction of parthenium weed habitat suitability within Australia using a generalized additive model (Mainali *et al.*, 2015) with occurrence data obtained from Australian Virtual Herbarium (http://avh. ala.org.au/occurrences/search?taxa =parthenium+hysterophorus#tab_mapView) and from Dhileepan and McFadyen (2012).

year, falling mainly in January to March (summer), with very little rain in the rest of the year. Temperatures are high in summer, with daily maxima exceeding 40°C for several days each year, but frosts may occur in winter. Annual rainfall is highest near the coast and is very variable from year to year (Anonymous, 2014). Development of grazing lands in this region increased from the 1960s, using large bulldozers to clear trees, and planting with improved tropical pastures developed by the Commonwealth Science and Industry Research Organisation (CSIRO) at their tropical research stations in Townsville and Rockhampton (Cook and Dias, 2006). Argentine beef was excluded from the USA in 1959, which greatly increased imports of Australian beef into the USA, and consequently US capital investment in Australian cattle properties. By 1974, King Ranch Development Company (owner R. Kleberg Jr of Texas, founder of the Santa Gertrudis cattle breed) owned 4 million ha in Australia, almost entirely in Queensland. Elgin Downs was one of their properties, consisting of 32,900 ha, of which 10,620 ha were cleared and sown to imported tropical pasture grasses by 1969. The carrying capacity of unimproved country was one beast to 6 ha, which rose to 1 beast to 3 ha after improvement, at a cost of AUS$80 per beast per area (equal to AUS$910 in 2015; Anonymous, 2016; see also Pearson and Lennon, 2010).

It was in this context of the rapid development of huge areas of sparsely populated

Table 11.1. Current status of parthenium weed in Australia and Pacific. (McFadyen, Kunjithapatham and Day.)

Country/ region	Earliest record	Source of new infestations	Current status	References
Queensland	1955	Machinery and pasture seed from USA	Widespread, largely under biological control	Everist (1976), Haseler (1976)
New South Wales	1982	Machinery, vehicles and stock feed from Queensland	Successful eradication programme	Blackmore and Johnson (2010)
Northern Territory	1990	Cattle and vehicles from Queensland	Several small infestations; all eradicated	Navie *et al.* (1996a), Australian Weeds Committee (2012)
Western Australia	2011	Vehicles from Queensland	Single infestation in Pilbara; under eradication	Penna and MacFarlane (2012)
Victoria, South Australia, Tasmania			No occurrences recorded	Australian Weeds Committee (2012)
Cocos Island			No occurrences recorded	Dodd and Reeves (2012)
Christmas Island	2006	Unknown; probably machinery from Queensland	Several small areas; under eradication	Dodd and Reeves (2012)
Papua New Guinea	2001	Vehicles from Queensland	Eradicated; not seen since 2006	Kawi and Orapa (2010)
New Caledonia	1881	Unknown	Present but not major weed	Gargominy *et al.* (1996)
Hawaii		Unknown	Present on several islands	PIER (2013)
Tahiti	2006		Only on Tubuai Island	Florence *et al.* (2007)
Vanuatu	1971	Machinery and vehicles from Queensland	Widespread on Efate and Tanna; eradicated from Aniwa; biological control underway	Vanuatu Forestry Service; S. Bule, Biosecurity Vanuatu, pers. comm. (2014), Day and Bule (2016)

country that the initial spread of parthenium weed occurred. Seed of the African pasture grass buffel grass (*Cenchrus ciliaris* L.), sourced from Texas, was imported and sown in a 3000-ha paddock on Elgin Downs. The imported seed was known to contain a low level of parthenium weed seed, but at the time parthenium weed was not a prohibited weed and therefore there were no quarantine barriers to its importation. Rainfall in Queensland in the 1960s was very low, in the driest 10% of years, but this was followed by exceptionally high rainfall (wettest 10% of years) in the 1973/74 summer (Anonymous, 2014). The presence of parthenium weed was noted in 1964, when the first herbarium specimen from that region was collected, but, as it was not a notifiable weed,

no action was taken. By 1973, after the first good rains for a decade, it was already spreading south 'at a frightening rate' (Everist, 1976). At the end of the 1974/75 summer, it was present as scattered infestations along roads 500 km south of the main infestation (Fig. 11.2), and spread continued at an increasing rate for several years (Auld *et al.*, 1983). In March 1975, F.J. Simmonds, Director of the Commonwealth Institute for Biological Control (CIBC), who lived in India, visited Brisbane and held discussions with W.H. Haseler, Director of the Biological Branch of the Queensland Lands Department, in which he alerted Haseler to the rapidly increasing problem of parthenium weed in India, specifically the problem caused by allergic reactions to the weed (W.H. Haseler

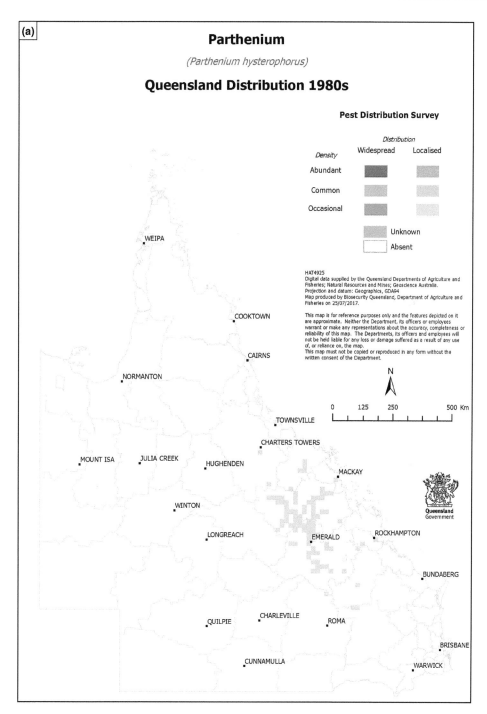

Fig. 11.2. Spread of parthenium weed in Queensland, Australia: from (a) 1980, (b) 2003 to (c) 2013. For 11.2A, black lines refer to survey roads/areas and the green lines refer to roads/area with parthenium infestations. (a: Queensland Department of Agriculture and Fisheries archive. b,c: Pest Info Queensland, https://www.daf.qld.gov.au/plants/weeds-pest-animals-ants/pest-mapping/annual-pest-distribution-maps)

continued

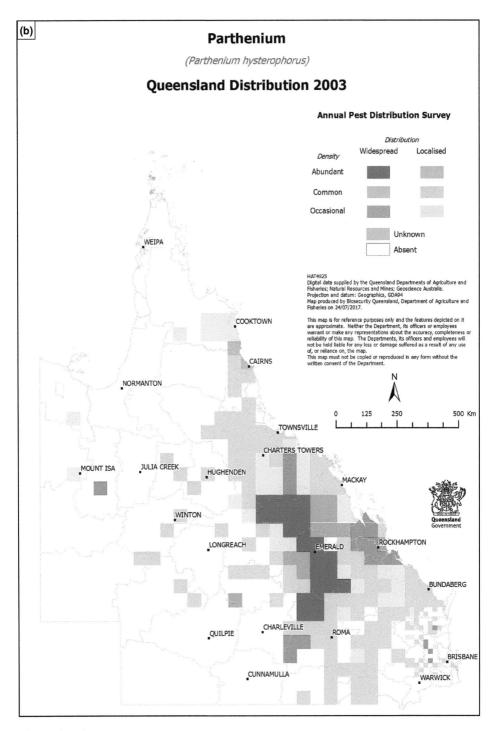

11.2   *continued*

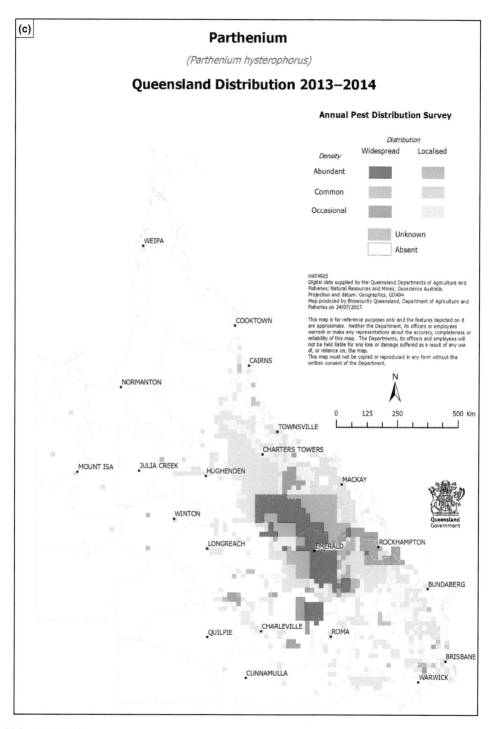

11.2 *continued*

letter, 1 April 1975). Allergic dermatitis among workers in Queensland was recorded by early 1976 (Armstrong, 1978), possibly in individuals already sensitized to *Xanthium* spp., which were common weeds in central and eastern Queensland at the time (Parsons and Cuthbertson, 1992).

In 1975, there was very little published information on parthenium weed as a weed. King (1966), in his book *Weeds of the World*, listed parthenium weed as a minor hayfever plant in the USA, but that is all; there was no indication that it was becoming a problem in India or elsewhere. In the USA, it was recorded as a minor weed in cultivation (Haseler, 1976), and was known to occur in the Caribbean and South America. Reports of the spread of parthenium weed in India had been published (Maheshwari, 1966; Vartak, 1968; Parker, 1968; Chandras and Vartak, 1970; Lonkar *et al.*, 1974), but few of these publications were readily available to Queensland scientists at the time. Consequently, the first response was to undertake a survey to establish the area infested, determine the main spread pathways and undertake control trials using herbicides registered for use in Queensland at the time (Armstrong, 1978). The approximate area of parthenium infestation by 1980 was estimated to be about 29,000 km^2, all restricted to central Queensland (Fig. 11.2a). The areas of parthenium infestation increased drastically over the next two decades, reaching about 520,522 km^2 in summer 2003 (Fig. 11.2b). There has been no significant increase in the area of parthenium infestation since then, and in fact there was a decline in the area infested, to about 368,586 km^2 in 2013 (Fig. 11.2c). The declining trend in the area with parthenium infestation since the mid-1990s was likely to be due to the effectiveness of biological control agents and improved roadside management and grazing management practices.

11.2.2 Legislative and political background

In 1974, Queensland had a long-term Liberal/National Party Government, in power from 1957 to 1989. Agriculture was a key industry in the state; mining was limited in size and impact, and governments saw support for agriculture as essential. More than 80% of Queensland was owned by the state and leased on long-term pastoral leases (McFadyen, 1991). Under the Stock Routes and Rural Lands Protection Acts 1944 (replaced by the Rural Lands Protection Act 1985, by the Land Protection – Pest and Stock Route Management – Act 2002, and by the Biosecurity Act in 2016), 'all landholders are required to destroy all noxious plants on their holdings and to keep the holding free of such plants at their own cost and expense' (Anonymous, 1980). Control of noxious weeds was overseen by the Queensland Lands Department, and research into management of these weeds was undertaken by the Biological Branch of the Lands Department, where W.H. Haseler was Director from 1970 to 1981. However, research into agricultural production, including development of improved pastures, was the responsibility of the Department of Primary Industries (DPI) through research and extension groups based in north and central Queensland. Plant identification was undertaken by the Queensland Herbarium, within the DPI. Pasture research was also undertaken by CSIRO from their laboratories in Townsville and Rockhampton in Queensland.

In practice, control of widespread weeds was not economically feasible on low-value extensive grazing lands, and historically the Queensland Government usually assisted landholders with a variety of incentive payments, as well as providing management recommendations and undertaking local area coordination (Green, 1965). Biological control of widespread weeds of pastoral country was undertaken by the Department of Lands, funded through various Government levies. The enormously successful campaign against the prickly pears (*Opuntia* species) in the 1920s and 1930s (Dodd, 1940) was still remembered by people influential in the Government, resulting in a willingness to provide financial support for biological control programmes.

11.2.3 Management response

From 1973 on, it had become clear that a major spread path for this new weed was along the roads. By the end of 1975, the weed was already east and south of Rolleston, 450 km by road from the original infestation in central Queensland (Everist, 1976). At the time, most roads in north and central Queensland were unsealed earth or gravel roads, maintained by graders and with wide earth shoulders. Vehicles used the shoulders to pass larger trucks or to park, and after rain the road shoulders became infested by dense stands of parthenium weed up to 2 m tall (Fig. 11.3a,b). As the plants dried out, seeds readily became trapped in the radiator mesh, on tyres and in mud on the bodywork, and were then dropped when the vehicles stopped, sometimes hundreds of kilometres away. Initial infestations were often in cattle yards on properties, where cattle trucks had loaded and unloaded, and could often be traced to transport of cattle from the original infested properties in central Queensland (Auld *et al.*, 1983).

The initial response by the Queensland Government was therefore to declare parthenium weed as a 'noxious weed' in October 1975 (Navie *et al.*, 1996a), which brought it under the Stock Routes and Rural Lands Protection Act. Landholders were required to destroy the weed wherever it occurred and to keep their land free of the weed. Free herbicide was given to landholders to control the weed in property roads and cattle yards, and Land Inspectors started a programme of regular property inspections in the affected council (local government) areas. At the same time, a programme was established to control the weed along the thousands of kilometres of affected roads, in order to reduce the continuing spread south and onto new properties. Ten metres either side of affected roads was sprayed two or three times each summer with herbicide. Affected councils were provided with free herbicide and additional financial and labour assistance.

Research trials were established to determine the most cost-effective method of control (Armstrong, 1978) and publicity

Fig. 11.3. Parthenium weed infestations in central Queensland (a, b) in 1981 before implementation of biological control. (Queensland Department of Agriculture and Fisheries archive.)

campaigns undertaken to make landholders aware of the plant and their legal obligations for control. It was quickly determined that the very large seed bank (>44,700 m^2; Joshi, 1991; Navie *et al.*, 1996a) and rapid germination after rain was a key feature of the plant, and the residual herbicide atrazine was widely used for continued control of its seedlings, usually in combination with a knock-down herbicide, usually 2,4-D, during this period. In badly infested areas, three treatments of the roadsides were conducted each year, by a contract operator. In Bely-ando Shire alone, 671 km of road were sprayed each year, 399 km sprayed to 5 m and 272 km to 10 m on the third occasion (W.H. Haseler, May 1978 Memo to Stock Routes Board).

At the same time, an ecologist was appointed to evaluate the impact of grazing cover on parthenium weed persistence (Dale *et al.*, 1978), but unfortunately his work was never completed or published. DPI agrono-mists based in central Queensland under-took trials of grazing management and the use of different pasture species (White, 1994). Studies at the University of Queensland established that parthenium weed was capable of spreading throughout the temperate humid and subhumid areas of Australia (Doley, 1977).

11.2.4 Costs

The management programme established in 1975 was expensive. The on-ground and roadside spraying programme cost AUD$250,000 each year from 1976 onwards (equivalent to AUD$1.5 million in 2015; Anonymous, 2016). Research costs were covered from consolidated revenue and not separately costed. Two staff spent most of their time on herbicide research and AUD$8500 (equivalent to AUD$42,000 in 2015) was budgeted for their material and travel costs (Stock Routes and Rural Lands Board 1978). By comparison, Government expenditure on the on-ground control of harrisia cactus (*Harrisia martinii* L.), another major weed of grazing lands, was AUD$618,000 in 1976 (McFadyen, 1991). In the 1990/91 financial year, Queensland Government expenditure on parthenium control was AUD$390,000 (equivalent to AUD$708,000 in 2015) on on-ground costs plus AUD$350,000 (AUD$636,000) on research (Chippendale and Panetta, 1994).

The programme was initially funded entirely by the Queensland Government, with support from levies from Queensland primary producers through the Rural Lands Protection Fund. At the time, there was no Australia-wide national weed legislation or system of cost-sharing, despite the obvious threat to cropping and grazing country in New South Wales (NSW). The Grains Research and Development Corporation funded research on herbicide trials, basic plant biology and other studies, largely at the University of Queensland. Two Coopera-tive Research Centres (CRCs), set up with Australian Government funding, supported various research and outreach programmes between 1991 (CRC for Tropical Pest Man-agement) and 2008 (CRC for Australian Weed Management).

Only two evaluations of the cost of parthenium weed in Australia have been published. Chippendale and Panetta (1994) estimated the annual cost to the Queensland cattle industry in 1991 to be AUD$16.5 mil-lion (equivalent to AUD$30 million in 2015), which could rise to AUD$109–129 million if spread continued unchecked. Page and Lacey (2006) estimated that the biological control programme cost a total of AUD$11 million, and produced a net positive value of AUD$33 million. Major costs due to parthenium weed were the loss of pasture, especially native pastures, the cost of establishing and using vehicle and machinery wash-down facilities, and medical costs for the approxi-mately 10% of exposed workers who devel-oped allergic reactions to the weed.

11.2.5 Health impacts

Allergic effects on the human population were evident immediately, with hay fever and dermatitis resulting from exposure to the plant reported by 1976 (Haseler, 1976; Armstrong, 1978). In 1991, a survey of workers in parthenium weed-affected areas indicated that 10% had developed allergic reactions to parthenium weed (McFadyen, 1995). Individuals were reported to be sell-ing farms and moving out of the area because of the health impacts on one or more family members. Total population in the major parthenium weed areas of Queensland was only 142,000 in 1995, so the number of people affected was relatively small (McFadyen, 1995). However, increased activity due to mining and gas extraction has resulted in a significant increase in transient workers in the region, many based in the coastal towns of Mackay and Townsville, with the regional population reaching about 400,000 in 2011. Parthenium weed is also spreading south into the Darling Downs regional area, with a human population of 250,000. Nevertheless, reports of allergic problems from parthenium weed have reduced since 2000, probably as a result of the reduced abundance of the weed.

11.2.6 Biological control programme

The first moves for biological control of parthenium were made by the CIBC in India, whose Director F.J. Simmonds was aware of increasing infestations and resulting human health issues around Bangalore in Karnataka State. Preliminary surveys of potential biocontrol agents were undertaken by CIBC staff in Trinidad in 1970 (Bennett and Cruttwell, 1970). A proposal for detailed surveys in the Americas was submitted to the Indian Council of Agricultural Research, but rejected as too expensive (Letter from T. Sankaran, Bangalore, 3 September 1974).

In Queensland, the possibility of biological control was raised as soon as the threat from parthenium weed was recognized. W.H. Haseler, Director of the Biological Branch from 1970 to 1981, was an entomologist who strongly supported biological control. Prospects for the biological control of parthenium weed were discussed with Simmonds on his visit to Brisbane in March 1975 (W.H. Haseler, 1 April 1975 letter), and biological control scientists based in South America and working for the Queensland Lands Department were asked to undertake preliminary surveys of insects attacking parthenium weed (McFadyen, 1979). Funding for preliminary surveys in North America was approved in May 1976 and the surveys undertaken the same (northern) summer (Bennett, 1976).

By 1976, the biological control programme against harrisia cactus was already

giving good results in the field (McFadyen, 2012), so there was political support for a similar programme against parthenium weed. A new quarantine insectary completed in 1975 allowed host-testing to be conducted in Brisbane rather than overseas. As a result, the Queensland Government undertook a major biological control programme against parthenium weed, commencing in 1977 and continuing until 2002. Two full-time scientists were based overseas, one in Mexico from 1978 to 1983, and one in southern Brazil from 1977 to 1980, and from 1978 a third scientist was based in Brisbane with responsibility for the detailed host-testing in quarantine, laboratory rearing and field releases of all suitable agents discovered (Dhileepan and McFadyen, 2012; see Dhileepan et al., Chapter 7, this volume). From 1995 to 1998, Meat and Livestock Australia (then called the Meat and Livestock Development Corporation) funded research on pathogens attacking parthenium weed in Mexico, and also field evaluation studies on the agents already released; their funds came equally from the Australian Government and from levies on sales of meat animals. Between 1980 and 2000, nine insect species and two rusts from North and South America were introduced and established in north and central Queensland (detailed information presented in Chapter 7, this volume).

11.3 National (Australia-wide) Management

11.3.1 National spread

Early studies had demonstrated that parthenium weed was capable of growing through much of NSW and northern Victoria (Doley, 1977). It was declared 'noxious' in NSW in 1978, but was first found in NSW in 1982 on a road north of Coonamble in north-west NSW (Table 11.1), where seeds had come off harvesting machinery originating from central Queensland; a major infestation was found over an area of 250 ha where the harvester had worked (I. Kelly, 1993 Report to

Castlereagh Macquarie County Council; Blackmore and Johnson, 2010). An eradication programme was immediately started (White, 1994). In contrast to the Queensland programme, on-ground eradication treatment was undertaken by Government staff at no cost to the landholder (other than lost production). A 4-year programme on a single property cost AUD\$105,400 (equivalent to AUD\$186,000 in 2015) from 1989 to 1993 (White, 1994). In April–June 1983, M. McMillan from NSW visited Queensland and his report emphasized that harvesting machinery was a particular problem; 'parthenium [in Queensland] was seen flowering in wheat (*Triticum aestivum* L.), sorghum (*Sorghum bicolor* (L.) Moench) and sunflower (*Helianthus annuus* L.) crops'. Movement of stock by road (in trucks) was also a source of spread. He said 'roadside control in Queensland appears to have been very effective in containing parthenium (weed)' and lists shires, such as Banana Shire, which had remained free of parthenium weed despite being adjacent to other heavily infested shires (M. McMillan, NSW Department of Agriculture unpublished report June 1983).

Machinery cleaning facilities were set up on roads crossing from Queensland into NSW, and a system of inspection developed with a permit issued to allow entry into NSW. A seed certification scheme was introduced by Queensland DPI in 1989; this cost growers AUD\$70/t but was widely used. Sale prices of cattle properties in central Queensland infested with parthenium weed were discounted by up to 30% during this period (I. Kelly, 1993 Report).

Subsequently grain from Queensland shipped south (mostly for use as stock fodder) was found to be a major source of new infestations (Blackmore, 2000). A 'Parthenium Weed Taskforce' was established in 1994 to help coordinate a strategic response to the parthenium weed threat (Blackmore, 2000). NSW Department of Agriculture records showed that 13,000 bales of hay were brought from Queensland into NSW in the 1993/94 year and 20,000 in the 1994/95 year, plus two million head of cattle in 1994/95. The Queensland sorghum harvest that year was 670,000 t, much of which was sent to NSW as stock feed. Sunflower seed was also a source; the seed was crushed for oil and the hulls, which would have contained intact parthenium seed, fed to cattle. In the 27 years from 1982, 70% of all parthenium weed infestations discovered in NSW were in the roadside corridor, i.e. probably originated from seed falling from grain trucks or off vehicles. Twenty-four per cent were on private property, and of these, 60% resulted from contaminated grain headers and a further 14% from other machinery, with 20% from cattle feedstock and 6% of unknown origin (Blackmore and Johnson, 2010). Numbers of newly discovered infestations in NSW fluctuated from year to year, ranging from 10 in 1986 to 86 in 1989, which probably reflected weather conditions and the movement of grain and stock feed. Over the two decades to 2004, there were on average 27 new infestations per year (Blackmore and Johnson, 2010).

By 1995, Indonesia, Malaysia and New Zealand had nil tolerance for parthenium weed seed in wheat shipments from Australia. Within Australia, up to 20 parthenium weed seeds were permitted in 500 ml of a grain sample. Wheat for milling would not have resulted in weed spread, but lower-grade wheat used for cattle and poultry feed was a problem.

11.3.2 National management

In the 1970s, there was no Australia-wide system of weed management. Each state and territory was responsible for its own weed management strategy, and had its own list of 'noxious' weeds, with separate legislation covering their control. National coordination was achieved through the Australian Weeds Committee (AWC), a committee of top bureaucrats from each state, with an observer from CSIRO but no input from the university sector. The AWC had no decision-making powers but served to inform state governments about new issues and to coordinate responses. The Council of Australian Weed Science Societies (CAWSS) held conferences every 3 years from 1978 and every

2 years from 2002. These conferences, attended by staff from universities and major chemical companies as well as government research groups, kept researchers in contact with new and emerging issues and problems. The first papers on parthenium weed in Queensland were published in 1976 in the *Queensland Agricultural Journal* and in *PANS*, and two papers were presented at the 1st CAWS Weeds Conference in 1978. Subsequent conferences contained many papers dealing with different aspects of parthenium weed biology and management.

The Landcare movement was established in Victoria in 1986 and nationally in 1989, initially with AUD$320 million to establish groups, employ coordinators and fund on-ground actions. Landcare brought farmers and conservationists together to encourage and develop integrated management of productive farmland and other privately owned land. Under the Landcare banner, a Parthenium Action Group was established in central Queensland.

The first National Weeds Strategy was developed in the late 1990s and published in 1997. As part of the Strategy, the Australian Government designated 20 Weeds of National Significance (WONS), one of which was parthenium weed, and allocated funding over 5 years for actions to meet national goals for each of these 20 WONS. A national coordinator was appointed for parthenium weed, and a National Parthenium Weed Management Group established, with members from affected communities, the university sector, and local and state government organizations. A Parthenium Weed Strategic Plan was developed in 2001 and reviewed in 2012 (Australian Weeds Committee, 2012), with one of the prime goals being to limit spread into new and previously uninfested areas. As a result, wash-down facilities were built by many local governments and regulations were developed to make the use of these mandatory for all contractors moving vehicles or machinery onto grazing or cropping land. Costs were high; construction or upgrading of 30 facilities by state and local governments over 10 years cost AUS$3.6 million (Australian Weeds Committee, 2012), and the time required to clean

machinery and vehicles is a continuing burden on contractors (Austin, 2013).

11.4 Current Situation in Australia

11.4.1 Spread

Fifty-five years after the initial introduction into central Queensland, parthenium weed infestations are still largely limited to Queensland (Fig. 11.2). There are no known infestations in Victoria, South Australia or Tasmania (Table 11.1). Spread into the very hot and seasonally dry areas of north-west Queensland, the Northern Territory and the north of Western Australia (WA) has been very limited (Table 11.1), and it is likely that these areas are not suitable for parthenium weed, except possibly along rivers and in irrigation areas. The two infestations in the Northern Territory near the Queensland border have been eradicated. An infestation in the Pilbara in the north of WA is currently being eradicated (Penna and MacFarlane, 2012). However, the limited spread into the north and west may be an artefact of the major spread routes, as there is relatively little movement of cattle and feed from eastern or central Queensland into these areas; movement is chiefly out of these areas to ports along the eastern coast.

In NSW, there is a continuing eradication programme, during which many small infestations and a few larger ones have been eradicated at an annual cost of AUS$150,000 (Blackmore and Johnson, 2010; A. McConnachie, NSW DAFF, 2015, personal communication). Wash-down facilities are now available in many areas throughout Queensland and NSW, and their use by contractors (e.g. involved with harvesting, stock movements, mining, seismic surveys, electricity transmission lines) who move vehicles and machinery onto grazing and cropping land is now mandatory (Austin, 2013). Information on clean-down procedures and requirements is available on the Queensland Government websites (for example, https://www.daf.qld.gov.au/business-priorities/plants/weeds-pest-animals-ants/weeds/preventing-weed-spread/ cleandown). Clearly these procedures are very expensive, both to local governments who build and maintain the facilities and to the industries which have to use them. Equally, though originally developed to reduce and prevent the movement of parthenium weed into clean properties in Queensland and into NSW, clean-down procedures also reduce the spread of many other weeds, particularly grass weeds such as the needle grasses *Nasella* spp. (Austin, 2013; Khan *et al.*, 2013).

In Queensland, parthenium weed continues to spread both to the south, north and, to a lesser extent, the west. The spread has greatly slowed from the early years, as can be seen by comparing the extent in the maps in Fig. 11.2. However, eradication of outlying infestations has not generally been successful, and where soils and climate are suitable the weed continues to spread. Published climate models predict that parthenium weed should thrive in large areas of southern and western Australia as well as in the east (McConnachie *et al.*, 2010; Mainali *et al.*, 2015). However on-ground evidence suggests that parthenium weed is not competitive in climatic zones with a predominantly winter rainfall pattern and hot dry summers. At the time of writing, all field occurrences in Australia have been in areas with predominantly summer rainfall (Fig. 11.2).

11.4.2 Biological control

Detailed information on the biological control agents released in Queensland is presented elsewhere (Chapter 7, this volume). No new biological agents have been introduced since 2000, but existing agents continue to be released into new areas, particularly in south and south-west Queensland where parthenium weed continues to expand its range and not all agents have successfully spread on their own (Callander and Dhileepan, 2016). Surveys of the agents present are made in autumn

every year, and these surveys have resulted in the discovery of viable field populations of agents previously believed to have failed (McFadyen, 1992). The galling weevil *Conotrachelus albocinereus* Fiedler, the stem-feeding moth *Platphalonidia mystica* (Razowski & Becker) and the sap-sucking bug *Stobaera concinna* (Stal) have now been found in several locations in central and southern Queensland (K. Dhileepan, 2016, Brisbane, unpublished data). The crown-boring moth *Carmenta* sp. nr *ithacae* (Beutenmuller), the stem-boring weevil *Listronotus setosipennis* (Hustache) and the seed-feeding weevil *Smicronyx lutulentus* Dietz are now present in most locations in central and north Queensland, and are progressively becoming established in southern Queensland. The leaf-feeding beetle *Zygogramma bicolorata* Pallister is present over much of central and southern Queensland but has not established in north Queensland. The stem-galling moth *Epiblema strenuana* Walker and the leaf-feeding moth *Bucculatrix parthenica* Bradley spread rapidly after release (McFadyen, 1992) and both moths are now present in all parthenium-infested areas. The two rust diseases, the winter rust *Puccinia abrupta* Diet. & Holw. var. *partheniicola* (Jackson) Parmelee and the summer rust *P. xanthii* var. *partheniihysterophorae* Seier, Evans & Romero, are much more limited in their distribution, though the winter rust is widespread in south-east Queensland, and distribution of both rusts into the southern areas is continuing.

The only detailed evaluation of the impact of the biological control agents, other than observational data, was undertaken in two grazing properties in north and central Queensland during 1996–2000, at which time only three agents were present. The insecticide carbofuran and the fungicide mancozeb were used to exclude biological control agents from sample plots. In the control plots, with the agents present, parthenium weed density was reduced by up to 90%, soil seed bank by up to 70% and grass production increased by up to 52% (Dhileepan, 2001). However, impacts varied greatly from year to year, with the greatest effect in years with good summer rainfall (Dhileepan, 2003; Dhileepan and McFadyen, 2012). This is borne out by observational data, that dry conditions in summer, especially early in the summer, greatly reduce the populations, and therefore the impact, of the various biological control agents. Conversely, wet summers favour the agents over the parthenium weed. Rainfall in the 2010/11 year was extremely high throughout most of Queensland, equalling that of 1973/74 (Anonymous, 2014), the year in which parthenium weed spread rapidly and successfully south from the original introduction site. Nevertheless, no resurgence of parthenium growth or spread has been reported after the 2011 floods subsided (M. Vitelli, 2016, AgForce Queensland, personal communication), and this is attributed to the control achieved and maintained by the biological control agents established in the previous decades.

11.4.3 Impact and management

Throughout the main area in central and north Queensland, the density and size of parthenium plants is greatly reduced compared to the situation in the 1980s and 1990s. The soil seed bank has reduced from 37,000 viable seeds m^2 in the 1990s to 6000 m^2 or less (Osunkoya et al., 2014). Nguyen (2011) examined seeds in sludge taken from five wash-down facilities in central Queensland in 2007, 2008 and 2009. On average, she found 6500 viable seeds per 100 kg of sludge, but only 2% of this seed was parthenium weed. The maximum amount of parthenium weed was 12% in samples taken in winter (when parthenium plants would have been dry and climatic conditions not suitable for germination) and the minimum was 0.3% in summer (when plants would have been green, with less ripe seed available to adhere onto vehicles). In 2009, which was a dry year, there was less seed overall but no change in the proportion of parthenium weed seed. That is,

parthenium weed is no longer a significant component of the weed seed load being transported by machinery and vehicles, even in the central Queensland infestation areas. As a result, new infestations recorded in NSW have dropped from the previous average of 27 per year (Blackmore and Johnson, 2010) to only 3 per year between 2006 and 2015 (S. Johnson, 2016, NSW DAF, personal communication).

Parthenium weed is no longer included in lists of the major weeds of crops in central and north Queensland (Osten et al., 2007; Widderick et al., 2014). The Area Pest Management Plan 2015–2020 for central Queensland, covering all the shires with major parthenium weed infestations, listed parthenium weed as number 20 in priority out of 26 pest plants (Central Highlands Regional Council 2015). As part of the Plan, a weed survey of all declared species was undertaken in 2015/16 and listed parthenium weed as 'low priority (category 3 out of 4)'. Grazing properties in central and northern Queensland no longer regard parthenium weed as a major weed, not does it require any special management (M. Vitelli, 2016, AgForce Policy Officer, personal communication). The Fitzroy Basin Association, a Regional Natural Resource Management group covering much of the central Queensland parthenium weed area, has a series of five modules on 'Grazing Best Management Practices', available through their website, in which weeds are not mentioned as an issue (Fitzroy Basin Association, 2013). Listed examples of 'managing pests' and 'managing run-down pastures' do not mention parthenium weed. Despite 2010/11 being the wettest year in Queensland since 1973/74, there were no research or discussion papers on management of parthenium weed in the 18th Australasian Weeds Conference in 2012, nor in the 19th in 2014, nor in the 12th Queensland Weeds Symposium in 2012. The conclusion is that, in most situations, parthenium weed is now successfully controlled by the biological control agents, and is no longer causing significant problems for landholders and farmers in Australia.

11.5 Pacific Region and Australian Indian Ocean Islands

Parthenium weed was first reported in Papua New Guinea in 2001 at Ela Motors Bond yard, Lae (Table 11.1), where it was thought to have arrived with used cars imported from Queensland several years earlier (Kawi and Orapa, 2010). In 2003, parthenium weed was reported 10 km away at Buambub Coconut (*Cocos nucifera* L.) Plantation, along the Highlands Highway, presumably having spread from the first site. Both sites were treated from 2001 to 2006 with paraquat and glyphosate as plants appeared, by staff from the National Agricultural Quarantine and Inspection Authority and the National Agriculture Research Institute, with funding of equipment and herbicide provided by the Secretariat of the Pacific Community. Parthenium weed was reported as eradicated in 2006 and no plants were seen during yearly checks to 2009 (Kawi and Orapa, 2010). The sites continue to be checked every year and no plants have been found to date (A. Kawi, Papua New Guinea, 2014, personal communication). This would seem to be one of the very few successful eradications of parthenium weed world-wide.

Parthenium weed is also reported in Hawaii, Tahiti and New Caledonia (Table 11.1; Gargominy *et al.*, 1996; Waterhouse, 1997; PIER, 2013). In Hawaii, it is found on the islands of Hawaii, Kauai, Maui, Molokai and Oahu (PIER, 2013). It was first reported in New Caledonia in 1881 (Gargominy *et al.*, 1996) and it is suspected that it spread from New Caledonia to Tahiti, where it was reported on Tubuai Island in February 2006 (Florence *et al.*, 2007), and to Vanuatu (Meyer *et al.*, 2006).

Vanuatu Forestry Service reported that parthenium weed was first seen in Vanuatu in 1971 (Table 11.1). Sylverio Bule, Biosecurity Vanuatu (Vanuatu Forestry Service, 2014, personal communication), believes it was introduced in 2009, brought in on machinery from Australia, as although parthenium weed is present in New Caledonia, machinery is not imported into Vanuatu from New Caledonia. Parthenium weed spread to Espiritu Santo the same year but was eradicated through hand-pulling young plants. It also spread from Efate to the southern islands of Tanna in 2011 and Aniwa in 2012. The biggest infestations are on Efate and Tanna islands, where there are some substantial infestations in paddocks. During surveys on Efate in October 2014 (Day and Bule, 2016), parthenium weed was seen at 41 sites, mainly as scattered plants along the sides of roads, particularly along the southern areas around Port Vila and around various villages on the north-western and northern parts of Efate.

Biological control for parthenium weed was initiated in Vanuatu in 2014, with the introduction of *Z. bicolorata* Pallister from Australia under an AusAID-funded project, managed by the Queensland Government. The beetle has been released on Efate and Tanna islands but establishment is not confirmed. In October 2014, *E. strenuana* Walker was seen at three sites on Efate. It is not known how it arrived in the country (Day and Bule, 2016). The moth was not seen when parthenium weed was surveyed in 2012 and 2013. As populations were still low, it may be a very recent introduction, possibly through larvae or pupae in plant material on imported machinery from Queensland, where the moth is seasonally common and abundant (Dhileepan and McFadyen, 2012).

The Australian Territories Christmas and Cocos Islands in the Indian Ocean are included here for completeness. Parthenium has not been recorded on the Cocos Islands but is present on Christmas Island (Table 11.1), where it is subject to an eradication programme managed by Western Australia (Dodd and Reeves, 2012).

11.6 Discussion

11.6.1 Impact of management programme

In the 40 years since its inception in 1975, the management programme in Queensland has been outstandingly successful. The initial spread south of parthenium weed was dramatically reduced, largely as a result of the roadside spraying programme in Queensland, coupled with controls on movement of contaminated feed and pasture seed and a major public awareness programme. Biosecurity legislation, supported by information and media campaigns, resulted in major efforts by oil and gas exploration firms, among others, to clean their vehicles and machinery when moving between and within properties (Austin, 2013). Spread into NSW was halted by these measures, together with the strictly enforced cleaning programme for harvesters and other agricultural machinery entering NSW, and the continuing search and eradication programme for all new infestations within NSW (Blackmore and Johnson, 2010; A. McConnachie 2015, NSW DAFF, personal communication). Procedures introduced initially to control the spread of parthenium weed have been adopted throughout Australia, and the first goal of the Australian Weeds Strategy is to reduce the spread of new weeds into, but also within, Australia (Noble, 2014).

At the same time, the increasing impact of the biological control programme resulted in greatly reduced seed production (Osunkoya et al., 2014), even in years of high rainfall such as 2010/11 (Anonymous, 2014), which in turn reduced the seed load heading south in produce and machinery. By 2009, in the heart of the parthenium weed-infested areas in central Queensland, parthenium weed was less than 2% of all viable seeds collected in the sludge from vehicle wash-down facilities (Nguyen, 2013). New infestations recorded in NSW were down to three per year from the previous average of 27 per year.

11.6.2 Current situation

The on-ground impact of parthenium weed in the main parthenium weed areas in Queensland has also been greatly reduced by the increasing effect of the biological control agents. In 1985, prior to biological control, parthenium weed was in the top ten weeds

of cropping lands in central Queensland (V. Pope, 1989, QDPI, unpublished report), but by 2007 it was no longer listed as a major weed of crops anywhere in Queensland (Osten et al., 2007; Widderick et al., 2014). Soil seed banks in heavily infested areas had dropped to 5000–6000/m² (Osunkoya et al., 2014) from the original level of 33,000/m² (Navie et al., 2004). Parthenium weed is not seen as a major weed of grazing lands in central and northern Queensland. Even after the very wet 2010/11 summer, graziers did not report major outbreaks of parthenium weed or problems in controlling the weed (M. Vitelli, 2016, AgForce, personal communication).

In conclusion, it is evident that the management programme in place from 1975 to the present day, along with the simultaneous development of on-ground methods to reduce spread and a well-funded biological control programme to reduce the impact of existing infestations, has proved highly effective. Parthenium weed is no longer a major threat to farmers and landholders in Australia, and no longer rates as one of the top weeds, even in the worst-affected areas.

Acknowledgements

We thank Bradly Gray (Biosecurity Queensland) for the historical ArcGIS maps of parthenium distribution and Asad Shabbir for creating the distribution map. We are also grateful to Warea Orapa and Anna Kawi, National Agriculture and Quarantine Inspection Authority and the late Sylverio Bule, Biosecurity Vanuatu for up-to-date information regarding the status of parthenium in Papua New Guinea and Vanuatu, respectively.

References

Anonymous (1980) *Parthenium Weed and Its Control.* Queensland Government, Brisbane, Australia.

Anonymous (2014) *Australia's Variable Rainfall: April to March Annual Australian Rainfall*

Relative to Historical Records 1890–2012. Department of Science, Information, Technology, Innovation and the Arts (DSITIA), Brisbane, Australia.

Anonymous (2016) *Inflation Calculator.* Reserve Bank of Australia, Sydney, Australia

Armstrong, T.R. (1978) Herbicidal control of *Parthenium hysterophorus* L. In: *Proceedings of the 1st Conference of the Council of Australian Weed Science Societies.* CAWSS, Melbourne, Australia, pp. 157–164.

Auld, B.A., Hosking, J. and McFadyen, R.E. (1983) An analysis of the spread of tiger pear and parthenium in Australia. *Australian Weeds* 2, 56–60.

Austin, P. (2013) Weed seed spread: It's everyone's business. In: O'Brien, M., Vitelli, J. and Thornby, D. (eds) *Proceedings of the 12th Queensland Weed Symposium,* Hervey Bay, Queensland. Weed Society of Queensland, Clifford Gardens, Australia, pp. 7–10.

Australian Weeds Committee (2012) *Parthenium (Parthenium hysterophorus L.) Strategic Plan 2012–17,* Weeds of National Significance. Australian Government Department of Agriculture, Fisheries and Forestry, Canberra.

Bennett, F.D. (1976) A preliminary survey of the insects and diseases attacking *Parthenium hysterophorus* L. (Compositae) in Mexico and the USA to evaluate the possibilities of its biological control in Australia. Unpublished report, CIBC West Indian Station, Curepe, Trinidad.

Bennett, F.D. and Cruttwell, R.E. (1970) Memorandum on the possibilities of biological control of *Parthenium hysterophorus* (L.) Compositae. Unpublished report, CIBC West Indian Station, Curepe, Trinidad.

Blackmore, P.J. (2000) Parthenium weed in New South Wales. In: Chamberlain, J., Dearden, S.W., Leitch, A. and Moran, A. (eds) *Parthenium Weed Best Management Practice.* Queensland Department of Primary Industries, Brisbane, Australia, p. 16.

Blackmore, P.J. and Johnson, S.B. (2010) Continuing successful eradication of parthenium weed (*Parthenium hysterophorus*) from New South Wales, Australia. In: Zydenbos, S.M. (ed.) *Proceedings of the 17th Australasian Weeds Conference,* Christchurch, New Zealand. New Zealand Plant Protection Society, Hastings, New Zealand, pp. 382–385.

Callander, J.T. and Dhileepan, K. (2016) Biological control of parthenium weed: Field collection and redistribution of established biological control agents. In: Randall, R., Lloyd, S. and Borger, C. (eds) *Proceedings of the 20th Australasian*

Weeds Conference. Weeds Society of Western Australia, Perth, pp. 242–245.

Central Highlands Regional Council (2015) *Area Pest Management Plan 2015–2020.* Available at: http://www.centralhighlands.qld.gov.au/water-waste-land-use/pests-nuisances-2/ (accessed June 2018).

Chandras, G.S. and Vartak, V.D. (1970) Symposium on problems caused by *Parthenium hysterophorus* in Mararashtra Region, India. *PANS* 16, 212–214.

Chippendale, J.F. and Panetta, F.D. (1994) The cost of parthenium weed to the Queensland cattle industry. *Plant Protection Quarterly* 9, 73–76.

Cook, G.D. and Dias, L. (2006) It was no accident: Deliberate plant introductions by Australian government agencies during the 20th century. *Australian Journal of Botany* 54, 601–625.

Dale, I.J., Jacobsen, C.N. and Tucker, R.J. (1978) An assessment of parthenium weed (*Parthenium hysterophorus*) in grazing lands – preliminary results. In: *Proceedings First Conference Council of the Australian Weed Science Society,* Melbourne, Australia, pp. 154–156.

Day, M.D. and Bule, S. (2016) The status of weed biological control in Vanuatu. *NeoBiota* 30, 151–166.

Dhileepan, K. (2001) Effectiveness of introduced biocontrol insects on the weed *Parthenium hysterophorus* (Asteraceae) in Australia. *Bulletin of Entomological Research* 91, 167–176.

Dhileepan, K. (2003) Seasonal variation in the effectiveness of leaf-feeding beetle *Zygogramma biocolorata* (Coleoptera), the (Coleoptera: Chrysomelidae) and stem-galling moth *Epiblema strenuana* (Lepidoptera: Tortricidae) as biological control agents on the weed *Parthenium hysterophorus* (Asteraceae) in Australia. *Bulletin of Entomological Research* 93, 393–401.

Dhileepan, K. and McFadyen, R.E. (2012) *Parthenium hysterophorus* L. – parthenium. In: Julien, M., McFadyen, R.E. and Cullen, J. (eds) *Biological Control of Weeds in Australia: 1960 to 2010.* CSIRO Publishing, Collingwood, Australia, pp. 448–462.

Dodd, A.P. (1940) *The Biological Campaign Against Prickly Pear.* Commonwealth Prickly Pear Board Bulletin. Government Printer, Brisbane, Australia.

Dodd, J. and Reeves, A.W. (2012) Overcoming weed eradication challenges in the Indian Ocean Territories. In: Eldershaw, V. (ed.) *Proceedings of the Eighteenth Australasian Weeds Conference.* Weed Society of Victoria, Melbourne, Australia, pp. 263–266.

Doley, D. (1977) Parthenium weed (*Parthenium hysterophorus* L.): Gas exchange characteristics as a basis for prediction of its geographical distribution. *Australian Journal of Agricultural Research* 28, 449–460.

Everist, S.L. (1976) Parthenium weed. *Queensland Agricultural Journal* 102(2), cover page.

Fitzroy Basin Association (2013) Grazing Best Management Practices. Fitzroy Basin Association, Rockhampton, Australia.

Florence, J., Chevillotte, H., Ollier, C. and Meyer, J.-Y. (2007) *Parthenium hysterophorus*. Base de données botaniques NIDEAUD de la flore de Polynésie française (INPN).

Gargominy, O., Bouchet, P., Pascal, M., Jaffre, T. and Tourneu, J.C. (1996) Conséquences des introductions d'espèces animales et végétales sur la biodiversité en Nouvelle-Calédonie. *Review Ecology (Terre Vie)* 51, 375–401.

Green, K.R. (1965) Noxious weed legislation. In: Anonymous (ed.) *Proceedings of the Third Australian Weeds Conference*, Vol. 2. CSIRO Publishing, Collingwood, Australia, pp. 70–174.

Haseler, W.H. (1976) *Parthenium hysterophorus* L. in Australia. *PANS* 22, 515–517.

Joshi, S. (1991) Biocontrol of *Parthenium hysterophorus* L. *Crop Protection* 10, 429–431.

Kawi, A. and Orapa, W. (2010) Status of parthenium weed in Papua New Guinea. *International Parthenium News* 2, 2–3.

Khan, I., O'Donnell, N., Navie, S., George, D., Nguyen, T. and Adkins, S. (2013) Weed seed spread by vehicles. In: O'Brien, M., Vitelli, J. and Thornby, D. (eds) *Proceedings of the 12th Queensland Weed Symposium*. Hervey Bay, Queensland. Weed Society of Queensland, Clifford Gardens, Australia, pp. 94–97.

King, L.J. (1966) *Weeds of the World: Biology and Control*. Leonard Hill, London.

Lonkar, A., Mitchell, J.C. and Calnan, C.B. (1974) Contact dermatitis from *Parthenium hysterophorus*. *Transactions of St John's Dermatological Society* 60, 43–53.

Maheshwari, J.K. (1966) *Parthenium hysterophorus*. *Current Science* 35, 181–183.

Mainali, K.P., Warren, D.L., Dhileepan, K., McConnachie, A., Strathie, L., *et al.* (2015) Projecting future expansion of invasive species: comparing and improving methodologies for species distribution modeling. *Global Change Biology* 21, 4464–4480.

McConnachie, A.J., Strathie, L.W., Mersie, W., Gebrehiwot, L., Zewdie, K., *et al.* (2010) Current and potential geographical distribution of the invasive plant *Parthenium hysterophorus* (Asteraceae) in eastern and southern Africa. *Weed Research* 51, 71–84.

McFadyen, P.J. (1979) A survey of insects attacking *Parthenium hysterophorus* L. (Compositae) in Argentina and Brazil. *Dusenia* 11, 42–45.

McFadyen, R.E. (1991) The Harrisia cactus eradication scheme: Policy-making by technocrats. MPubAdmin thesis, University of Queensland, St Lucia, Australia.

McFadyen, R.E. (1992) Biological control against parthenium weed in Australia. *Crop Protection* 11, 400–407.

McFadyen, R.E. (1995) Parthenium weed and human health in Queensland. *Australian Family Physician* 24, 1455–1459.

McFadyen, R.E.C. (2012) *Harrisia martinii* – Harrisia cactus: *Acanthocereus tetragonus* – sword pear. In: Julien, M., McFadyen, R. and Cullen, J. (eds) *Biological Control of Weeds in Australia*. CSIRO Publishing, Collingwood, Australia, pp. 274–281

Meyer, J.-Y., Loope, L., Sheppard, A., Munzinger, J. and Jaffre, T. (2006) Les plantes envahissantes et potentiellement envahissantes dans l'archipel néo-calédonien: Première évaluation et recommandations de gestion. In Beauvais, M.-L. *et al.* (eds) *Les Espèces Envahissantes dans L'archipel Néo-Calédonie*. IRD Éditions, Paris.

Navie, S.C., McFadyen, R.E., Panetta, F.D. and Adkins, S.W. (1996a) The biology of Australian weeds. 27. *Parthenium hysterophorus* L. *Plant Protection Quarterly* 11, 76–88.

Navie, S.C., McFadyen, R.E., Panetta, F.D. and Adkins, S.W. (1996b) A comparison of the growth and phenology of two introduced biotypes of *Parthenium hysterophorus*. In Shepherd, R.C.H. (ed.) *Proceedings of the Eleventh Australasian Weeds Conference*. Weed Society of Victoria, Melbourne, Australia, pp. 313–316.

Navie, S.C., Panetta, F.D., McFadyen, R.E. and Adkins, S.W. (2004) Germinable soil seed banks of central Queensland rangelands invaded by the exotic weed *Parthenium hysterophorus* L. *Weed Biology and Management* 4, 154–167.

Nguyen, T.L.T. (2011) Biology of parthenium weed (*Parthenium hysterophorus* L.) in Australia. PhD thesis, The University of Queensland, Brisbane, Australia.

Noble, M. (2014) Towards improved effectiveness: Weed hygiene regulation within Australia. In Baker, M. (ed.) *Proceedings of the Nineteenth Australasian Weeds Conference*. Tasmanian Weed Society, Hobart, Australia, pp. 170–173.

Osten, V.A., Walker, S.R., Storrie, A., Widderick, M., Moylan, P., Robinson, G.R. and Galea, K. (2007)

Survey of weed flora and management relative to cropping practices in the north-eastern grain region of Australia. *Australian Journal of Experimental Agriculture* 47, 57–70.

Osunkoya, O.O., Ali, S., Nguyen, T., Shabbir, A., Navie, S., *et al.* (2014) Soil seed bank dynamics in response to an extreme flood event in a riparian habitat. *Ecological Research* 29, 1115–1129.

Page, A.R. and Lacey, K.L. (2006) *Economic Impact Assessment of Australian Weed Biological Control.* Technical Series 10. CRC for Australian Weed Management, Adelaide, Australia.

PIER (2013) *Plant Threats to Pacific Ecosystems. Pacific Island Ecosystems at Risk.* Available at: http://www.hear.org/pier/species.htm (Accessed November 2014).

Parker, C. (1968) Weed problems in India, West Pakistan and Ceylon. *PANS* 14, 217–228.

Parsons, W.T. and Cuthbertson, E.G. (1992) *Noxious Weeds of Australia.* Inkata Press, Melbourne, Australia.

Pearson, M. and Lennon, J. (2010) *Pastoral Australia: Fortunes, Failures and Hard Yakka: A Historical Overview 1788–1967.* CSIRO Publishing, Collingwood, Australia.

Penna, A.M. and MacFarlane, M. (2012) Parthenium incident in the Pilbara, Western Australia: How is this 'a good news' story? In Eldershaw, V. (ed.) *Proceedings of the Eighteenth Australasian Weeds Conference.* Weed Society of Victoria, Melbourne, Australia, pp. 13–16.

Stock Routes and Rural Lands Board (1978) *Annual Report.* Queensland Government, Brisbane, Australia.

Vartak, V.D. (1968) Weed that threatens crops and grasslands in Maharashtra. *Indian Farming* 18, 23–24.

Waterhouse, D.F. (1997) *The Major Vertebrate Pests and Weeds of Agriculture and Plantation Forestry in the Southern and Western Pacific.* ACIAR Monograph 44. Australian Centre for International Agricultural Research, Canberra.

White, G.G. (1994) *Workshop Report: Parthenium Weed.* Cooperative Research Centre for Tropical Pest Management, Brisbane, Australia.

Widderick, M., Cook, T., McLean, A., Churchett, J., Keenan, M., Miller, B. and Davidson, B. (2014) Improved management of key northern region weeds: Diverse problems, diverse solutions. In Baker, M. (ed.) *Proceedings of the Nineteenth Australasian Weeds Conference.* Tasmanian Weed Society, Hobart, Australia, pp. 312–315.

12 History and Management – Southern Asia

Asad Shabbir,[1]* Bharat B. Shrestha,[2] Muhammad H. Ali[3] and Steve W. Adkins[4]

[1]University of the Punjab, Lahore, Pakistan; current affiliation: Plant Breeding Institute, the University of Sydney, Narrabri, New South Wales, Australia; [2]Tribhuvan University, Kathmandu, Nepal; [3]First Capital University of Bangladesh, Chuadanga, Bangladesh; [4]The University of Queensland, Gatton, Queensland, Australia

12.1 History of Invasion and Spread in Southern Asia

In Southern Asia, parthenium weed is now present in most countries, including India (Yaduraju et al., 2005), Pakistan (Shabbir and Bajwa, 2006), Sri Lanka (Jayasuriya, 1999), Nepal (Shrestha et al., 2015), Bangladesh (Karim, 2009) and Bhutan (Biswas and Das, 2007). Although there are no confirmed reports of its presence in Afghanistan or Maldives, it is highly likely that it is present in Afghanistan. Khan et al. (2014) has reported anecdotal evidence that the weed may be present in the southern parts of Afghanistan due to the movement of trucks and other vehicles across the border from Pakistan on a daily basis.

12.1.1 India

The weed was first recorded in India in 1810 (Paul, 2010), but the fate of that original population is unknown. The current infestation in India is thought to be from a contaminated wheat (Triticum aestivum L.) food aid shipment (lot Public Law 480) coming from the USA in the 1950s (Kapoor, 2012) (Public Law 480 was passed in 1954 to give food grain to developing countries to eliminate starvation and malnutrition). It was first observed in Pune, Maharashtra in 1955. Since then, the weed has spread to all the states in India (Fig. 12.1), covering about 5 million ha by 1975 and around 35 million ha by 2010 (Sushilkumar and Varsheny, 2010).

Until the 1980s, parthenium weed infestations were confined mainly to uncultivable land. At that time it was not considered to be a problem in Rajasthan, Gujrat, Jammu & Kashmir, Kerala, Orissa, West Bengal and the north-eastern states, while its presence in the Andaman and Nicobar Islands, Arunachal Pradesh, Daman and Diu, Goa, Kerala, Manipur, Mizoram, Mehgalaya, Nagaland, Pondicherry and Sikkim was considered to be a minor problem. However, in the last 30 years, parthenium weed has become distributed widely throughout the whole of India with no state able to say that it is free of the weed (Fig. 12.1). The arid areas in Rajasthan and the hilly areas in Uttarakhand and Himachal Pradesh have also become heavily infested with the weed in recent years.

In India, parthenium weed has invaded all kinds of habitats, from the coastal areas up to the higher mountainous ranges.

* asad.shabbir@sydney.edu.au

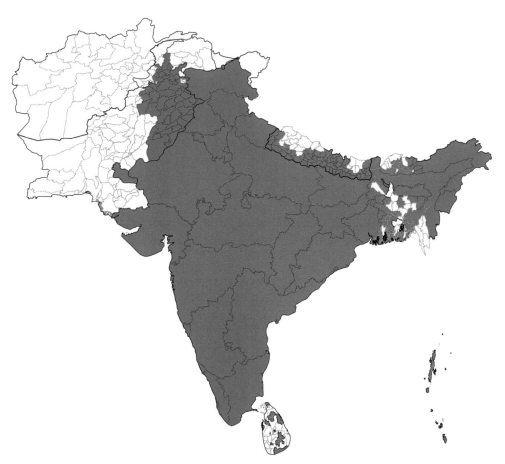

Fig. 12.1. Distribution of parthenium weed in Southern Asia. Red indicates the districts or areas where parthenium weed is present in Pakistan, India, Nepal, Bhutan, Bangladesh and Sri Lanka. (A. Shabbir.)

Initially it was reported as a weed of wastelands and roadsides, but it has now become a problematic weed of agricultural lands, pastures, national parks and forests (Sushilkumar, 2014). Parthenium weed has achieved the status of India's 'worst weed' owing to its negative effects on agricultural crop production and its harmful effects on human and animal health (Towers and Subba Rao, 1992; Narasimhan *et al.*, 1977). Parthenium weed has been reported as a problematic weed in many cereal, oil, fibre, fodder and vegetable crops. Some of the major cereal crops infested by the weed are wheat (*T. aestivum* L.), rice (*Oryza sativa* L.), sorghum (*Sorghum bicolor* L. Moench), and maize (*Zeya mays* L.) (see Bajwa *et al.*,

Chapter 4, this volume; Adkins and Shabbir, 2014). Parthenium weed has invaded several grasslands in India, where it has reduced the percentage cover of pasture species by as much as 90% (Jayachandra, 1971). It has also invaded several forests and national parks in India, where it is reported to reduce the growth of forest trees through the release of allelochemicals (see Witt and Belgeri, Chapter 5, this volume; Towers *et al.*, 1977; Swaminathan *et al.*, 1990).

Among many ill-effects of parthenium weed, the one of most concern is its negative effect upon human and animal health. It has been reported that more than 18 million people suffer from parthenium weed allergy and hay fever each year (see Allan *et al.*,

Chapter 6, this volume). Another survey carried out in Bangalore has shown that over 7% of the population suffer from allergenic rhinitis and about 40% of the population are sensitive to its pollen (Towers and Subba Rao, 1992). In India, it has been reported that allergy due to parthenium weed is so severe that it has led to a number of cases of suicide (Kololgi *et al.*, 1997). Farm workers are one group of people who are commonly affected due to their exposure to the weed in their fields.

According to Sushilkumar (2014), the rapid spread of parthenium weed in India was assisted by the Public Distribution System, a food security system established by the Government of India in the 1950s. Another suggested cause for its rapid spread is through its seeds being carried by motor vehicles, farm machinery, trains and in packaging materials. There is no effective legislation in place in India to stop the spread of this weed. Karnataka in 1969 became the first and only state to declare this weed an agricultural pest, but unfortunately this legislation has not been enforced and has proven to be ineffective in stopping the spread of weed within and outside of Karnataka (Mahadevappa, 2009; Adkins and Shabbir, 2014).

The economic losses caused by the weed in India are huge, but no comprehensive study has been undertaken at a national level to document this. Sushilkumar and Varshney (2010) reported that about 35 million ha are now infested with the weed and they estimate that the cost of manual and chemical control (Table 12.1) would have been about 3.0 billion Indian Rupees (Rs) if proper management had been implemented since its arrival in 1956. Furthermore, they estimated the costs for the treatment of parthenium weed-induced health problems as Rs 88 million. It is quite clear that parthenium weed, if not better managed, will expand its range to further states and its population's size will increase. The growing problem in India therefore warrants the development of an effective national strategy for the management of this weed and for its immediate implementation.

12.1.2 Pakistan

The first published reference to the occurrence of parthenium weed in Pakistan dates back to the 1980s, in the Gujarat district of the Punjab Province (Razaq *et al.*, 1994). Following a 20-year period of slow spread, in last 10 years the weed has extended its range rapidly and is now to be found in most districts of the Punjab, in Khyber Pakhtunkhwa

Table 12.1. Herbicides that are used in Southern Asia and are effective against parthenium weed, and the various situations where they have been used. (Asad Shabbir.)

Herbicide	Rate (kg/ha)	Situation	Reference
Fernoxone	1.5	Crops	Leela (1987)
Glyphosate	0.84	Experimental field	Reddy *et al.* (2007)
	1.0	Non-cropped area	Mishra and Bhan (1995)
Glyphosate	3.6	Experimental field	Singh *et al.* (2004)
Glufosinate	0.48		Reddy *et al.* (2007)
Chlorimuron ethyl	0.20	Non-cropped area	Mishra and Bhan (1995)
Metsulfuron methyl	0.004		
Oyflurofen	0.15	Onion crop	Dalavai *et al.* (2008)
Norflurazon	2.3	Glasshouse crop	Reddy *et al.* (2007)
Clomazone	1.1	Glasshouse crop	Reddy *et al.* (2007)
Bentazon	1.5	Soybean crop	Mishra and Bhan (1996)
Atrazine	3.0	Experimental field	Khan *et al.* (2012b)
Isoproturon	2.0		Shabbir (2014)
Metribuzin	2.0		Khan *et al.* (2012b)

(KP), the Federally Administered Tribal Areas (FATA), Azad Jammu & Kashmir and more recently Sind Province (Shabbir *et al.*, 2012; Fig. 12.2). During a 1985/86 ecological survey of several locations in Islamabad and Rawalpindi, it was reported that parthenium weed was absent; however, 20 years later no area within these cities is free of the weed (Awan *et al.*, 1987; Shabbir and Bajwa, 2006; Kumar *et al.*, 2013).

Parthenium weed is not only a dominant weed in wastelands, it is becoming a problematic weed in irrigated and rain-fed cropping systems, range lands, forests and protected areas (Shabbir *et al.*, 2012). Studies conducted prior to 2000 indicated that parthenium weed was only infesting the northern districts of the Punjab, but a recent

survey carried out in 2009/10 revealed the weed to have spread rapidly from the northern to the southern districts of the Punjab and into Sind. In the districts of Okara, Pakpattan, Sahiwal, Khanewal, Multan and Bahawalpur, it is now threatening the cotton (*Gossypim hirsutum* L.) and dairy industries (Fig. 12.2) (Shabbir *et al.*, 2012). Parthenium weed is an emerging weed in several other crops, including cereals, vegetables and fodder crops, where it inflicts significant yield losses. In maize, for example, up to 50% yield losses have been reported (Shabbir, 2006; Anwar *et al.*, 2012; Khan *et al.*, 2013; Safdar *et al.*, 2014; Tanveer *et al.*, 2015).

Many protected areas (national parks, reserves) have been invaded by parthenium weed in Pakistan, where the weed is now

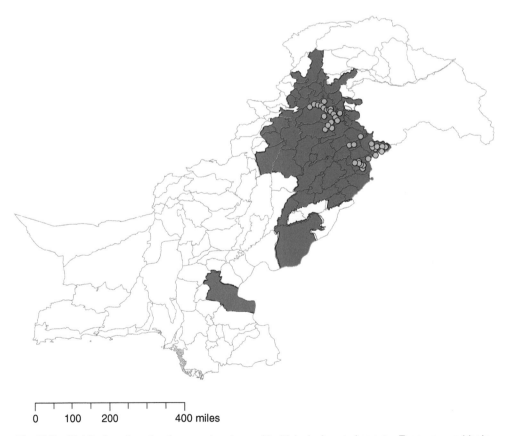

0   100   200   400 miles

Fig. 12.2. Distribution of parthenium weed and one of its biological control agents, *Zygogramma bicolorata*, in 2010, Pakistan. Red indicates the districts where parthenium weed is present, green dots indicate the presence of *Z. bicolorata*. (A. Shabbir.)

threatening native plant diversity and the environment (Mujahid, 2015; Khan, 2015). Mujahid (2015) investigated the impact of this weed on the composition and structure of the above-ground vegetation and the soil seed bank of the protected Jhok reserve forest near Lahore. Parthenium weed has significantly decreased the composition of both the above-ground vegetation and the soil seed bank of this reserve forest and its further invasion poses a serious threat to the native plant biodiversity of this protected area. In the Changa Manga plantation near the district of Lahore (Punjab), parthenium weed has emerged as one of the most dominant plant species of the understorey vegetation.

In the Punjab, parthenium weed is spreading to agricultural farms mainly through the extensive canal network used to irrigate the land, but other vectors such as vehicles and farm machinery are also contributing to the longer distance spread within the country (Shabbir et al., 2012). The horticultural trade is also considered as one of the important sources of spread of this weed through the movement of ornamental plants potted in soil that is contaminated with seeds of parthenium weed. Naveed (2015) investigated the seed content of soil collected from 30 plant nurseries located in Lahore, Kasur and Islamabad. The study found parthenium weed seed to be present in most samples. Floral bouquet makers, particularly those in the Punjab, are using the flower-bearing branches of parthenium weed as fillers in their garlands. These garlands are not only a health hazard but also act as a vector for further spread of this weed to new locations.

Parthenium weed is causing significant yield losses in some of the major cereal crops in Khyber Pakhtunkhwa (KP) province. Up to 40% yield reductions have been reported due to parthenium weed in both sorghum and maize crops (Khan et al., 2013). It is also having negative effects on the health and wellbeing of the people living in districts where parthenium weed is present. Both human and animal health is negatively affected by the weed. In a recent survey, Khan et al. (2013) reported that parthenium weed is becoming a serious weed of range lands in Charsada district (KP) and that the milk produced by animals in that area had become bitter and of poor quality.

A recent analysis suggests that there are still vast areas in the Punjab, KP and Sindh Provinces that are climatically suitable for parthenium weed invasion (Shabbir, 2012). This growing problem in Pakistan therefore warrants the development of an effective national parthenium weed management strategy and its immediate implementation.

12.1.3 Nepal

Although parthenium weed was known to the scientific community of Nepal fairly early in the 1960s (Tiwari et al., 2005), rapid spread of this weed did not occur until the 1990s (Shrestha et al., 2015). From the locations of these early infestations and the current distribution pattern of the weed within Nepal (Shrestha et al., 2015), it is evident that the weed entered the country from India. Due to the numerous open border crossings and the number of vehicles that pass every day, it is highly likely that the weed entered on a number of occasions.

At present, parthenium weed is a dominant species in the grasslands, fallow lands and roadside vegetation of several major districts in Nepal, including Kathmandu, Biratnagar, Birgunj, Hetaunda, Narayangargh, Butwal, Pokhara, Dang and Surkhet. In most places, invasion is now considered to be at the high colonization stage and the average stem density has grown to 402 stems/m^2. A recent study revealed that this weed is expanding along the road network, spreading from the Tarai plains in the south to the mid hills to their north by seed attached to vehicles and transportation of agricultural products such as grains and fodder (Fig.12.3) (Shrestha et al., 2015). The weed has been observed up to 1935 m asl.

Recently, this weed has been spreading rapidly in urban areas and gradually invading agricultural land and forested areas. The weed has invaded several economically important crops such as maize, sugarcane

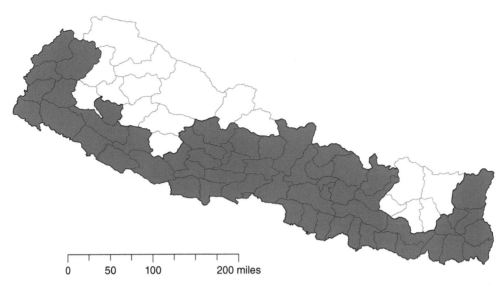

| 0 | 50 | 100 | | 200 miles |

Fig. 12.3. Distribution of parthenium weed in Nepal. Red indicates the districts where parthenium weed is present. (A. Shabbir.)

(*Saccharum officinarum* L.), potatoes (*Solanum tuberosum* L.) and mustard (*Brassica compestris* L.). After agricultural land has been abandoned for 1 to 2 years, parthenium weed becomes a dominant weed. In the upland cropping system, parthenium weed forms a seasonal dense cover in fallow fields that come under cultivation for upland rice. The high density of weeds during the early summer fallow period poses a great threat to the upcoming rice crop (Shrestha *et al.*, 2015). In India, parthenium weed has already become a problematic weed in upland rice (Paradkar *et al.*, 1997).

The rapid range expansion of parthenium weed is a new environmental problem in urban and peri-urban areas of Nepal, particularly in the southern Tarai plains, in the Siwalik range and the mid hills region (Shrestha, 2008). In peri-urban areas, this weed is mainly found in the grasslands used for livestock grazing and in abandoned agricultural lands. It has been found less frequently in cultivated agricultural lands, but invades some of the important crops such as potato (*Solanum tuberosum* L.) and wheat. Parthenium weed has been reported from community forests and protected areas, including Chitwan National Park, a world

heritage site. The weed is found inside the park and its surrounding buffer zone. Once it moves further into the National Park, the weed will have undesired ecological impacts, suppressing native biodiversity and altering the nutrient supply and species composition (Timisna *et al.*, 2011).

The impact on human and animal health has not been studied systematically but there are several reports of allergic responses to this weed. Farmers involved in fodder harvesting have developed allergic reactions on exposed skin (Karki, 2009). Similarly, lesions on the skin of livestock have been observed when parthenium weed was used as a bedding material. Bitterness in the milk of buffalos fed on fodder from parthenium weed-infested grasslands has been reported (Shrestha *et al.*, 2015).

Parthenium weed is spreading from the low altitudes of the south of Nepal to the higher lands. If it is not managed effectively, it is expected to expand its range, mainly through movement along road corridors, to most parts of Nepal below *c.*2500m asl. There is an urgent need to assess the economic impact of this weed on livelihoods and to ensure the policy makers and other stakeholders take immediate action.

12.1.4 Sri Lanka

Parthenium weed was first recorded in Sri Lanka in 1999. These early populations were in the northern district of Vavuniya, but later found in Pallekele in Kandy district, Central Province and then in Trincomalee in Eastern Province in May 2000 (Jayasuriya, 2005). Subsequently more districts in central, north and north-western parts of the country were invaded, including Kilinochchi, Matale and Jaffna (Fig. 12.4). The weed is thought to have arrived in Sri Lanka as a contaminant of mustard seed brought in by an Indian Peace Keeping Force (IPKF) to their camps in northern Sri Lanka during the civil war in the 1980s. However, another second view is that the weed was introduced to Sri Lanka, by the IPKF, with goats brought from India in 1987 (Jayasuriya, 2005). It is commonly known as 'Indian weed' in Sri Lanka, and may have been introduced recurrently through seed lots of mustard, chilli (*Capsicum annum* L.) and onion (*Allium cepa* L.). Thus, it would seem that parthenium weed has been introduced to Sri Lanka on a number of occasions since the 1980s.

In Vavuniya, parthenium weed initially occurred in large populations on the bunds of the irrigation canals but then spread rapidly into the irrigated paddy rice fields. From there it has spread to upland fields used for other crops during the Yala season (May to September). Parthenium weed has become a problematic weed in vegetable crops of Northern Province, including tomato (*Solanum lycopersicum* L.) in the Jaffna peninsula where it has been shown to reduce yields significantly (Nishanthan *et al.*, 2013). Plant dispersal is thought to involve seed transported by human and animal movements, wind and water, and in river sand taken from the Mahawali River for construction purposes in Kandaketiya district (Jayasuria, 2005). Studies carried out on the impact of the weed on plant species diversity in Jaffana have shown that the weed can completely suppress all natural vegetation and form monocultures (Nandakumar *et al.*, 2001; Sukanthan *et al.*, 2001). The effects of parthenium weed on human and animal health have not been well studied in Sri Lanka. However, Jayasuria (2005) described a personal observation of one severe allergy case.

After its initial invasion, an awareness programme started in Sri Lanka and the weed was declared a Weed of National Significance by the Sri Lanka Council for Agricultural Research Policy in its National Weed Strategy for 2009–2014 (Rajapakse *et al.*, 2012). However, knowledge on the spread of parthenium weed in Sri Lanka and the weed's management remain poor.

12.1.5 Bangladesh

It is suspected that the initial source of parthenium in Bangladesh was seed attached to vehicles from India, where this weed has been widespread for 30 years (Akter and Zuberi, 2009). This possibility is supported by the presence of parthenium weed in the vicinity of vehicle service stations near the border with India. The first report of the weed in Bangladesh was in 1994 (Azam, 1999). However, 20 years later the weed has spread to at least 35 districts, including Jessore, Faridpur, Norail, Magura, Rajshahi, Natore, Sirajgonj, Manikgonj, Dhaka and Mymensingh (Fig. 12.5). The present infestations are mostly along the roadsides, but in a 2011 survey it was found that eight different crops (rice, wheat, onion, cucurbit, field pea, bean, mustard and banana) had already been invaded by this weed (S.W. Adkins, Australia, 2011, personal observation). Although there are quarantine centres in Bangladesh for detecting invasive species, informal trade of food crops within and across the border is likely to be adding to the weed's spread. Although studies on the impact of parthenium weed on crop production have been undertaken in other Asian countries, very few studies have been done in Bangladesh, which has to import huge amounts of food grain from India to sustain its population. Although the weed has now invaded most parts of Bangladesh, very little is known about its impacts upon human or animal health, agriculture and the environment.

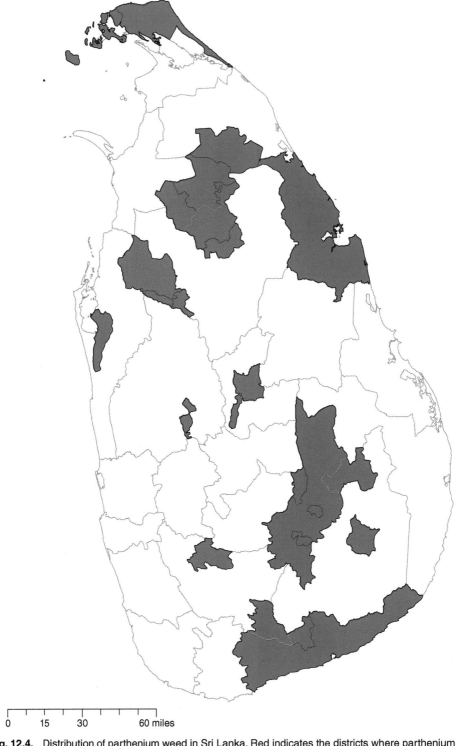

Fig. 12.4.   Distribution of parthenium weed in Sri Lanka. Red indicates the districts where parthenium weed is present. (A. Shabbir.)

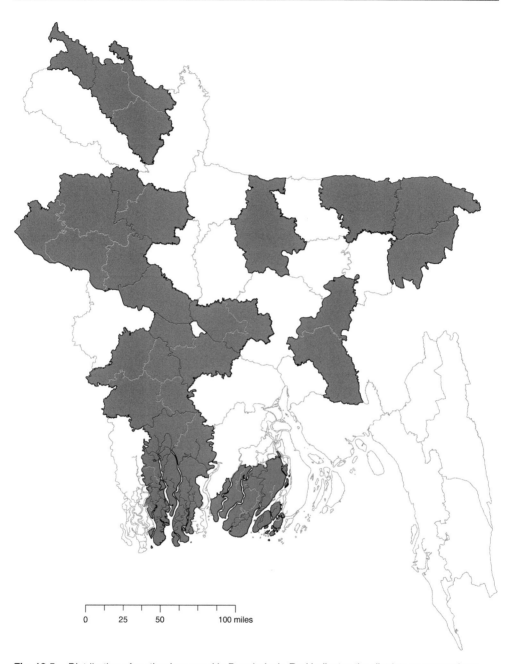

Fig. 12.5.   Distribution of parthenium weed in Bangladesh. Red indicates the districts or areas where parthenium weed is present. (A. Shabbir.)

12.1.6 Bhutan

Parthenium weed was introduced to Bhutan through India (Parker, 1992) in the 1980s. At present, it is mainly a weed of roadsides, wastelands and some crops (Tshering and Adkins, 2012). It is a dry land weed that mostly occurs below 1700 m asl (districts of Chhukha, Samdrupjongkhar, Lhuentse, Trashigang, Monggar, Zhemgang, Trongsa,

Sarpang, Tsirang, Dagana, Chhukha, Samtse, Punakha), but it has been found up to 2320 m asl in Thimpu (Fig. 12.6; Tshering and Adkins, 2012). A survey that covered Sarpang (250 m asl) to Thimphu (2320 m asl), including Tsirang and Wangdue districts, has shown that between Sarpang and Tsirang the weed was absent but that it started to appear in patches especially near the road settlements between Tsirang and Wangdue. Interestingly, local people do recognize it as a problem weed (Biswas and Das, 2007; Tshering and Adkins, 2012). In a national newspaper article it was reported that the weed was causing hay fever and dermatitis in students and teachers in Gyalposhing High School in Monggar district.

Until now, the negative impacts of parthenium weed have gone unnoticed in Bhutan due to limited scientific research the country has undertaken. Further research on its impacts on crop production, a national strategy for the weed's management and awareness on the health aspects of this weed are all needed.

12.2 Management of Parthenium Weed

The rapid spread of parthenium weed throughout the Southern Asian countries is now massively impacting agricultural production, reducing natural ecosystem biodiversity and affecting the health of up to 40 million people. In most countries an effort is underway to manage the weed using a number of methods. However, no single approach appears to be satisfactory, as each method suffers from one or more limitations, including inefficiency, high cost, impracticability and environmental safety.

12.2.1 Cultural control

Among Southern Asian countries, Sri Lanka is one of two countries to develop legislation for parthenium weed. Under the Plant Protection Ordinance No. 35 of 1999, parthenium weed is recognized as a noxious plant. It is given the status of a 'Weed of National

0   20   40   80 miles

Fig. 12.6. Distribution of parthenium weed in Bhutan. Red indicates the districts or areas where parthenium weed is present. (A. Shabbir.)

Significance' by the Sri Lanka Council for Agricultural Research Policy in its National Weed Strategy for 2009–2014 (Rajapakse et al., 2012). As it is also a plant on the National Priorities in Plant Protection Research 2011–2013 register (Marambe et al., 2011), this weed has a high level of recognition as a weed requiring management. In India, the State of Karnataka issued a notification on 23 October 1975 declaring parthenium weed as noxious plant under the Karnataka Agricultural Pests and Disease Act 1968. Under this Act, in the 1980s the Government of Karnataka issued notices to Bangalore to eradicate parthenium weed, but these notices were not upheld. Thus, in spite of this comprehensive Act, the management of the weed in Bangalore and elsewhere in India did not improve (Mahadevappa, 2009). Yaduraju et al. (2005) emphasized the importance of greater public and political involvement to help tackle the problem of parthenium weed in India and suggested that simply declaring the weed noxious will serve little purpose unless action is implemented effectively. In other countries in Southern Asia where parthenium weed has become a problem, legislative measures have not been developed.

As for most weeds, prevention of parthenium weed coming into a new area is much cheaper than trying to eradicate it once it has arrived. Since vehicles, livestock and seed and feed lots are known risks, management of seed movement through these pathways could be something Southern Asia could introduce as a cultural control approach.

12.2.2 Physical control

Manual uprooting of parthenium weed before flowering is an effective control approach that can easily be undertaken when the soil is moist. However, this method is only effective on small and isolated populations, such as those found in residential zones or in small agricultural fields. It is not economical or practical over the much larger areas where the weed most often appears

(Yaduraju et al., 2005). The hand pulling strategy is more commonly used in Southern Asian countries than in other locations around the world because labour is cheap and most people are unaware of the associated health risks of the plant. However, in India, manual methods of removal are gradually being reduced as they are becoming more expensive to apply (Kandasamy, 2005). Hand pulling of individual plants must be performed carefully and protective clothing should be worn and subsequently washed to prevent the possibility of an allergic reaction developing (Gupta and Sharma, 1977; Parsons and Cuthbertson, 1992). Another disadvantage of this method is the rapid regeneration of new populations from the seed bank after manual removal, in addition to plants regrowing from the remains of plants with a root system left behind after pulling (Gupta and Sharma, 1977; Shrestha et al., 2015).

Ploughing before plants reach the flowering growth stage and establishing a pasture or another crop can be effective in the management of the weed. However, such approaches are not often used in Southern Asia. To be effective using this method, seeds need to be buried by more than 5 cm (Tamado et al., 2002). In Southern Asian countries where rototilling is financially viable, future management could be undertaken by this method. Mechanical treatments on their own are not considered efficient in managing the weed: the weed often sets some seed before these methods can be carried out and these methods may also aid the spread of parthenium weed seeds (Haseler, 1976).

Cutting and slashing is a common method of parthenium weed management in developing countries of Southern Asia. In Nepal, municipalities and village committees often deploy labourers to clean up parthenium weed growing along roadsides (Shrestha et al., 2015). In Pakistan, slashing the weed on the bunds of watercourses is a common practice, especially during the rainy season when the growth of the weed is luxuriant. Unfortunately, most of the slashed plants are not destroyed properly and are allowed to remain by the watercourses to dry. This practice allows seed to become part of the soil seed bank from which the weed will regrow. Mowing can be effective but can also result in the rapid regeneration of new plants from uncut lateral shoots that are close to the ground (Gupta and Sharma, 1977).

In implementing physical control, the people involved should be made aware of the possible health hazard and the precautions needed to minimize contact with airborne plant parts (see Allan *et al.*, Chapter 6, this volume). In one clinical patch study involving 300 workers employed to manually uproot parthenium weed in Bangalore, 56% became sensitized to the weed and 4% developed contact skin dermatitis (Rao *et al.*, 1977).

The role of burning in the management of parthenium weed in Southern Asia has not been studied in any great detail and needs further investigation. In wheat crops in the Punjab Province of Pakistan, the residues are burnt after harvest; it was noticed that parthenium weed seedlings emerged rapidly from the patches that had escaped the fire (Fig. 12.7), but not from the inter-row areas that were burnt (Shabbir, 2007).

12.2.3 Chemical control

Various herbicides have been reported to be effective, over the shorter term, in the management of parthenium weed in a number of situations in Southern Asia (Table 12.1). In certain urban areas of India, parthenium weed is managed successfully with sodium chloride (15% solution) (Banerjee *et al.*, 1977). However, in agricultural regions post-emergence application of synthetic herbicides is more common. Fernoxone (80%, 2,4-D sodium salt) used as a post-emergence application could control parthenium weed within 20 days after application (Leela, 1987). Mishra and Bhan (1995) evaluated the efficacy of two sulfonylurea herbicides (chlorimuron ethyl and metsulfuron methyl), glyphosate and atrazine against parthenium weed in a non-cropped area of Jabalpur. Most treatments were found to be effective in controlling parthenium weed, but atrazine could only reduce flower numbers and seed set per plant. Field experiments conducted in a non-cropped situation in Haryana State to control parthenium weed revealed that new formulations of glyphosate (MON 8793 and MON 8794) provided excellent control of the weed (Singh *et al.*, 2004).

Fig. 12.7. Emergence of parthenium weed plants through the residues after the burning of the trash from a wheat crop. (A. Shabbir.)

In a non-cropped broad-acre field study in Pakistan, Khan *et al.* (2012a) evaluated a number of herbicides, alone or in combination (e.g. bromoxynil + MCPA @ 0.80; glyphosate @ 4.00; atrazine @ 1.0; 2,4-D @ 1.00; metribuzin @ 2.00; S-metolachlor @ 1.92; triasulfuron + terbutryn @ 0.30; pendimethethalin @ 1.50; atrazine + S-metolachlor @ 1.50 kg active ingredient/ha) to control parthenium weed at two growth stages (vegetative and flowering stages). At both stages, maximum parthenium weed mortality was recorded with glyphosate and metribuzin treatments and at 4 weeks after treatment. In another field study, Shabbir (2014) tested the full recommended and a lower dose of two herbicides (glyphosate and isoproturon) and found both were effective in killing parthenium weed.

In Southern Asia, chemical control of parthenium weed is financially viable in some high-value crops and in other important circumstances, such as along roadsides, in public parks or on private properties. However, it is not cost effective to control this weed chemically over the vast areas of wastelands, rangelands, lower value field crops or within forests where the weed is also commonly found (Shrestha *et al.*, 2015). Moreover, the use of chemicals may be an environmental concern because non-target plants can also be damaged or killed. With any long-term chemical management approach, the potential for herbicide resistance within the weed community is always possible. Parthenium weed has already developed resistance to herbicides in other parts of the world (e.g. Njoroge, 1991; Vila-Aiub *et al.*, 2008), but not yet in Southern Asia.

12.3 Biological Control

12.3.1 Classical biological control

Biological control, an approach that uses insect herbivores or plant pathogens to suppress plant growth, is one of the most important tactics used for managing invasive weeds around the world, however it is poorly adopted in Southern Asia. In this region only India has so far officially released biological control agents against parthenium weed. Sri Lanka, Pakistan and Nepal have agents that have arrived there accidentally (see Dhileepan *et al.*, Chapter 7, this volume; Adkins and Shabbir, 2014).

India

Due to the absence of effective natural enemies in India, efforts were made to import herbivorous insects that specifically attack parthenium weed from the weed's native range. The introduction into quarantine of *Smicronyx lutulentus* Dietz. (Coleoptera: Curculionidae), *Epiblema strenuana* (Walker) (Lepidoptera: Tortricidae) and *Zygogramma bicolorata* Pallister (Coleoptera: Chrysomelidae) was undertaken on permits issued by the Plant Protection Advisor to the Government of India in 1983 (Table 12.2). The culture of *S. lutulentus* could not be established and the culture of *E. strenuana* was destroyed because this agent was shown to lay eggs and develop on niger (*Guizotia abyssinica* L.), a cultivated oilseed crop grown in India. The chrysomelid leaf-feeding beetle, *Z. bicolorata*, imported from Mexico, was released after host specificity testing in 1984 (Jayanth and Nagarkatti, 1987). These field releases were carried out in Bangalore and since then the beetle has established widely across the country (Jayanth and Bali, 1994; Dhiman and Bhargava, 2005). Both the larvae and adults feed on parthenium weed leaves and can completely defoliate parthenium weed plants and reduce flower production by 98% in a few days (Dhileepan *et al.*, 2000; Jayanth and Bali, 1994).

With the help of mass-rearing and release efforts by the Indian Council of Agricultural Research, it has been estimated that the beetle has travelled and multiplied to cover *c.*5 million ha within 15 states. These populations have the potential to spread further under ideal climatic conditions (Jayanth and Visalakshy, 1994; Viraktamath *et al.*, 2004). In a recent study, Dhileepan and Wilmot Senaratne (2009) mapped the incidence of the beetle in India based on metadata and found three distinct clusters

Table 12.2. The classical biological control agents of parthenium weed that have been introduced or arrived into Southern Asian countries.

Country	Agent	Source	Year of release/report	Status
India	*Smicronyx lutulentus* Dietz	Mexico	1983	Could not be reared in quarantine
	Zygogramma bicolorata Pallister	Mexico	1983	First released in Bangalore and then several releases to other states. Established widely, aestivates in dry period
	Epiblema strenuana Walker	Mexico via Australia	1985	Since insect was able to develop on niger crop (*Guziotia abyssinica* L.), the stock was destroyed in quarantine
	Puccinia abrupta var. *partheniicola* (Jackson) Parmelee	unknown	unknown	Very localized at higher elevations, impact minimal
Pakistan	*Z. bicolorata*	India	2002	Entered from India. Well established in central and northern Punjab, Islamabad Capital Territory and parts of Khyber Pakhtunkhwa
Nepal	*Z. bicolorata*	India	2004 to 2006	Entered from India. Well established in southern part of Nepal
	P. abrupta var. *parthenicola*	unknown	2011	Some areas in Kathmandu Valley
Sri Lanka	*P. xanthii*	unknown	2005?	Released after host-range testing
	E. strenuana			

of establishment which were in the north, south and eastern regions, but covering most states in India.

Parthenium plants are reported to produce an average of 5925 flowers per plant (Joshi, 1991), which release enormous quantities of pollen (624 million pollen grains per plant) into the atmosphere (Towers *et al.*, 1977). The defoliation of parthenium weed plants by *Z. bicolorata* in and around Bangalore has caused an overall reduction in the pollen density in the atmosphere (Jayanth, 1996) and a subsequent reduction in the incidence of allergies. In response to defoliation of parthenium weed by *Z. bicolorata*, 40 different plant species, including eight grasses, were observed to grow in its place (Jayanth and Visalakshy, 1996).

As reported elsewhere (Jayanth and Visalakshy, 1994), the effectiveness of the leaf-feeding beetle as a biological control agent for parthenium weed is limited to a period of activity during the summer and autumn months. During the remainder of the year it undergoes diapause in the soil. As parthenium weed can germinate and flower throughout the whole year in some regions (provided there is adequate soil moisture), the cooler season generation of this weed easily escapes the impact of the agent.

Another biological control agent, the winter rust *Puccinia abrupta* Diet and Holw. var. *partheniicola* (Jackson) Parmelee, is also present in India. This agent is highly host specific, but it was not released by authorities and its source is unknown. Although the rust is established in India, its impact on parthenium weed plant health is minimal (Parker *et al.*, 1994; Bagyanarayana and Manoharachary, 1997).

Nepal

The fortuitous arrival of *Z. bicolorata* into Nepal in the period 2004–2006 (Shrestha *et al.*, 2010) has led to a degree of weed suppression since 2009 in the areas of Hetaunda, Bharatpur, Butawal and Bhairahawa in the Tarai region (Shrestha *et al.*, 2010). It is speculated that the leaf-feeding beetle entered Nepal in the same way as parthenium weed, that is along the road corridors from India, where it had been first released in Bangalore in 1984 and in the northern states of India (e.g. West Bengal, Uttar Pradesh, Bihar, Uttarakhand) in 2000 (Sushilkumar, 2009). The populations of the beetle are expanding naturally into new areas, including those within the Kathmandu Valley (Shrestha *et al.*, 2011). Until now, the mass rearing and release of the beetle into new locations has not been possible, because the national authorities (e.g. Departments of Agriculture, Forest, Public Health) have not acknowledged that parthenium weed is a 'serious' threat to health, agricultural production and the environment (Shrestha *et al.*, 2015). In Nepal, the leaf-feeding beetle is only an active leaf feeder for about 4 months (June to September), during the monsoon season. The current known distribution of the beetle in Nepal lies well within a CLIMEX projection (a projection created using modified

parameters first established by Dhileepan and Senaratne, 2009) of its potential distribution (Fig. 12.7). This projection also shows that there are many other areas in Nepal that are climatically suitable for the development of the beetle outside of its present range. According to the CLIMEX projections, the most southerly regions (Tarai, Siwalik and Hill) in central and western Nepal are slightly less suitable for the leaf-feeding beetle, while the northern (Middle and High Mountains) sector is not favourable (Fig. 12.8).

Another biological control agent, *P. abrupta* var. *partheniicola*, the winter rust, is also present in Nepal (Table 12.2). This rust was first encountered at Kirtipur, in the Kathmandu Valley, in May 2011 (Shrestha, 2012). However, its distribution has remained localized to just a few places in the Valley and the damage it has caused to parthenium weed has been minimal. The rust has been reported on parthenium weed plants at high elevations (930 m asl; Kumar

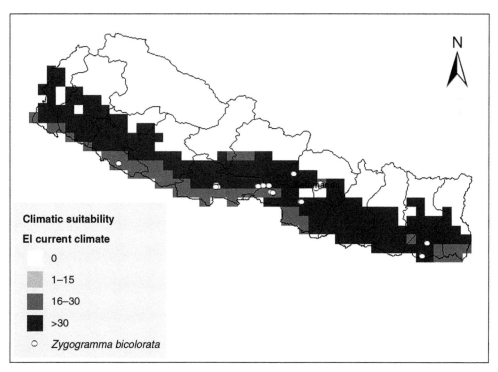

Fig. 12.8. The current distribution and climatic suitability for leaf-feeding beetle (*Zygogramma bicolorata*) in Nepal, modelled using CLIMEX software. (Reproduced with permission from Shrestha *et al.*, 2015.)

et al., 2008). It is highly host specific. It is found infecting parthenium weed in Central and South America (Parmelee, 1967), where the strains are more aggressive than those present in India (Kumar and Evans, 2005). This rust has also been reported in other Asian countries (e.g. India and China). However, as with Nepal, this agent was not released intentionally and its pathway of introduction into these countries remains unknown, but could have been as a seed-born pathogen on the initially introduced seed.

Pakistan

In Pakistan, the leaf-feeding beetle *Z. bicolorata* was first found at the Changa Manga forest reserves near the district Lahore (Javaid and Shabbir, 2007). Presumably the agent arrived in Pakistan from India with the introduction of the weed. Shabbir *et al.* (2012) collected data on the distribution of the leaf-feeding beetle and parthenium weed in the Punjab Province of Pakistan and found that two major clusters of establishment existed: one cluster in the eastern districts (Lahore, Narowal, Gujrat and Sialkot) on the Indian border and one in north-western districts (Attock, Chakwal). The beetle was not recorded from central and southern districts of Punjab Province (Fig. 12.2).

12.3.2 Mycoherbicide/inundated biological control

In India, a number of fungal pathogens (such as *Colletotrichum gloeosporioides, C. capsici, Exserohilum rostratum, Oidium parthenii, Sclerotium rolfsii, Fusarium pallidoroseum, F. moniliforme, F. oxysporum, F. solani, Myrothecium roridum, Alternaria alternata, A. dianthi, A. macrosporus, Cladosporium cladosporioides, Aspergillus fumigatus* and *Stachybotrys* sp.) have been isolated from various parts of the parthenium plant (Kumar and Rao, 1977; Pandey *et al.*, 1992) for possible use in a mycoherbicide formulation. Similarly, bacteria such as *Pseudomonas solanacearum, Xanthomonas campestris* pv.

parthenii nov. and *X. campestris* pv. *Phaseoli* have also been isolated from parthenium weed (Kishun and Chand, 1987; Ovies and Larrinaga, 1988; Chand *et al.*, 1995) and considered as possible components to a new weed-management formulation. Research on mycoherbicides under a Department for International Development-sponsored project resulted in the identification of several new pathogens on parthenium weed in India and a number were evaluated as possible new mycoherbicides (Evans *et al.*, 2000). Based on the early results, *F. pallidoroseum* [isolate WF (Ph) 30] was selected for the development of a mycoherbicide for parthenium weed, but since then no commercial product has been developed. Populations of the weed from different Indian states were found to be susceptible to this isolate, but formulations did not prove to give satisfactory control in a field trial conducted in Bangalore (Kumar, 2005).

12.3.3 Suppressive plants

India

Early work undertaken on plants that could compete or suppress the growth of parthenium weed has been undertaken in India. Parthenium weed growth has been shown to be suppressed by a number of companion planted crops (fodder plants, sorghum, sunflower and maize) and by self-perpetuating plant species such as oneleaf senna (*Senna uniflora* (Mill.) Irwin & Barneby), sicklepod (*Senna tora* (L.) Roxb.), Indian marigold (*Tagetes erecta* L.), Indian mallow (*Abutilon indicum* (Link) Sweet), ban tolsi (*Croton bonplandianus* Baill.) and matura tea tree (*Cassia auriculata* (L.) Roxb.) in non-crop areas.

Among the various plants screened, oneleaf senna, a leguminous undershrub, was found to be the most suitable plant to suppress the growth of parthenium weed and to reduce its population size (Singh, 1983). Joshi (1991) revealed that seedlings of oneleaf senna could outcompete the seedlings of parthenium weed, causing a reduction in their height, their dry weight and the number of flowers they produced. It was able to replace 90% of a 4800 m² parthenium weed population in a period of 5 years. Another species, matura tea tree, was also reported to suppress the growth of parthenium weed in another region of India (Joshi and Mahadevappa, 1986). A field study carried out by Kandasamy and Sankaran (1997) revealed Indian mallow, a naturally occurring plant in wastelands of India, could reduce the growth of parthenium weed populations by 50%. In certain parts of India, crop rotation using Indian marigold during the rainy season instead of the usual crop is effective in reducing parthenium weed infestation in cultivated areas (Kumari, 2014).

Some of the species used in India for suppressing the growth of parthenium weed are themselves considered to be weeds and the use of such plants in a management approach should be considered carefully. Moreover, the basis for growth suppression is unknown as in the field it is not clear whether the effects are due to allelopathy or competition, or both (Khan et al., 2012a). Out of all the studies undertaken so far, none have focused on the effect of suppressive plants on the production of parthenium weed seed, or on the subsequent size of its soil seed bank. Moreover, no attempt has been made to look at the combined effect of other management strategies with suppressive plants on parthenium weed growth and reproduction.

Pakistan

In Pakistan, native and exotic grasses such as Congo grass (*Imperata cylindrica* (L.) Beauv.), halfa grass (*Desmostachya bipinnata* (L.) Stapf.), blue-stem (*Dichanthium annulatum* (Forssk.) Stapf.), slender buffel grass (*Cenchrus pennisetiformis* Hochst. & Steud.) and Johnson grass (*Sorghum halepense* (L.) Pers.) have all been found to suppress the growth of parthenium weed by a mechanism that was thought to involve allelopathy (Anjum et al., 2005; Anjum and Bajwa, 2005; Javaid et al., 2005; Javaid and Anjum, 2006). Interestingly, most of these plants have been listed as noxious weeds in other countries (Holm et al., 1977; Parsons and Cuthbertson, 1992). More recently, Khan

et al. (2014) tested a number of fodder species at two contrasting locations in Pakistan. Most of the fodder species suppressed the growth of parthenium weed, with Columbus grass (*Sorghum almum* Parodi), buffel grass (*Cenchrus ciliaris* L.) and Rhodes grass (*Chloris gayana* Kunth) all suppressing growth by >73% and producing at least 600 g/m² of dry fodder biomass. Sowing of selected pasture plants in parthenium weed-infested areas is therefore thought to be a useful way to suppress the growth of the weed and to provide improved fodder for stock. The use of pasture suppression is an easy approach to apply, sustainable over time, profitable under a wide range of environmental conditions and can promote native plant biodiversity (Adkins and Shabbir, 2014).

The testing and release of classical biological control agents against parthenium weed in Southern Asian countries would seem to be a long and costly process. Therefore the use of suppressive plants could be seen as stand-alone or complementary strategies to that of the classical biological control approach.

12.4 Conclusions

Parthenium weed, arguably the most important invasive plant species in Southern Asia, is well adapted to thrive in the wide-ranging habitats of this region, making management extremely difficult. One conclusion reached is that only an integrated management approach, which involves a suite of strategies (cultural, physical, biological and chemical), will be effective in managing this weed on a long-term and sustainable basis in this region. Management has to be viewed in the long term and needs to involve collaboration across states, regions and countries, as there seems to be no 'silver bullet' solution to this challenging weed.

Acknowledgements

The authors are grateful to K. Dhileepan (Australia), P. Saravanane (India), Rezaul

Karim and Ilias Hossain (Bangladesh), Buddhi Marambe (Sri Lanka) and Karma Chophyll (Bhutan) for providing information on parthenium weed distribution maps.

References

Adkins, S. and Shabbir, A. (2014) Biology, ecology and management of the invasive parthenium weed (*Parthenium hysterophorus* L.). *Pest Management Science* 70, 1023–1029.

Akter, A. and Zuberi, M.I. (2009) Invasive alien species in Northern Bangladesh: Identification, inventory and impacts. *International Journal of Biodiversity Conservation* 15, 129–134.

Anjum, T. and Bajwa, R. (2005) Biocontrol potential of grasses against parthenium weed. In: Prasad, T.V.R., Nanjappa, H.V., Devendra, R., Manjunath, A., Subramanya, S.C., Chandrashekar *et al.* (eds) *Proceedings of the Second International Conference on Parthenium Management*, University of Agricultural Sciences, Bangalore, India, pp. 143–146.

Anjum, T., Bajwa, R. and Javaid, A. (2005) Biological control of parthenium 1: Effect of *Imperata cyclindrica* on distribution, germination and seedling growth of *Parthenium hysterophorus* L. *International Journal of Agriculture and Biology* 7, 448–450.

Anwar, W., Khan, S.N., Tahira, J.J. and Suliman, R. (2012) *Parthenium hysterophorus*: An emerging threat for *Curcuma longa* fields of Kasur District, Punjab, Pakistan. *Pakistan Journal of Weed Science Research* 18, 91–97.

Awan, M.R., Niazi, B.H. and Khattak, Z. (1987) Impact of soil on vegetation in Islamabad and Rawalpindi areas. *Pakistan Journal of Agricultural Research* 13, 368–372.

Azam, M.A. (1999) Early warning systems for forest invasive species in Bangladesh. Available at: http://www.apfisn.net/sites/default/files/Bangladesh.pdf (accessed 24 May 2018).

Bagyanarayana, G. and Manoharachary, C. (1997) Studies on *Puccinia abrupta* var. *partheniicola* potential mycoherbicide. In: Mahadevappa, M. and Patil, V.C. (eds) *Proceedings of the First International Conference on Parthenium Management*, Vol. II. University of Agricultural Sciences, Dharwad, India, pp. 95–96.

Banerjee, S.N., Paharia, K.D. and Rajak, R.L. (1977) Control of *Parthenium hysterophorus* L. Paper presented at Weed Science Conference and Workshop, Hyderabad, India. Paper no. 69, pp. 107–108.

Biswas, R. and Das, A.P. (2007) A note on the distribution of *Parthenium hysterophorus* L. in the Trongsa District of Bhutan. *Pleione* 2, 58.

Chand, R., Singh, B.D., Singh, D. and Singh, P.N. (1995) *Xanthomonas campestris* pv. *parthenii* pathovar nov. incitant of leaf blight of parthenium. *Antonie van Leeuwenhoek* 68, 161–164.

Dalavai, B.L., Kandasamy, O.S., Hanumanthappa, M. and Arasumallaiah, L. (2008) Effect of herbicides and their application techniques on weed flora in onion at Coimbatore. *Environment and Ecology* 26, 2157–2160.

Dhileepan, K. and Wilmot Senaratne, K. (2009) How widespread is *Parthenium hysterophorus* and its biological control agent *Zygogramma bicolorata* in South Asia? *Weed Research* 49, 557–562.

Dhileepan, K., Setter, S.D. and McFadyen, R.E. (2000a) Impact of defoliation by the biocontrol agent *Zygogramma bicolorata* on the weed *Parthenium hysterophorus* in Australia. *Biocontrol* 45, 501–512.

Dhiman, S.C. and Bhargava, M.L. (2005) Seasonal occurrence and bio-control efficacy of *Zygogramma bicolorata* Ballister (Coleoptera: Chrysomelidae) on *Parthenium hysterophorus*. *Annals of Plant Protection Sciences* 13, 81–84.

Evans, H.C., Seier, M., Harvey, J., Djeddour, D., Aneja, K.R., *et al.* (2000) *Final Technical Report: Developing Strategies for the Control of Parthenium Weed in India Using Fungal Pathogens*. Submitted to the Department for International Development, United Kingdom.

Gupta, O.P. and Sharma, J.J. (1977) Parthenium menace in India and possible control measures. *FAO Plant Protection Bulletin* 25, 112–117.

Haseler, W.H. (1976) *Parthenium hysterophorus* L. in Australia. *PANS* 22, 515–517.

Holm, L., Plucknett, D., Pancho, J. and Herberger, J. (1977) *The World's Worst Weeds: Distribution and Biology*. University of Hawaii Press, Honolulu, Hawaii.

Javaid, A. and Anjum, T. (2006) Control of *Parthenium hysterophorus* L. by aqueous extracts of allelopathic grasses. *Pakistan Journal of Botany* 38, 139–146.

Javaid, A. and Shabbir, A. (2007) First report of biological control of *Parthenium hysterophorus* by *Zygogramma bicolorata* in Pakistan. *Pakistan Journal of Phytopathology* 18, 99–200.

Javaid, A., Anjum, T. and Bajwa, R. (2005) Biological control of parthenium II: Allelopathic effect of *Desmostachya bipinnata* on distribution and early seedling growth of *Parthenium hysterophorus* L. *International Journal of Biology and Biotechnology* 2, 459–463.

Jayachandra (1971) Parthenium weed in Mysore State and its control. *Current Science* 40, 568–569.

Jayanth, K.P. (1996) Status of biological control trials against *Parthenium hysterophorus* by *Zygogramma bicolorata* in India. *Madras Agricultural Journal* 83, 672–678.

Jayanth, K.P. and Bali, G. (1994) Life table of the parthenium beetle, *Zygogramma bicolorata* Pallister (Coleoptera: Chrysomelidae) in Bangalore, India. *Insect Science and its Application* 15, 19–23.

Jayanth, K.P. and Nagarkatti, S. (1987) Investigations on the host-specificity and damage potential of *Zygogramma bicolorata* Pallister (Coleoptera: Chrysomelidae) introduced into India for the biological control of *Parthenium hysterophorus*. *Entomology* 12, 141–145.

Jayanth, K.P. and Visalakshy, P.N.G. (1994) Dispersal of the parthenium beetle *Zygogramma bicolorata* (Chrysomelidae) in India. *Biocontrol Science and Technology* 4, 363–365.

Jayanth, K.P. and Visalakshy, P.N.G. (1996) Succession of vegetation after suppression of parthenium weed by *Zygogramma bicolorata* in Bangalore, India. *Biological Agriculture & Horticulture* 12, 303–309.

Jayasuriya, A.H.M. (1999) Weed alert. *Parthenium hysterophorus* in Sri Lanka. *Plant Protection Newsletter* 3(2). Postgraduate Institute of Agriculture, University of Peradeniya, Sri Lanka.

Jayasuriya, A.H.M. (2005) Parthenium weed – status and management in Sri Lanka. In: Prasad, T.V.R., Nanjappa, H.V., Devendra, R., Manjunath, A., Subramanya, S.C., Chandrashekar et al. (eds) *Proceedings of the Second International Conference on Parthenium Management*. University of Agricultural Sciences, Bangalore, India, pp. 36–43.

Joshi, S. (1991) Biocontrol of *Parthenium hysterophorus* L. *Crop Protection* 10, 429–431.

Joshi, S. and Mahadevappa, M. (1986) *Cassia sericea* SW. to fight *Parthenium hysterophorus*. *Current Science* 55, 261–262.

Kandasamy, O.S. (2005) Parthenium weed: Status and prospects of chemical control in India. In: Ramachandra Prasad, T.V., Nanjappa, H.V., Devendra, R., Manjunath, A., Subramanya, S.C., Chandrashekar et al. (eds) *Proceedings of the Second International Conference on Parthenium Management*. University of Agricultural Sciences, Bangalore, India, pp. 134–142.

Kandasamy, O.S. and Sankaran, S. (1997) Biological suppression of parthenium weed using competitive crops and plants. In: Mahadevappa, M. and Patil, V.C. (eds) *Proceedings of the First International Conference on Parthenium Management*. University of Agricultural Sciences, Dharwad, India, pp. 33–36.

Kapoor, R.T. (2012) Awareness related survey of an invasive alien weed, *Parthenium hysterophorus* L. in Gautam Budh Nagar District, Uttar Pradesh, India. *Journal of Agricultural Technology* 8, 1129–1140.

Karim, S.M.R. (2009) Parthenium weed: A new introduction to Bangladesh. *Abstracts, 21st Bangladesh Science Conference* Bangladesh Agricultural Research Institute, Gazipur, pp. 43–44.

Karki, D. (2009) Ecological and socio-economic impact of *Parthenium hysterophorus* L. Invasion in two urban cities of south-central Nepal. MSc thesis, Tribhuvan University, Kathmandu.

Khan, H., Marwat, K.B., Hassan, G. and Khan, M.A. (2012b) Chemical control of *Parthenium hysterophorus* L. at different growth stages in non-cropped area. *Pakistan Journal of Botany* 44, 1721–1726.

Khan, H., Marwat, K.B., Hassan, G. and Khan, M.A. (2013) Socio-economic impacts of parthenium weed in Peshawar valley, Khyber Pakhtunkhwa. *Pakistan Journal of Weed Science Research* 19, 275–293.

Khan, I (2015) The impact of *Lantana camara* invasion on the plant biodiversity of a reserve forest. MPhil thesis, Department of Botany University of the Punjab, Lahore, Pakistan.

Khan, N., O'Donnell, C., George, D. and Adkins, S.W. (2012a) Suppressive ability of selected fodder plants on the growth of parthenium weed. *Weed Research* 53, 61–68.

Kishun, R. and Chand, R. (1987) New collateral hosts for *Pseudomonas solanacearum*. *Indian Journal of Mycology and Plant Pathology* 17, 237.

Kololgi, P.D., Kololgi, S.D. and Kololgi, N.P. (1997) Dermatologic hazards of Parthenium in human beings. In: Mahadevappa, M. and Patil, V.C. (eds) *Proceedings of the First International Conference on Parthenium Management*. University of Agricultural Sciences, Dharwad, India, pp. 18–19.

Kumar, C. and Rao, A. (1977) Two new leaf spot diseases. *Indian Phytopathology* 30, 118–120.

Kumar, D., Ahmed, J. and Singh, S. (2013) Distribution and effect of *Parthenium hysterophorus* L. in Mehari sub-watershed of Rajouri Forest Range, J&K. *International Journal of Scientific Research* 2, 304–306.

Kumar, P.S. (2005) Scope of fungal pathogens in weed control in India. In: Rabindra, R.J., Hussaini, S.S. and Ramanujam, B. (eds) *Proceedings of the ICAR-CABI Workshop on*

Microbial Biopesticide Formulations and Application, 9–13 December 2002. Technical Document No. 55. Project Directorate of Biological Control, Bangalore, India, pp. 203–211.

Kumar, P.S. and Evans, H.C. (2005) The mycobiota of *Parthenium hysterophorus* in its native and exotic ranges: Opportunities for biological control in India. In: Prasad, T.V.R., Nanjappa, H.V., Devendra, R., Manjunath, A., Subramanya, S.C., *et al.* (eds) *Proceedings of the Second International Conference on Parthenium Management.* University of Agricultural Sciences, Bangalore, India, pp. 107–113

Kumar, P.S., Rabindra, R.J. and Ellison, C.A. (2008) Expanding classical biological control of weeds with pathogens in India: The way forward. In: *Proceedings of the XII International Symposium on Biological Control of Weeds. La Grande Motte, France, 22–27 April 2007.* CAB International, Wallingford, UK, pp 165–172.

Kumari, M. (2014) *Parthenium hysterophorus* L.: A noxious and rapidly spreading weed of India. *Journal of Chemical, Biological and Physical Sciences* 4, 1620–1628.

Leela, D. (1987) Control of *Parthenium hysterophorus* L. by herbicides. *Final Technical Report* AICRP-WC (ICAR/PL-480). IIHR, Bangalore, Inida.

Mahadevappa, M. (2009) *Parthenium: Insight into its Menace and Management.* Stadium Press, New Delhi, India.

Marambe, B., Jayaskera, S., Samarasinghe, J.D. and Wimal Kumara, P. (2011) *National Priorities in Plant Protection Research – Strategic Approach 2011–2013.* National Plant Protection Committee, Sri Lanka Council for Agricultural Research Policy.

Mishra, J.S. and Bhan, V.M. (1995) Efficacy of sulfonylurea herbicides against *Parthenium hysterophorus* L. *Indian Journal of Weed Science Research* 27, 45–48.

Mishra, J.S. and Bhan, V.M. (1996) Chemical control of carrot grass (*Parthenium hysterophorus*) and associated weeds in soybean (*Glycine max*). *Indian Journal of Agricultural Sciences* 66, 518–521.

Mujahid, I. (2015) Effect of *Parthenium hysterophorus* L. invasion on above ground vegetation and soilseed bank of Jhok Reserve Forest of Lahore. Master of Philosophy thesis, Department of Botany, University of the Punjab Lahore, Pakistan.

Nandakumar, J., Sukanthan, K. and Kugathasan, K.S. (2001) A study of the effects of *Parthenium hysterophorus* on the species diversity of local habitats. In: *Proceedings of the 57th Annual Sessions, Part I – Abstracts.* Sri Lanka Association for the Advancement of Science (SLAAS), Colombo, Sri Lanka, p. 167.

Narasimhan, T.R., Ananth, M., Narayana, M.S., Rajendra, M.B., Mangala, A. and Suba Rao, P.V. (1980) Toxicity of *Parthenium hysterophorus* L.: Partheniosis in cattle and buffaloes. *Indian Journal of Animal Science* 50, 173–178.

Naveed, M. (2015) Introduction and spread of exotic plant species through horticulture trade in Pakistan: A case study. MSc thesis, Department of Botany, University of the Punjab, Lahore, Pakistan.

Nishanthan, K., Sivachandiran, S. and Maram, B. (2013) Control of *Parthenium hysterophorus* L. and its impact on yield performance of tomato (*Solanum lycopersicum* L.) in the Northern Province of Sri Lanka. *Tropical Agricultural Research* 25, 56–68.

Njoroge, J.M. (1991) Tolerance of *Bidens pilosa* L and *Parthenium hysterophorus* L to paraquat (Gramoxone) in Kenya coffee. *Kenya Coffee* 56, 999–1001.

Ovies, J. and Larrinaga, L. (1988) Transmission of *Xanthomonas campestris* pv. *phaseoli* by a wild host. *Ciencia y Tecnica en la Agricultura, Proteccion de Plantas* 11, 23–30.

Pandey, A.K., Mishra, J., Rajak, R.C. and Hasija, S.K. (1992) Possibility of managing *Parthenium hysterophorus* L. through *Sclerotium rolfsii* Sacc. *National Academy Science Letters* 15, 111–112.

Parker, A., Holden, A.N.G. and Tomley, A.J. (1994) Host specificity testing and assessment of the pathogenicity of the rust, *Puccinia abrupta* var. *partheniicola,* as a biological control agent of parthenium weed (*Parthenium hysterophorus*). *Plant Pathology* 43, 1–16.

Parker, C. (1992) *Weeds of Bhutan.* Royal Government of Bhutan, Thimpu.

Parmelee, J.A. (1967) The autoecious species of *Puccinia* on Heliantheae in North America. *Canadian Journal of Botany* 45(12), 2267–2327.

Parsons, W.T. and Cuthbertson, E.G. (2001) *Noxious Weeds of Australia,* 2nd edn. CSIRO Publishing, Collingwood, Australia.

Paul, T.K. (2010) The earliest record of *Parthenium hysterophorus* L. (Asteraceae) in India. *Current Science* 98, 1272.

Rajapakse, R., Chandrasena, N., Marambe, B. and Amarasinghe, L. (2012) Planning for effective weed management: lessons from Sri Lanka. *Pakistan Journal of Weed Science Research* 18, 843–853.

Rao, P.V.S., Mangala, A., Rao, S.S.S. and Prakash, K.M.. (1977) Clinical and immunological studies on persons exposed to *Parthenium hysterophorus* L. *Experientia* 33, 1387–1388.

Razaq, Z.A., Vahidy, A.A. and Ali, S.I. (1994) Chromosome numbers in Compositae from Pakistan. *Annals of the Missouri Botanical Garden* 81, 800–808.

Reddy, K.N., Bryson, C.T. and Burke, I.C. (2007) Ragweed parthenium (*Parthenium hysterophorus*) control with pre-emergence and post-emergence herbicides. *Weed Technology* 21, 982–986.

Safdar, M.E., Tanveer, A., Khaliq, A. and Naeem, M.S. (2014) Allelopathic action of parthenium and its rhizospheric soil on maize as influenced by growing conditions. *Planta Daninha* 32, 243–253.

Shabbir, A. (2006) Parthenium weed: a threatening weed for agricultural lands. *Parthenium News* 1.

Shabbir, A. (2007) Burning crop residues reduces the parthenium weed emergence: A study. *Parthenium News* 2, 2.

Shabbir, A. (2012) Towards the improved management of parthenium weed: Complementing biological control with plant suppression. PhD thesis, The University of Queensland, Brisbane, Australia.

Shabbir, A. (2014) Chemical control of *Parthenium hysterophorus* L. *Pakistan Journal of Weed Science Research* 20, 1–10.

Shabbir, A. and Bajwa, R. (2006) Distribution of parthenium weed (*Parthenium hysterophorus* L.), an alien invasive weed species threatening the biodiversity of Islamabad. *Weed Biology and Management* 6, 89–95.

Shabbir, A., Dhileepan, K. and Adkins, S.W. (2012) Spread of parthenium weed and its biological control agent in the Punjab, Pakistan. *Pakistan Journal of Weed Science Research* 18, 581–588.

Shrestha, B.B. (2008) Phailedai Hanikarak Banaspati [Nepali]. *Kantipur National News* 17 May, p. 10.

Shrestha, B.B. (2012) *Puccinia abrupta* var. *partheniicola*: a biocontrol agent of *Parthenium hysterophorus* new to Nepal. *Biocontrol News and Information* 33, 2N.

Shrestha, B.B., Poudel, A., Karki, J., Gautam, R.D. and Jha, P.K. (2010) Fortuitous biological control of *Parthenium hysterophorus* by *Zygogramma bicolorata* in Nepal. *Journal of Natural History Museum* 25, 332–337.

Shrestha, B.B., Thapa-Magar, K.B., Poudel, A. and Shrestha, U.B. (2011) Beetle on the battle: Defoliation of *Parthenium hysterophorus* by *Zygogramma bicolorata* in Kathmandu valley, Nepal. *Botanica Orientalis* 8, 100–104.

Shrestha, B.B., Shabbir, A. and Adkins, S.W. (2015) *Parthenium hysterophorus* in Nepal: A review of its weed status and possibilities for management. *Weed Research* 55, 132–144.

Singh, N.P. (1983) Potential biological control of *Parthenium hysterophorus* L. *Current Science* 52, 644.

Singh, S.A., Yadav, S.B., Rajender, R.K., Malik and Singh, M. (2004) Control of ragweed parthenium (*Parthenium hysterophorus*) and associated weeds. *Weed Technology* 18, 658–664.

Sukanthan, K., Nandakumar, J. and Kugathasan, K.S. (2001) Spread of *Parthenium hysterophorus* in Valikamum. *Proceedings of Jaffna Science Association* 9, 1.

Sushilkumar (2009) Biological control of Parthenium in India: status and prospects. *Indian Journal of Weed Science* 41(1&2), 1–18.

Sushilkumar (2014) Spread, menace and management of *Parthenium*. *Indian Journal of Weed Science* 46, 205–219.

Sushilkumar and Varshney, J.G. (2010) Parthenium infestation and its estimated cost management in India. *Indian Journal of Weed Science* 42, 73–77.

Swaminathan, C., Vinayaand, R.S., Suresh, K.K. (1990) Allelopathic effects of *Parthenium hysterophorus* on germination and seedling growth of a few multi-purpose trees and arable crops. *The International Tree Crops Journal* 6, 143–150.

Tamado, T., Ohlander, L. and Milberg, P. (2002) Interference by the weed *Parthenium hysterophorus* L. with grain sorghum: Influence of weed density and duration of competition. *International Journal of Pest Management* 48, 183–188.

Tanveer, A., Khaliq, A., Ali, H.H., Mahajan, G. and Chauhan, B.S. (2015) Interference and management of *Parthenium*: The world's most important invasive weed. *Crop Protection* 68, 49–59.

Timsina, B., Shrestha, B.B., Rokaya, M.B. and Munzbergova, Z. (2011) Impact of *Parthenium hysterophorus* L. invasion on plant species composition and soil properties of grassland communities in Nepal. *Flora (Jena)* 206, 233–240.

Tiwari, S., Adhikari, B., Siwakoti, M. and Subedi, K. (2005) *An Inventory and Assessment of Invasive Alien Plant Species of Nepal*. IUCN – The World Conservation Union, Nepal.

Towers, G.H.N., Wat, C.K., Graham, E., Bandoni, R.J., Chan, G.F.Q., Mitchell, J.C. and Lam, J. (1977) Ultraviolet mediated antibiotic activity of species of Compositae caused by polyactylenic compounds. *Lloydia* 40, 487–498.

Towers, G.H.N. and Subba Rao, P.V. (1992) Impact of the pan-tropical weed, *Parthenium hysterophorus* L. on human affairs. In: Richardson,

R.G. (ed.) *Proceedings of the 1st International Weed Control Congress.* Weed Science Society of Victoria, Melbourne, Australia, pp. 134–138.

Tshering, C. and Adkins, S. (2012) Parthenium weed in Bhutan. *International Parthenium Weed Newsletter* 5, 9–10.

Vila-Aiub, M.M., Vidal, R.A., Balbi, M.C., Gundel, P.E., Trucco, F. and Ghersa, C.M. (2008) Glyphosate-resistant weeds of South America cropping systems: An overview. *Pest Management Science* 64, 366–371.

Viraktamath, C.A., Bhumannavar, B. and Patel, V. (2004) Biology and ecology of *Zygogramma bicolorata* Pallister. In: Jolivet, P., Santiago-Blay, J.A. and Schmitt, M. (eds) *New Developments in the Biology of Chrysomelidae*, SPB Academic, The Hague, The Netherlands, pp. 767–777.

Yaduraju, N.T., Sushilkumar, Prasad Babu, M.B.B. and Gogoi, A.K, (2005) *Parthenium hysterophorus* – distribution, problems and management strategies in India. In: Prasad, T.V.R., Nanjappa, H.V., Devendra, R., Manjunath, A., Subramanya, S.C., *et al.* (eds) *Proceedings of the Second International Conference on Parthenium Management.* University of Agricultural Sciences, Bangalore, India, pp. 6–10.

13 History and Management in East and South-east Asia

Boyang Shi,[1]* Saichun Tang,[2] Nguyen Thi Lan Thi[3] and Kunjithapatham Dhileepan[1]

[1]Biosecurity Queensland, Department of Agriculture and Fisheries, Brisbane, Queensland, Australia; [2]Guangxi Institute of Botany, Chinese Academy of Sciences, Guilin, China; [3]University of Sciences, Ho Chi Minh City, Vietnam

13.1 History of Parthenium Weed in East and South-east Asia

Parthenium weed (*Parthenium hysterophorus* L.) has become a dominant weed in a number of regions around the world, especially within eastern and southern Africa, Australia and the Pacific, and East, South and South-east Asia (Mainali *et al.*, 2015). In countries that have been badly infested, many natural ecosystems have also come under threat from its presence and persistence (Bajwa *et al.*, 2016). This weed has serious adverse impacts upon the biodiversity of natural communities, the productivity of rangelands and cropping systems as well as human and animal health (Chippendale and Panetta, 1994; Navie *et al.*, 2005; Adkins and Shabbir, 2014).

In East and South-east Asia, parthenium weed has invaded many countries, including Vietnam in 1922 (Arenes *et al.*, 1922), Southern China in 1926 (Li and Xie, 2002), Japan in 1972 (Tachikake and Nakamura, 2007), Taiwan in 1988 (Peng *et al.*, 1988), Korea in 1993 (Kil *et al.*, 2004; Kim, 2013), a new second introduction has been identified in Northern China (Shandong Province) in 2004, and it has been found in peninsular Malaysia and Sarawak in 2013 (Karim, 2013) and Thailand in 2016

(Fig. 13.1). So far, there has been no report of parthenium weed from Indonesia, the Philippines, Singapore or Timor-Leste (East Timor). However, Dr Siriporn Zungsontiporn (M. Day, Australia, personal communication) has discovered parthenium weed in Thailand in 2016 and her team members are working to eradicate it. In addition, there is also no report of parthenium weed in Myanmar, although it may be present in the north of the country (regions bordering with India and Bangladesh), or Brunei, where there is a good probability of it invading from Sarawak. Parthenium weed may also be present in Laos because this country is very close to the northern regions of Vietnam where the weed is common and has been present for many years. It is the view of many scientists that parthenium weed is still spreading rapidly in the East and South-east Asian region and its full invasion range is still to be realized (Tang *et al.*, 2008; McConnachie *et al.*, 2011; Karim, 2013; Kim, 2013; Li *et al.*, 2014).

Parthenium weed continues to be introduced to new regions, then from those points of introduction it spreads rapidly out into neighbouring areas that have a suitable climate to support its growth (Shi *et al.*, 2015). For example, after its introduction into southern China, it has rapidly spread from small isolated patches on roadsides

* Boyang.Shi@daf.qld.gov.au

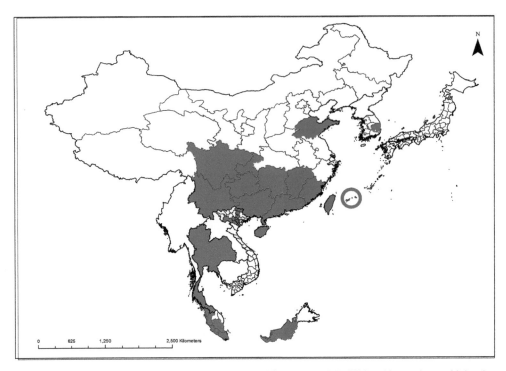

Fig. 13.1. Distribution of parthenium weed in East and South-east Asia (China, Korea, Japan, Malaysia, Vietnam and Thailand). Red shading or circled areas indicate provinces, municipalities or districts where parthenium weed is present. (A. Shabbir.)

and wasteland into field crops and is now commonly found in maize (*Zea mays* L.), sugarcane (*Saccharum officinarum* L.), groundnut (*Arachis hypogaea* L.) and vegetable crops, as well as in grasslands, open forested areas and along roadsides. It is in these new locations where it has its most damaging effects, including reducing crop yield, causing environmental damage and hindering traffic on both local and national roads (Tang *et al.*, 2008) (Fig. 13.2). In the early days of its invasion, farmers and government officials did not pay much attention to the management of this weed, which is likely to be a major factor behind its rapid spread (Sun, 2010; Wang and Xu, 2012). As in many other regions of the world, in East and South-east Asian countries parthenium weed has a long reproductive phase, which allows for the continuous production of seed over many months, which in turn results in massive stands of seedlings, then plants (Tang *et al.*, 2012).

13.2 China

13.2.1 History

The first Chinese parthenium weed specimen was collected in Yunnan Province in 1926 (Li and Xie, 2002), just after it had been reported in northern Vietnam. Since then, parthenium weed has slowly spread to the east and north-east, arriving in the provinces of Guangxi by 1935, Guangdong by 1932, Hainan by 1948, Guizhou by 1960 and Fujian by 1977, and in Hong Kong by 2002 and Macau by 2004. More recently, the weed spread into the provinces of Sichuan by 2008 (Zhou *et al.*, 2008), Hunan by 2012 (Xie and Zhang, 2012), Jiangxi by 2012 (Zeng and Qiu, 2012) and Chongqing Municipality by September 2015 (Chen *et al.*, 2016) (Fig. 13.1). In 2004 a newly introduced parthenium weed population was discovered in the northern province of Shandong (Wang and Hou, 2004). As in all

Fig. 13.2. Parthenium weed in a number of different habitats in southern China and Vietnam. China: (a) within grassland; (b) in low-density forest; (c) around a farm house. Vietnam: (d) a waterway in Huong Son Forest Protection; (e) Tam Dao National Park; (e) urban land around Hanoi. (a,b,c: Saichun Tang; d,e,f: Nguyen Thi Lan Thi.)

other locations around the world, spread of the weed is thought to be by seed entrapped on vehicles, in the transportation of seed crops, in soil around tree seedlings or in other agricultural products. Other natural methods of spread include wind and water, including irrigation, flood and river water (Sun, 2010). The recently discovered population in Shandong Province has been shown to be genetically different to that in southern China and therefore is believed to represent a separate introduction (Tang et al., 2012). The growth characteristics of this northern population are quite different from the southern population, with the northern population producing plants of greater stature (with larger stem diameters, longer internode and inflorescences, greater leaf area) and appearing to be much more invasive (Tang et al., 2012).

13.2.2 Biology

Parthenium weed grows from spring to autumn (March to October) and produces seed for most of this time, but with a peak in flower and seed production in the autumn (September to October). It grows well in areas where the annual rainfall is greater than 500 mm and when rain falls predominantly in the summer months. Its most dominant habitats in China are grasslands, sparse forests and roadsides. The population density, flower number, seed production and 1000-seed weight can all be modified by local environmental conditions. The population densities of parthenium weed in maize crops, sparse forested land, grasslands and roadside habitats were about 10 ± 2, 19 ± 3, 73 ± 25 and 80 ± 24 plants/m^2, respectively. The flower numbers/m^2 in these habitats were $10,298\pm1745$, 2618 ± 343, $14,539\pm3296$ and 5324 ± 1208, respectively. Seed production was highest in grasslands ($c.60,361\pm14,661$ seeds/m^2), and $42,085\pm6418$ seeds/m^2 in maize, $10,805\pm1489$ seeds/m^2 in sparse forested land and $10,308\pm1636$ seeds/m^2 in roadside habitats (Pu *et al.*, 2010). Seeds were heavier and larger in plants growing along roadsides than in all other habitats (the 1000-seed weight was 0.46 ± 0.01 g in roadside habitats and 0.34 ± 0.02 g in the other three habitats).

In China, parthenium weed can grow on a wide range of soil types with different nutrient contents. Plant height and seed weight increased with increasing soil phosphorus level. Stem, branch and inflorescence number, total biomass, flower mass and proportion of viable seed set all increased with increasing soil nitrogen content. In addition, plants increased their seed production under increased nitrogen levels, through enhanced branch and inflorescence production, greater flower mass and seed set; phosphorous-rich environments increased seed quality. The addition of nitrogen and phosphorus to soil could therefore promote the reproductive ability of parthenium weed and increase its invasiveness. Parthenium weed was able to acclimatize to different light intensities (22–100% light intensity) by changing certain growth parameters, such as total biomass and its allocation, total leaf area and specific leaf area (Pu *et al.*, 2010; Huang *et al.*, 2012). However, plants could not survive under very low light conditions (*c.*5% light intensity).

Parthenium weed produces a large number of white flower heads, and each flower head usually contains five fertile florets. However, flower heads can contain 6–9 ray florets and it is interesting to note that the only other common reports of this is in Vietnamese populations (S. Adkins, Queensland, personal observations) and may suggest a common ancestry.

13.2.3 Present distribution and crops infested

Parthenium weed is very commonly distributed in the southern provinces of China and here it is still spreading (Gao *et al.*, 2013). In addition, a new introduction is now found in the northern province of Shandong. In these two regions the weed can be found most commonly growing in wastelands, on vacant urban lands, in orchards, forested lands, floodplains, in agricultural areas, overgrazed pastures and along roadsides and railway tracks (Figs 13.2 and 13.3). Its reproductive traits and the strong adaptability of its seed germination to a wide range of environmental conditions are considered to be important factors that make it a successful invader in southern China.

In Shandong Province, parthenium weed was first observed at Shizi Road, Junan in 2004 (Wang and Hou, 2004) and was believed to have been introduced along with an imported soybean (*Glycine max* L.) seed lot from the USA. Wang and Xu (2011) have reported that the weed has spread dramatically in this province since 2004, with it now being found in Linmu County and in Hedong and Lanshan Regions in 2005. A survey in 2010 revealed parthenium weed covered 750 ha in four counties, 27 towns and 351 villages, appearing along roadsides, and on wastelands, river banks, cropping lands and land surrounding farm houses (Wang and Xu, 2011).

Fig. 13.3. Parthenium weed has been found in a wide range of cropping land in East and South-east Asian countries. Vietnam: (a) maize crop. China: (b) aubergine crop; (c) bank of paddy field; (d) mung-bean crop; (e) peanut crop; (f) maize crop. (a: Nguyen Thi Lan Thi; b,c,d: Saichun Tang; e,f: Boyang Shi.)

In the northwest of Guangxi, in the Zhuang Autonomous Region, parthenium weed was found in Longlin and Lingyun Counties and Nanning City during the 1950s and 1960s. By the 1970s, this weed had spread to Guigang City, Beihai City, Hepu County, Luocheng County, Pingqiang City, Ningming County, Tiane County, Laibin City and Donglan County. In the 1980s, the weed was also found in Yulin City and Leye County. It is interesting to note that the plant checklist of Guangxi published in 1971 did not record parthenium weed as being present, however this is likely to be an oversight. Parthenium weed has now spread throughout Guangxi Province (Fig. 13.1) and scientists have begun to realize that this weed is a serious problem. However, most people have remained unaware of the dangers of this weed, especially to their health. In Nanning, Guangxi Province, the weed has been found on farm land and in orchards, but the population densities are much lower than those on roadsides, riverbanks, wastelands and in forests. The view developed that these were the main sources of the infestations found on farm lands (Huang et al., 2012). Parthenium weed has significantly reduced vegetable production in these areas, where local farmers mainly depend on vegetable farming (Chen et al., 2016).

13.2.4 Management

Cultural weed control approaches can help reduce the spread of parthenium weed in the agricultural landscape, especially when the movement of machinery or vehicles is suspected to be one of the main vectors of spread. Cleaning of agricultural machinery to eliminate weed seed contamination, using appropriate sowing times for crops, the use of high-quality crop seed lots (without weed-seed contamination) and choosing an appropriate fertilizer regime to sustain effective crop growth will all help to reduce the parthenium weed threat (Parsons and Cuthbertson, 1992). However, such practices are rarely applied to the management of parthenium weed in China.

Physical weed control on the other hand plays an important role in the overall management approach to parthenium weed in China. Such physical approaches include mowing, slashing, smothering (using mulch) and ploughing, hoeing and hand pulling. All of these physical methods can effectively manage parthenium weed, but only over the short term (Kohli et al., 1997) because recruitment from the seed bank often follows and the management process has to be repeated. In China, because of cheap labour, physical approaches to weed control are among the most common and effective methods used to minimize parthenium weed populations (Sun, 2010). Manual uprooting or cutting of parthenium weed are very effective methods, but they need to be undertaken before flowering and repeated several times throughout the year when replenishment occurs from the seed bank (Gupta and Sharma, 1977; Khan, 2011).

Chemical weed control is another effective method to manage parthenium weed.

However, as different herbicides or different concentrations need to be applied at the different growth stages (Wang and Xu, 2011), such approaches become costly and difficult. In Shandong Province, a 10% glyphosate solution has been shown to give good management of adult plants (Sun, 2010), and in southern China either 2,4-D or glyphosate. Such chemical approaches can be cost effective when used in cropping land or on the large roadside populations, but not over huge areas of wastelands, in forest or grasslands where large populations of parthenium weed may be found (Khan, 2011). Farmers are aware that the best time to control parthenium weed chemically is when plants are small, and certainly before they flower (Huang et al., 2012).

There is no active biological control programme specifically targeting parthenium weed in any of the countries in East and South-east Asia, including China. However, biological agents are present in China because they were introduced to help manage closely related species such as ragweed (Ambrosia artemisiifolia L. and A. trifida L.). One biological control agent, the stem-galling moth (Epiblema strenuana Walker), introduced from Australia to China in 1990 for the management of ragweed (Wan and Ding, 1993), is now to be found on parthenium weed in Guangxi Province (S. Adkins, Queensland, personal observation in 2012). Another agent, winter rust (Puccinia abrupta v. partheniicola (Jackson) Parmelee) has been found in Baoxiu, Yunnan Province (S. Tang, Guangxi, personal observation in 2007) (Fig. 13.2). However, nine insects and two rust pathogens have been released as biological control agents for parthenium weed in other countries, such as Australia. So future introductions of some of these agents may add to the management of parthenium weed that has so far been by E. strenuana alone (Dhileepan, 2009).

Management approaches that involve the sowing of suppressive plant species into parthenium weed-infested areas have also proven to be useful in other countries (Wahab, 2005). However, this type of approach has not yet been used in China.

13.2.5 Health effects

Many people who have come into contact with parthenium weed in China have developed skin allergies (Chen *et al.*, 2016). One village doctor recalled seven of his patients had developed a very serious allergic reaction to parthenium weed and several others also had relatively mild allergic reactions (B. Shi, Queensland, personal observation). In the same village, 15% of the children also showed signs of allergy and these cases accounted for 2% of the total village population (Chang *et al.*, 2009; Sun, 2010). Allergic symptoms often appear as inflamed skin, accompanied by itching. Most people recover after treatment, but the problems often recur (Sun, 2010).

13.3 Vietnam

13.3.1 History

In Vietnam, parthenium weed has been present in Hanoi and its surrounding regions since 1922 (Arenes *et al.*, 1922). It was first officially identified as parthenium weed in 1924 (Li and Xie, 2002).

13.3.2 Biology

Several biological characteristics of the weed, including the density of plants, canopy diameter and seed production, were measured in populations growing in Cuc Phuong National Park (Ninh Binh Province) and Huong Son Protection Forest (Ha Tay, Hanoi) in October 2008 (Nguyen *et al.*, 2011; Fig. 13.2). The density of parthenium weed was *c.* 50 and 70 plants/m^2 in Cuc Phuong National Park and the Huong Son Protection Forest, respectively; the average canopy diameter was 30 and 70 cm, respectively. About 2500 seeds per plant were produced by plants at Cuc Phuong National Park; an average of 4100 seeds per plant were produced by plants in Huong Son Protection Forest. The study also indicated that the two biotypes present in both locations were similar in morphological appearance, but were different from Australian biotypes.

13.3.3 Present distribution

In an attempt to gain further information on the spread of parthenium weed in Vietnam, a rapid survey technique was used to determine the presence of the weed along roadsides, in fallow and cropping land, from the north to the south of Vietnam, and in more detail around Hanoi and several protected areas in the north of Vietnam, to create a distribution map of the weed (Nguyen *et al.*, 2011; Fig. 13.1). Parthenium weed was found in a number of provinces in the north, including Cao Bang, Bac Kan, Son La, Thai Nguyen, Phu Tho, Vinh Phuc, Bac Ninh, Hanoi Capital, Hung Yen, Hai Duong, Hai Phong, Ha Nam, Nam Dinh, Hoa Binh and Ninh Binh. However, no parthenium weed was present from Ho Chi Minh City to the Mekong Delta, or in other southern regions of Vietnam. It is thought that parthenium weed may be present in other northern regions, and possibly central Vietnam as well.

13.3.4 Management

Farmers often use mechanical methods to control the weed when it is present in cropping lands, but few farmers recognize this weed or know about its health impacts, so management is usually not practised. However, in 2011 the Vietnam Environment Administration of the Ministry of Natural Resources and Environment recognized parthenium weed as a significant threat and included it on their list of high-risk invasive weeds, and now encourage its better management.

13.3.5 Health effects

At present, there have been no records of human allergy or other health impacts due to parthenium weed reported from

Vietnam, although people do live permanently adjacent to dense stands of the weed in Huong Son Protection Forest and in Hanoi City. It is interesting to note that the local people living in the Red River Valley use parthenium weed for making a compost mixed with other plant materials and they have been doing this since the 1980s (T. Nguyen, Vietnam, personal observation). In addition, parthenium weed inflorescences are often used in bridal bouquets in Vinh Phuc Province.

13.4 Japan

Parthenium weed was first introduced accidentally into Okinawa, Japan around 1972 (Tachikake and Nakamura, 2007) and has spread very little, but it does now occur in other parts of southern Japan (Fig. 13.1).

13.5 South Korea

Parthenium weed was first observed in South Korea in 1993, then identified in 1995 (Kim, 2013). Its main distribution is around Masan City, in the south-eastern part of the Korean peninsula (Fig. 13.1). So far, there have been no reports of human allergy and crop production losses due to parthenium weed in South Korea. The weed was first found around Masan harbour. It may have come from Japan in the early 1990s because this harbour was within a free-export-zone receiving imported materials from Japan at that time. However, it is interesting to note that parthenium weed has not been reported in the region of Japan where these imports may have orginated.

13.6 Malaysia

One of the most recent countries where parthenium weed has been reported is Malaysia, in 2013 (Karim, 2013). However, it is thought from the size and frequency of the populations there that the weed must have been present in this country for a number of years before its identification. Now parthenium weed can be found in most districts of Malaysia, except for Kelantan and Terengganiu States (Fig. 13.1). In Malaysia, the weed has been found alongside roads and rivers, on wasteland, in farming and cropping lands, in parks and in the backyards of homes (Karim, 2013). It is thought that the spread of parthenium weed has been mainly by seed on vehicles and machinery, contamination of crop and pasture seed lots, livestock, hay and also in flood water (Karim, 2013). There is some evidence of allergy-induced mouth ulcers in goats in parthenium weed-infested areas (Karim, 2014). However, as most farmers do not recognize this weed, and fewer realize that it is toxic, few reports have linked parthenium weed to allergic reactions in people or their animals.

13.7 Recommendations

13.7.1 Cultural control

In East and South-east Asian countries, preventing the further spread of parthenium weed should be considered to be the most important management approach and should be implemented immediately. However, since the weed is poorly known, an effort should first be made to inform the public about this weed and its impacts. Governments should be encouraged to pay more attention to its invasive potential, establish policies and then develop strategies to manage this weed. Methods to enhance public awareness such as television reports, lectures, newspapers and distribution of technical information should be considered (Karim, 2014), so that people can recognize the weed, understand its harmful effects and know how to take action against it (Sun, 2010; Wang and Xu, 2011). More effort should also be made to examine seed and food grains as possible introduction routes. This will require governments to impose strict guidelines on the movement of seed, vegetables, food, including imported and exported agricultural commodities and domestic transportation of food and crops (Sun, 2010). In addition, governments should be encouraged to strengthen their scientific research and to create and improve the early warning systems used to detect new invasions so that management can be swiftly employed (Wang and Xu, 2011).

13.7.2 Integrated weed management

A wide range of approaches have been employed to attempt to manage parthenium weed in East and South-east Asia. However, as in many other parts of the world, the utilization of a single management approach has not been proven effective. An integrated approach seems to be the only successful approach that can be used for its effective management. The integrated management approaches developed are often site-specific, but usually contain some or all of the major control approaches. A weed best practice management manual should be developed to provide basic information about the ecology and biology, reproduction and spread, current distribution and potential threat of this weed, and also information on parthenium weed management and control aspects, including spread minimization, pasture management, chemical and biological control and health aspects.

13.8 Conclusions

Only a limited amount of information is available on the incidence of parthenium weed in East and South-east Asian countries. More survey work is needed to determine the full extent of the parthenium weed invasion in these countries. Although some countries have reported the presence of the weed and have some knowledge of its spread, legislative frameworks to help prevent further invasion and spread of parthenium weed are still needed. Educational tools need to be developed to raise public awareness. Methods to prevent further spread need to be implemented, including restrictions on the movement of seed lots and food

products from infested to non-infested weed-free regions, and the potential for future biological control investigated.

Acknowledgements

We are grateful to Asad Shabbir for creating the distribution map and Steve Adkins for editing this chapter. Thanks to Le Bach Mai, Dang Quoc Quan and Le Van Cuong for their assistance with field work and data collection in Vietnam. Special thanks also to Nguyen Phi Nga and Tohru Tominaga, who contributed information on the weed in Vietnam and Japan, respectively. We also thank Chunqiang Wei for providing photographic images.

References

Adkins, S. and Shabbir, A. (2014) Biology, ecology and management of the invasive parthenium weed (*Parthenium hysterophorus* L.). *Pest Management Science* 70, 1023–1029.

Arenes, J., Bonati, D.P., Dop, P., Gagnepain, F., Guillaumin, L.H., Pellegrin, F. and Pitard, J. (1922) Capripoliacees a Apocynacees. In: Lecomte, H. and Humbert, H. *Flore Generale de L'Indo-chine.* Tome III. Masson et Cie, Paris, pp. 589–590.

Bajwa, A.A., Chauhan, B.S., Farooq, M., Shabbir, A. and Adkins, S.W. (2016) What do we really know about alien plant invasion? A review of the invasion mechanism of one of the world's worst weeds. *Planta* 244, 1–19.

Chang, Z., Zhang, D., Yuan, Y., Zhang, J., Xu, D. and Jin, Y. (2009) The premilinary research on intergrated management of parthenium weed and its occurrence regulation. *China Plant Protection* 29, 26–27.

Chen, Y., Liu, Z., Zhang, J., Jin, J., Han, R. and Lin, M. (2016) A newly recorded genus of vicious alien invasive plant in Chongqing: *Parthenium* L. *Guizhou Agricultural Sciences* 44, 153–155.

Chippendale, J. and Panetta, F. (1994) The cost of parthenium weed to the Queensland cattle industry. *Plant Protection Quarterly* 9, 73–76.

Dhileepan, K. (2009) Managing parthenium weed across diverse landscapes: Prospects and limitations. In: Indergit (ed.) *Management of Invasive Weeds.* Springer Science, the Netherlands, pp. 227–259.

Gao, X., Li, M., Gao, Z., Zhang, J., Liu, Y. and Cao, A. (2013) Seed germination characteristics and clone reproductive capacity of *Parthenium hysterophorus* L. *Ecology and Environmental Sciences* 22, 100–104.

Gupta, O. and Sharma, J. (1977) Parthenium menace in India and possible control measures. *FAO Plant Protection Bulletin* 25, 112–117.

Huang, Y., Liu, X., Tang, W. and Zeng, D. (2012) Study on population size and occurrence reason of *Parthenium hysterophorus* in different habitats. *Weed Science (China)* 30, 37–39.

Karim, S.M.R. (2013) Obnoxious environmental pollutant, the parthenium weed: A possibility of ecological disaster in Malaysia. In: *Joint Seminar on Natural Disasters: Experiences in Malaysia and* Indonesia. Faculty of Earth Science, University Malaysia Kelantan, Jeli, Malaysia.

Karim, S.M.R. (2014) Malaysia invaded: weed it out before it is too late. *International Parthenium News* 9, 1.

Khan, N. (2011) Long term, sustainable management of parthenium weed (*Parthenium hysterophorus* L.) using suppressive pasture plants. PhD thesis, University of Queensland, Brisbane, Australia.

Kil, J., Shim, K., Park, S., Koh, K., Suh, M., *et al.* (2004) Distributions of naturalized alien plants in South Korea. *Weed Technology* 18, 1493–1495.

Kim, J.W. (2013) Parthenium weed in Korea. *International Parthenium News* 7, 1.

Kohli, R., Batish, D. and Singh, H. (1997) Management of *Parthenium hysterophorus* L. through an integrated approach. In: Mahadevappa, M. and Patil, V.C. (eds) *Proceedings of First International Conference on Parthenium Management.* University of Agricultural Sciences, Dharwad, India, pp. 6–62.

Li, P., Shen, W., Wan, F. and An, R. (2014) Effects of global climate change on potential distribution of *Parthenium hysterophorus* in China. *Inner Mongolia Forestry Investigation and Design* 37, 112–116.

Li, Z. and Xie, Y. (2002) *Invasive Alien Species in China.* Forestry Publishing Company of China, Beijing, China.

Mainali, K.P., Warren, D.L., Dhileepan, K., McConnachie, A., Strathie, L., *et al.* (2015) Projecting future expansion of invasive species: comparing and improving methodologies for species distribution modeling. *Global Change Biology* 21, 4464–4480.

McConnachie, A.J., Strathie, L.W., Mersie, W., Gebrehiwot, L., Zewdie, K., *et al.* (2011) Current and potential geographical distribution of the invasive plant *Parthenium hysterophorus* (Asteraceae) in eastern and southern Africa. *Weed Research* 51, 71–84.

Navie, S.C., McFadyen, R.E., Panetta, F.D. and Adkins, S.W. (2005) The effect of CO_2 enrichment on the growth of a C3 weed (*Parthenium hysterophorus* L.) and its competitive interaction with a C4 grass (*Cenchrus ciliaris* L.). *Plant Protection Quarterly* 20, 61.

Nguyen, T., Nguyen, N. and Adkins, S. (2011) Parthenium weed (*Parthenium hysterophorus* L.) in Vietnam. In: *Proceedings of 23rd Asian–Pacific Weed Science Society Conference.* The Sebel Cairns, Weed Management in a Changing World, Cairns, Australia, pp. 401–402.

Parsons, W.T. and Cuthbertson, E.G. (1992) *Noxious Weeds of Australia.* Inkata Press, Melbourne, Australia.

Peng, C., Hu, L. and Kao, M. (1988) Unwelcome naturalisation of *Parthenium hysterophorus* (Asteraceae) in Taiwan. *Journal of Taiwan Museum* 41, 95–101.

Pu, G., Tang, S., Pan, Y., Wei, C. and Cen, Y. (2010) Phenotypic plasticity and modular biomass of invasive *Parthenium hysterophorus* in different habitats in south China. *Guihaia* 30, 641–646.

Shi, B., Aslam, Z. and Adkins, S. (2015) The invasive potential of parthenium weed: A role for allelopathy. In: Price, J.E. (ed.) *New Developments in Allelopathy Research.* Nova Science Publishers, New York.

Sun, D. (2010) Occurrence and management of dangerous parthenium weed in Shandong. *Plant Quarantine* 24, 61–62.

Tachikake, M. and Nakamura, S. (2007) *The Handbook of Naturalized Plants in Japan.* Hiba Society of Natural History, Hiroshima, Japan.

Tang, S., Lv, S., He, C., Li, X., Pan, Y. and Pu, G. (2008) Distribution and harmful effects of alien invasive plant *Parthenium hysterophorus* in Guangxi. *Guihaia* 28, 197–200.

Tang, S., Geng, Y., Zhang, Q., Liu, Z. and Lei, X. (2012) Comparison of growth characteristics of different geographic populations of invasive *Parthenium hysterophorus* under common garden conditions. *Journal of Guangxi Normal University: Natural Science Edition* 30, 257–262.

Wahab, S. (2005) Management of parthenium weed through an integrated approach initiatives, achievements and research opportunities in India. In: Prasad, T.V.R., Nanjappa, H.V., Devendra, R., Manjunath, A., Subramanya, S.C., *et al.* (eds) *Proceedings of the Second International Conference on Parthenium Management,* University of Agricultural Sciences, Bangalore, India, pp. 55–59.

Wan, F. and Ding, J. (1993) Host specificity of *Epiblema strenuana* (Lepidoptera: Tortricidae): A potential bio-control agent for *Ambrosia artemisiifolia* and *A. trifida* (Compositae). *Chinese Journal of Biological Control* 9, 69–75.

Wang, K. and Hou, Y. (2004) *Parthenium hysterophorus* L (Asteraceae) a newly naturalized record generus in Shandong province. *Journal of Qufu Normal University* 30, 83–84.

Wang, Y. and Xu, D. (2011) Parthenium weed situation and management in Linyi. *Modern Agriculture Science* 16, 170.

Wang, Y. and Xu, D. (2012) The potential invasiveness and management of alien parthenium weed. *Jiangsu Agriculture Science* 40, 109–110.

Xie, H. and Zhang, X. (2012) Research on invasive alien plant in Hunan Province. *Modern Agricultural Science and Technology* 5, 178–181.

Zeng, X. and Qiu, H. (2012) Parthenium L., a newly recorded genus of naturalized plant in Jiangxi province, China. *Guangdong Agriculture Science* 16, 46–47.

Zhou, X., Chen, Q., Zhang, H., Zheng, Y., Gao, H., *et al.* (2008) Invasive alien weeds species in farmland and forest in Sichuan Province. *Southwest China Journal of Agricultural Sciences* 21, 852–858.

14 History and Management – Southern Africa and Western Indian Ocean Islands

Lorraine W. Strathie[1]* and Andrew J. McConnachie[2]

[1]Agricultural Research Council – Plant Health and Protection, Hilton, South Africa; [2]Department of Primary Industries, Biosecurity and Food Safety, Orange, New South Wales, Australia

14.1 Introduction

In Africa, parthenium weed (*Parthenium hysterophorus* L.) is present in Egypt in North Africa and in the East African countries of Somalia, Tanzania, Kenya, Uganda, Rwanda, Eritrea and Ethiopia (McConnachie and Witt, Chapter 15, this volume). In Southern Africa, parthenium weed has invaded South Africa, Swaziland, Mozambique, Zimbabwe, Botswana and the western Indian Ocean islands of Mauritius, Réunion, Madagascar and the Seychelles, among others (Nath, 1988; McConnachie *et al.*, 2011). On mainland Southern Africa, the plant has not been detected or reported yet in Lesotho, Namibia, Angola, Malawi or Zambia. Ecoclimatic modelling indicates that some of these countries are climatically suitable for the growth of parthenium weed (McConnachie *et al.*, 2011), so new invasions are inevitable in the future, if not already present at undetected levels. The considerable trans-border and long-distance movement of vehicles, people and goods within Southern Africa, with limited controls on the movement of plant material and agricultural produce across borders, increases the potential for translocation of parthenium weed within the region. Extensive infestations of parthenium weed occur along many of the national road networks that link South Africa, Swaziland and Mozambique, as well as at or near country border posts, increasing the risk of dispersal to new countries. The probability of invasion increases with time as infestations continue to expand.

Dense parthenium weed infestations occur in subsistence (Fig. 14.1A) and commercial (Fig. 14.1B) agricultural production land (cropping and livestock) (Fig. 14.1C), in fallow land, and along water courses and roadside verges (Fig. 14.1D) in affected Southern African countries. Homesteads (Fig. 14.1E), schools, clinics and towns (Fig. 14.1F), particularly in rural communities, are surrounded by parthenium weed. Dirt or gravel-surfaced road networks exacerbate the spread, particularly due to road construction and grading, and movement of seed in the mud carried on vehicles travelling such routes. Road verge maintenance (grass-cutting) along tarred road networks also exacerbates spread of parthenium weed seed. Travel by minibus public transport and off-road vehicles may move seed long distances, resulting in the formation of new, isolated populations. Infestations of parthenium weed occur around roadside minibus informal washing facilities. Parthenium weed has established in numerous national

* StrathieL@arc.agric.za

Fig. 14.1.   In Southern Africa, *Parthenium hysterophorus* invades (a) subsistence farming,
(b) commercial farming, (c) natural grazing for livestock production, (d) major and minor road networks,
(e) homesteads in rural communities, (f) pathways in or near towns, (g) key protected areas such as
Ndumo Nature Reserve in Northern KwaZulu-Natal and (h) the world-renowned Kruger National Park,
South Africa. (L. Strathie.)

parks (Fig. 14.1g,h) and smaller, private protected areas that have been set aside for the conservation of faunal and floral diversity. Subsistence farmers in rural communities are particularly affected by parthenium weed as they often lack the resources to use herbicide control and are regularly exposed to dense infestations of the weed. High cultural value placed on the ownership of cattle as a sign of wealth leads to high stocking rates and overgrazing, in unfenced, communal grazing systems. These scenarios create ideal circumstances for the incursion of parthenium weed, resulting in the establishment of extensive, dense stands of the weed.

Numerous factors at several levels, from authorities through to the general public, contribute to the expanding parthenium weed problem and that of other invasive plant species found in Southern Africa. These include a poor understanding of invasive plants, a greater focus on agricultural production rather than biodiversity conservation, a lack of policy and legislation to help control the movement of invasive plant species and a lack of management activities. In some cases, there is also a negative perception of the intervention methods that could be used, including chemical and biological control, and limited resources to invest in suitable management interventions. Despite considerable infestations of parthenium weed in the region, awareness of the problem and possible solutions remain limited.

14.2 History of Parthenium Weed Invasions and Current Distributions

14.2.1 Overview

Parthenium weed has been recorded in the Southern African countries of South Africa, Swaziland, Zimbabwe, Mozambique and most recently Botswana (Fig. 14.2), and the western Indian Ocean islands of Madagascar, Mauritius including Rodrigues, as well as Réunion, Seychelles, Comoros, Mayotte and Salomon islands (Fig. 14.3). Much of sub-Saharan Africa is considered to be eco-climatically suitable and therefore at risk of invasion by parthenium weed (McConnachie et al., 2011; Mainali et al., 2015).

14.2.2 South Africa

In South Africa, parthenium weed was first recorded in 1880 at Inanda, KwaZulu-Natal (Hilliard, 1977). Wood (1897) recorded parthenium weed growing in disturbed areas and along roadsides around Verulam in KwaZulu-Natal Province. By 1977, it was an uncommon weed of cultivated and disturbed areas along the KwaZulu-Natal coast near Durban, below about 600 m asl (Hilliard, 1977). It became more common after cyclone Demoina caused extensive flooding along the east coast of Southern Africa in 1984, after which it was locally named 'Demoina weed'. Other common local names include 'Maria-Maria' and more recently 'Famine weed' or 'Umbulalazwe' in isiZulu, highlighting the potentially devastating impacts of this plant on agricultural production and the natural environment. Parthenium weed occurs in the sub-tropical savanna biome of South Africa, in eastern subtropical KwaZulu-Natal Province, to the borders of Mozambique and Swaziland, northwards into Mpumalanga Province and to a lesser extent, although increasing, in North-west and Limpopo Provinces (Fig. 14.2; Henderson, 2001; McConnachie et al., 2011; Terblanche et al., 2016).

Parthenium weed was designated a declared weed or category 1 plant in South Africa, according to Regulation 15 of the Conservation of Agricultural Resources Act (Act No. 43 of 1983), under the authority of the Department of Agriculture, Forestry and Fisheries (DAFF). It was declared as a prohibited species that could not be allowed to occur on any land. Land users were to control it by means of any appropriate control methods that were effective and had minimal environmental impact. Furthermore, no person was permitted to sell, advertise, exhibit, transmit, send and deliver for sale, exchange or distribute it, and cause or

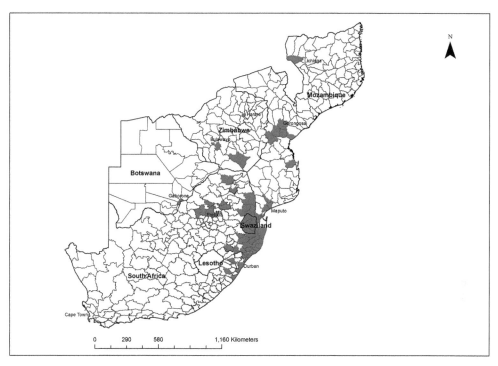

Fig. 14.2. Known distribution of *Parthenium hysterophorus* within southern Africa. Darkened areas indicate local municipalities of South Africa, and districts in Swaziland, Zimbabwe, Mozambique and Botswana, within which parthenium weed has been recorded. Distribution is over-represented graphically due to the use of administrative units rather than point localities.

permit its dispersal. Directives could be issued by the DAFF for the property to be cleared of the weed.

Currently, the National Environmental Biodiversity: Resource Act 10 of 2004 (NEMBA) provides the framework for the regulation of alien and invasive species. Alien and Invasive Species Regulations pertaining to the implementation of this Act were published in 2014. Parthenium weed has been listed as a Category 1b species. It must therefore be controlled, the landowner must notify the relevant authority of possession of the species and take steps to control and eradicate it, and prevent its spread. If an invasive species management programme has been developed for the species, then the landowner must manage the species in accordance with it. The Department of Environmental Affairs (DEA) is responsible for the management of this Act, and may enter a property to monitor,

assist with or implement management of the species.

Parthenium weed continues to increase in distribution and density in South Africa (McConnachie *et al.*, 2011; Terblanche *et al.*, 2016). Currently, it is recorded within at least 70 quarter-degree squares (*c.*25 × 25 km each) in South Africa, less than 44,000 km^2 (Henderson, 2017). It rapidly colonizes wasteland, roadsides, railway lines and sidings, watercourses, cultivated and fallow fields and overgrazed grasslands. It impacts on conservation and agricultural land, including sugarcane (*Saccharum officinarum* L.) and banana (*Musa* sp.) plantations as well as rural community land. Seed is spread through road construction and road maintenance machinery, building construction materials and watercourses, while long-distance dispersal occurs by vehicles, animals and farm machinery, especially in attached mud. Important national parks

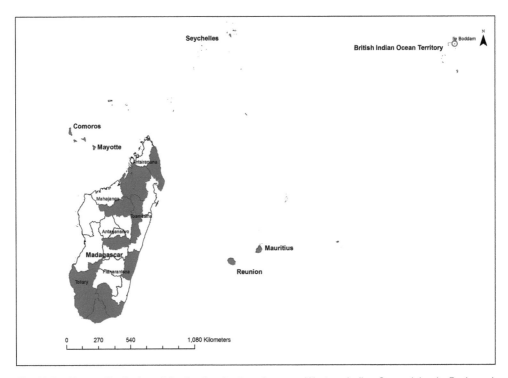

Fig. 14.3. Known distribution of *Parthenium hysterophorus* on Western Indian Ocean Islands. Darkened areas indicate regions of Madagascar within which parthenium weed has been recorded. Other islands on which parthenium weed is present are indicated in their entirety due to broad coverage or indefinite records. Distribution is over-represented graphically due to the use of administrative units rather than point localities.

that are invaded include the world-renowned Kruger National Park, Hluhluwe-iMfolozi Game Reserve, Isimangaliso Wetland Park (a UNESCO World Heritage Site), Phongolo, Ndumo, Tembe and Mkhuze Game Reserves, as well as numerous smaller private game reserves, particularly in KwaZulu-Natal and Mpumalanga Provinces. Within these reserves, parthenium weed occurs particularly along road verges, in game capture and holding pens, dump sites, around staff villages, alongside watercourses and in overgrazed veld. Privately owned lands, which have been converted from agricultural (crop or livestock) to game (wildlife) production, are more heavily invaded due to the historical disturbance of the land. Spread is noticeably increased during above-average rainfall periods and retarded by drought.

Despite the presence of parthenium weed just north of Durban for more than a century, the most severe infestations occur several hundred kilometres further north, in northern KwaZulu-Natal Province, closer to neighbouring Swaziland and Mozambique, and in Mpumalanga Province, possibly due to more suitable climatic conditions for its growth (McConnachie *et al.*, 2011), although separate introductions have not been ruled out. Geographically separate populations occur in the north of South Africa, near Rustenburg and Brits in North West Province and even further north near Tzaneen in Limpopo Province, with little to no parthenium weed in between for hundreds of kilometres. This may have been caused by long-distance dispersal of seed by vehicles. Agricultural land use within these regions that are most densely invaded by parthenium weed include commercial sugar cane and banana production, and to a lesser extent cotton (*Gossypium* sp.), as well as vegetable and fruit crops. Following the assessment of changes in distribution of alien plants in Southern Africa, parthenium weed was one of nine taxa reported by Henderson and Wilson (2017) to be of particular concern as a major increasing environmental threat. Up to the year 2000, parthenium weed had been recorded in 15 quarter-degree squares by the Southern African Plant Invaders Atlas (SAPIA), but by 2016 this figure had increased to 89 quarter-degree squares (Henderson and Wilson, 2017). The atlas region primarily covers South Africa, and to a much lesser extent, neighbouring countries (Henderson and Wilson, 2017).

14.2.3 Swaziland

Swaziland, a small, landlocked country, has about 4% of its area as protected land; nearly 70% is used for grazing, with an additional 14% for cultivated land for feeding livestock during the dry season (Keatimilwe and Mlangeni, 2003). Despite being classified as a lower middle income country (The World Food Program, 2015), Swaziland has economic problems more similar to those of low-income countries, with some of the widest disparities in household income and with 63% of the population living below the poverty line (Kariuki and Leigh, 2015). Overgrazing is common in the communal lands, but no effective management systems exist to address this issue (Keatimilwe and Mlangeni, 2003). Parthenium weed, locally named 'Indodengaziwa', has become one of the primary invaders in Swaziland. It was recorded as the most prolific of the more than 30 invasive plant species in the Mlawula/Ndzindza Reserve (International Union for Conservation of Nature, 1992). In a survey of three reserves of the Lubombo Conservancy in Swaziland, parthenium weed occurred in overgrazed areas, particularly in the south of Mbuluzi Game Reserve and along the western side of Mlawula Nature Reserve (Bowen, 2001). Roadside surveys of parthenium weed in 2006 and 2007 (McConnachie et al., 2011), together

with records from the Swaziland Alien Plants Database (Braun et al., 2004), indicated that the weed was widespread throughout the country, occurring in almost every quarter-degree square (Fig. 14.2). Roadsides, cultivated and fallow fields, grazing land, abandoned land, protected areas and surrounds of homesteads are invaded.

Despite awareness of the impact of parthenium weed in Swaziland, resources are limited, so negligible management intervention has taken place. The National Strategy for the Management of Famine Weed (*Parthenium hysterophorus*) in South Africa proposed international cooperation with neighbouring countries to improve management of this species in the whole of the Southern African region (Terblanche, 2014; Terblanche et al., 2016). Improvements to land management practices by reducing stocking densities of livestock and game would be beneficial, as demonstrated by a study in Mlawula Reserve, where reduced game-stocking levels decreased grazing pressure, leading to reduction in incidence of parthenium weed infestations (Bowen, 2001).

14.2.4 Zimbabwe

Parthenium weed was not recorded in Zimbabwe by Wild (1967) when reporting on the Flora Zambesiaca region (Botswana, Zambia, Zimbabwe, Malawi and Mozambique). However, herbarium specimens were collected 15 years later from near Rainham Dam in 1982 and 1983 (R.B. Drummond in Hyde et al., 2016). Some years later, parthenium weed populations were reported to be spreading along roadsides in Harare, the main Harare/Mutare highway near Mutare and in rhino enclosures at Chipangali Wildlife Sanctuary near Bulawayo (http://treesociety.mweb.co.zw/2002/jan.htm). The weed was observed in Harare (c.1500 m asl) in about 1990, along the roadside verges in Chancellor Avenue, but only collected later in 1993, by which time there had been intermittent spread northwards along Chancellor Avenue as far as the Veterinary Research establishment and southwards to about 100 m south of Josiah Tongogara Avenue (M.A. Hyde, Zimbabwe, 2016, personal communication). Additional plants have also appeared along the Fifth Street extension, to the north of Josiah Tongogara Avenue, but in general terms, spread around Harare has been surprisingly slow (Hyde et al., 2016). Parthenium weed occurs along roadsides and in disturbed sites in the west, centre and east of Zimbabwe (Hyde et al., 2016). It was listed in a checklist of Zimbabwean vascular plants (Mapaura and Timberlake, 2004). The Flora of Zimbabwe (Hyde et al., 2016) contains 14 records of parthenium weed from 1993 to mid-2017, at various sites around Harare and Esigodini, with many large stands reported in some cases (Fig. 14.2). A record at Old Runde Bridge, Mwenezi in January 2015 appears to be the first record from the south (Hyde et al., 2016). No specific intervention approaches against parthenium weed appear to have been used in Zimbabwe.

14.2.5 Mozambique

Wild (1967) recorded parthenium weed in Mozambique. One specimen of the weed was collected in 1961 from Ponta Vermelha, in current Maputo (then named Lourenço Marques), and is housed in the National Herbarium and Botanic Garden in Harare, Zimbabwe (Wild, 1967). Wild (1967) reported parthenium weed as an introduced weed in the Lourenço Marques region of Mozambique, common along roadsides and on disturbed ground. The Flora of Mozambique database (Hyde et al., 2016) contains 22 records of parthenium weed from 2006 to 2015 (Fig. 14.2). Sightings were recorded along roadsides and/or disturbed areas at Mt Gorongosa, Gorongosa National Park, Chimoio, Sussundenga, Manica, along Beira road at Nhamatanda, Chiluwa village and Inchope, and Gondola, Dondo and Pungwe River (Hyde et al., 2016).

Parthenium weed is common in southern Mozambique, in cultivated fields, wetlands and roadsides (S. Bandeira, 2007,

Universidade Eduardo Mondlane, Mozambique, personal communication). It occurs widely around Maputo, along the seafront esplanade and near the docks, the old fort area; also alongside the road to Bilene (north-east of Maputo); at Macia (Inkomati River system) and Chokwe (Limpopo River system), about 80–150 km north of Maputo; at Lichinga in northern Mozambique; and at Beira.

Bandeira et al. (2006) reported parthenium weed to be one of the most common invasive plant species within the Chibuto-Missavene Wetland Reserve, with extensive areas invaded in its drier areas. Most Chibuto inhabitants rely on agriculture and cattle production from the Missavene wetland and its surrounds (Bandeira et al., 2006). About 20–40% cover by parthenium weed was recorded within 10×10 m quadrats, and together with Egyptian riverhemp (Sesbania sesban (L.) Merr.) and cocklebur (Xanthium strumarium L.), up to 5 ha was invaded in the study area (Bandeira et al., 2006).

Zachariades (2017) reported parthenium weed as by far the most abundant and widespread of 17 invasive alien plant species recorded during a survey of southern Mozambique. It was ubiquitous at inland sites, common but less dense closer to the coast and absent from three very xeric sites. Its presence was recorded at multiple localities within the districts of Boane, Namaacha, Moamba, Magude and Manhica (Zachariades, 2017).

14.2.6 Botswana

Until recently, parthenium weed was not known to occur in Botswana. Surveys of selected areas considered to be likely possibilities for invasion did not yield any occurrences during the late 2000s (McConnachie et al., 2011). However, one or more plants were recorded for the first time east of Gaborone in early 2015, along a recently upgraded dirt road between Tlokweng and Ruretse (Sheila Gregory, United Kingdom, 2017, personal communication) (Fig. 14.1).

Although no further records are available, subsequent spread is likely to have occurred.

14.2.7 The Mascarene and Western Indian Ocean Islands

Parthenium weed has been present for some time in the three Mascarene islands (Mauritius, Réunion and Rodrigues; Hind *et al.*, 1993), where it is commonly found in drier cultivated areas, in fallow, scrub and wastelands and along roadsides (C. Baider, The Mauritius Herbarium, 2015, personal communication). The Mauritius Herbarium houses parthenium weed specimens from the Mascarene and western Indian Ocean islands – Mauritius, Rodrigues, Réunion, Chagos Archipelago and the Seychelles (Fig. 14.3) – although the collection is fairly limited (C. Baider, The Mauritius Herbarium, 2015, personal communication).

14.2.8 Mauritius

Parthenium weed was first recorded on Mauritius between 1752 and 1761 by Fusée-Aublet (1775), and was referred to as 'Herbe à mouton' or 'Herbe à Samson'. Bory de Saint-Vincent (1804) reported parthenium weed in the streets of Port-Louis; it was also grown in the Botanical Gardens in Europe (Guillemin, 1828), while Bojer (1837) called it 'Herbe blanche' and reported that it grew in 'all quarters of the island'. Bouton (1864) recorded that parthenium weed originated from Jamaica and was naturalized in Mauritius, where it was very common. No other records from that time report parthenium weed to be common anywhere else in Southern Africa or the western Indian Ocean islands. All parts of the plant were reported to be excessively bitter and the Creole-speaking people sometimes used it as an anthelmintic for eradication of intestinal worms (Bouton, 1864). Baker (1877) reported that parthenium weed was a common weed in Mauritius, it had been collected from Flat Island, a small island off the north coast of Mauritius, and had also established

on Bourbon (now known as Réunion). Later, Daruty (1886), referring to it as 'Camomille du pays', reported that it was used for medicinal (anti-emetic) purposes. In Mauritius parthenium weed is used to treat gynaecological problems (Pourchez, 2011), and it is also used as a tea in Creole medicine in Mauritius and Réunion for the treatment of fever (Autran, 2010). However, Autran (2010) commented that as parthenium weed can be allergenic and some studies indicate liver toxicity, its medicinal use should be limited.

Parthenium weed is not referred to in Rochecouste *et al.* (1981), nor in McIntyre's (1991) record of weeds affecting sugarcane in Mauritius, although Vaughan and Wiehe (1937) did record it as one of the annual species to be found in cane-field communities. However, they commented that such species are short-lived, and are replaced by the perennial Poaceae to form savannahs or plains, destroyed by cultivation of the land, or lead to a *Cordia* (Boraginaceae)-*Lantana* (Verbenaceae) scrub (Vaughan and Wiehe, 1937). Despite the duration of establishment, parthenium weed reportedly does not have any great important effect on agricultural production, and has not been recorded as one of the major problem plants of sugarcane in Mauritius in general (C. Baider, The Mauritius Herbarium, 2015, personal communication). However, it is reported more recently to be particularly important as a weed in sugarcane and vegetable fields in some specific regions in the north around the villages of Pamplemousses and Souvenir (A. Gaungoo, Mauritius Sugarcane Industry Research Institute, 2015, personal communication). The Mauritius Herbarium collections recorded parthenium weed from Réduit, and a wetlands survey recorded presence in Pereybere, Grand Baie (C. Baider, The Mauritius Herbarium, 2017, personal communication). It is rarely encountered in agricultural situations in other parts of the country, though it occasionally occurs on abandoned and fallow lands, and along roadsides which are cut regularly (C. Baider, The Mauritius Herbarium, 2015, personal communication). Detailed surveys on islets around Mauritius did not record parthenium weed, but a few coastal wet-lands in the north are known to have the weed (C. Baider, The Mauritius Herbarium, 2015, personal communication). The Weed Identification and Knowledge in the West-ern Indian Ocean (WIKWIO) portal (Le Bourgeois *et al.*, 2015) reports occurrence of parthenium weed in most Mauritian dis-tricts, including Port Louis, Pamplemousses, Plaines Wilhems, Riviere du Rempart, Riv-iere Noire, Grand Port and Flacq (Fig. 14.3). Le Bourgeois *et al.* (2008) report it as a weed of dry, rain-fed crops, fallow lands, vacant lots and roadsides.

14.2.9 Rodrigues

Wiehe (1938) did not list parthenium weed as a major weed on Rodrigues, a 108 km² autonomous outer island of the Republic of Mauritius. Two herbarium specimens of parthenium weed from Rodrigues are lodged in the Mauritius Herbarium – from culti-vated land in Port Mathurin, the capital city, and a roadside weed near the coast (C. Baider, The Mauritius Herbarium, 2015, personal communication). Parthenium weed is referred to as 'herbe blanche' on Rodrigues, where it is used to treat gynaecological and urinary infections (Pourchez, 2011).

14.2.10 Réunion

Parthenium weed was not recorded within the first full floral survey of 1156 species on Réunion, undertaken in 1895 (de Cordemoy, 1895). A single specimen of parthenium weed from Réunion is housed in the Mauri-tius Herbarium, found growing as a roadside weed in Gillot (around the airport) (C. Baider, The Mauritius Herbarium, 2015, personal communication). However, a later survey conducted by the Conservatoire National Botanique de Mascarin (Mascarine Cadetiana, 2016) records 158 observations of parthenium weed from all around the island, particularly from the western and south-western drier parts, to a lesser degree in the north and east, and mostly along the coast and up to 400–800 m asl (Fig. 14.3). In these locations the weed grows in aban-doned and open areas, in newly disturbed sites, shaded and slightly moist places, and particularly on clay soils (Le Bourgeois *et al.*, 2008). Some have reported it as a minor weed in about 15% of the cultivated areas and gardens (D. Strasberg, Université de la Réunion, 2015, personal communication). It sometimes occurs in young sugarcane fields without forming dense stands, but it is often more abundant in vegetable crops (Le Bourgeois *et al.*, 2008). It colonizes dis-turbed areas, roadsides and forest roads, and suppresses the growth of other species.

Parthenium weed is referred to as 'kamomyi' (chamomile) because it vaguely resembles European chamomile (*Matricaria chamomilla* L.), and was most likely named during the 18th century by European set-tlers to Réunion (Pourchez, 2011). It is also referred to as 'Camomille', Camomie', 'Cam-omille z'oiseaux', 'Camomille balais' and 'Herbe blanche' (Le Bourgeois *et al.*, 2008). 'Chamomile z'oiseaux' is also called 'bastard absinthe' because it looks like wormwood (*Artemisia absinthium* L.), a native to Eurasia and Northern Africa, that is used in the spirit absinthe. Parthenium weed is used in the treatment of gynaecological and urinary infections (Pourchez, 2011) in Réunion, as well as a remedy for skin inflammation, rashes, chicken pox and other superficial skin conditions, rheumatic pain, diarrhoea, dysentery, malaria and neuralgia. It is also used to treat an early childhood disease in Réunion called 'tanbav syndrome' ('that the doctor does not know'; Pourchez, 2011). The whole dried plant is used to prepare an infu-sion against fever.

14.2.11 Madagascar

Madagascar, a large island of 592,800 km², separated from the African mainland for more than 160 million years, has 11,220 plant species, with 82% endemism (Callmander *et al.*, 2011). A biodiversity hotspot of global significance, conservation of its unique but threatened flora and ecosystems is critical (Phillipson *et al.*, 2006). Until the 1920s, publications focused mainly on Madagascar's native vegetation with little mention of invasive species (Binggeli, 2003). Perrier de la Bâthie (1928, 1931) first documented non-native species and weeds on the islands (Kull *et al.*, 2014) but did not perceive introduced plants as a threat to the native biodiversity (Binggeli, 2003). Successful biological control of drooping prickly pear (*Opuntia monacantha* Haw.) during the 1920s in Madagascar created a widespread interest in weeds and their impacts (Binggeli, 2003). Perrier de la Bâthie (1931) reported parthenium weed around the villages on the east coast and at Tuléar. Humbert (1963) reported parthenium weed to be a ruderal weed, growing here and there, all year. Plants were reported in Antalaha in the east (Sakabato, Sakaleony) by Perrier de la Bâthie, in Moramanga in central Madagascar by d'Alleizette, as well as in the south of Madagascar in Mahatsinjo by Decary, in Ampandrandava by Segrig and Ambovombe, and by Decary, and in Tulear in the south-west by Poisson. Binggeli (2003) listed 38 flowering plant species as being highly invasive in Madagascar, or present and known to be invasive elsewhere, but parthenium weed was not among them, although species may have been under-recorded. Little awareness or knowledge existed on the distribution and impact of invasive plant species, possibly due to the more overriding impacts of deforestation, erosion and fire on biodiversity (Binggeli, 2003). Kull *et al.* (2012) provided an updated inventory of introduced plants, of which about 9% were reported as invasive, which is fairly high compared with other island systems, such as the Mascarenes, Galapagos, Hawaii and Polynesian islands. It was suggested that there had been numerous opportunities for plant transfers to have occurred during colonization, from trade and other historical introductions (Binggeli, 2003).

The Catalogue of the Plants of Madagascar (2017) currently records parthenium weed in the eastern and south-eastern regions of Madagascar. This catalogue is based on herbaria specimens, collected by botanists primarily interested in native flora, thereby under-representing the introduced plants (Kull *et al.*, 2012). The online database lists parthenium weed as being naturalized in Madagascar, occurring on anthropic lands, on dunes and strand, among forest vegetation, in dry, humid, sub-arid, and sub-humid climates, at elevations below 1000 m (Catalogue of the Plants of Madagascar, 2017). Records include the Provinces of Antananarivo (Vakinankaratra), Antsiranana (SAVA), Fianarantsoa (Vatovavy-Fitovinany) by H. Perrier de la Bâthie, Toamasina (Alaotra-Mangoro), Toliara (Atsimo-Adrefana) and in the protected areas of Andohahela and Tsimanampetsotsa (Fig. 14.3). Parthenium weed was observed in the north of Madagascar, at Berômba along the Route Nationale 6 between Anivorano Nord and Antsakoabe (L. Strathie, South Africa, 2009, personal observation).

Locally it is named 'Jamalanjirike' (Le Bourgeois *et al.*, 2008), Marorokamboa (Husson *et al.*, 2010), as well as 'Fausse camomille', 'Grande camomille', 'camomille balais' and 'camomille z'oiseaux'. Parthenium weed is a weed of 'tanety' (high and fairly level ground, uncultivated hillsides) and 'baiboho' (rain-fed flood recession crops) in the west and south-west of Madagascar, and on alluvial land, developing in the cool, dry season, in lowland areas (Le Bourgeois *et al.*, 2008; Husson *et al.*, 2010). It grows in disturbed areas, along edges of fields and wasteland, in shaded and slightly moist sites, but is not common. It occurs in fallow and crop loads, although it is scarce in crops in general (Husson *et al.*, 2010), yet it can be very competitive in vegetable crops and sugarcane, impacting strongly on yield (Le Bourgeois *et al.*, 2008). Control methods employed are manual control by hand-pulling and chemical control using 2,4-D (Husson *et al.*, 2010).

14.2.12 Seychelles

Parthenium weed is referred to as 'Herbe blanche' in the Seychelles. It occurs in clearings and abandoned places but is uncommon (Friedmann, 2011, and confirmed by

B. Senterre, Seychelles Natural History Museum, 2015, personal communication) or rarely abundant, with low impact (Le Bourgeois et al., 2008). It has not been observed to invade natural or semi-natural forests or inselbergs, occurring more in ruderal stands (B. Senterre, Seychelles Natural History Museum, 2015, personal communication). A herbarium specimen of parthenium weed from 1962 records it as a weed in coconut (Cocos nucifera L.) plantation in the La Passe village on Silhouette island (Seychelles Plant Gallery, 2013). Parthenium weed is also recorded from the largest island Mahé, and from Farquhar (Robertson, 1989) (Fig. 14.3).

Kueffer (2010) attributed the relative paucity of invasive plants in the Seychelles to (i) the dominance of cinnamon (Cinnamomum verum L.) since the 19th century in most forests, which may have excluded other invasive species; (ii) nutrient-poor soils, which particularly lack phosphorus; (iii) unlike many other islands, absence of large feral animals which assist plant invasion through soil and vegetation disturbance; and (iv) intensive foreign trade and travel only as a recent phenomenon.

The coral islands of the western Indian Ocean comprise: the islands of the Mozambique Channel, including Europa; the Aldabra group (Aldabra, Assumption, Cosmoledo and Astove); the Farquhar group (Farquhar, St Pierre and Providence); the Amirantes (African Banks, Remire, D'Arros, St Joseph, Desroches, Poivre, Etoile, Boudeuse, Marie-Louise and Desnoeufs); Bird and Dennis Islands, northern Seychelles Bank; Cargados Carajos; and isolated islands, including Gloriosa, Agalega, Tromelin, Coetivy and Alphonse (Stoddart, 1970). These islands are either elevated reef-limestone islands, or sand cays on sea-level reefs (Stoddart, 1970). Early accounts of fauna and flora on these coral islands are generally lacking; ecological accounts are only recorded sporadically, from two major expeditions in 1882, and subsequent surveys in the 1930s and 1960s (Stoddart, 1970). Farquhar Atoll, with an area of 170 km², is part of the Farquhar Group of islands that are part of the Outer Islands and in the most

southerly part of the Seychelles. It lies 770 km south-west of the capital, Victoria on Mahé Island, and 285 km north-east of Madagascar (Stoddart and Poore, 1970). Limited ecological knowledge of Farquhar existed until the 1960s, when several accounts were published (Stoddart and Poore, 1970). Flora and vegetation of Farquhar differ considerably from those of elevated limestone islands in the Aldabra group, and are more similar to those of the sand cays of the Amirantes and the central Indian Ocean (Fosberg and Renvoize, 1970). The islands of Farquhar are simple sand cays with dunes, but a long history of human interference has resulted in a strong gradient in number of introduced species southwards from the North Island settlement (Fosberg and Renvoize, 1970). Coconut woodland is the primary vegetation type on both islands, but the smaller North Island has many more introduced herbs and grasses (Fosberg and Renvoize, 1970). Parthenium weed was recorded on Farquhar by Stoddart and Poore (1970).

14.2.13 Comoros

In the Comoros, parthenium weed is present in regions of low to medium altitude up to 500 m asl, in recently cleared fields (Le Bourgeois et al., 2008). It is uncommon but locally abundant (Le Bourgeois et al., 2008) (Fig. 14.3).

14.2.14 Mayotte

Mayotte is a group of French territory islands, 374 km² in area, situated in the northern Mozambique Channel between Madagascar and Mozambique, with Comoros to the north-west. Volcanic rock results in rich soils on the Mayotte islands. Humbert (1963) listed parthenium weed to be present on Mayotte and Pamanzi (or Petite Terre), an island just off the main island Grand Terre (or Maore) (Fig. 14.3). Parthenium weed is invasive mainly in disturbed areas of Mayotte (UICN, 2017). It is locally named 'Malandi voa' by the Bushi (ShiBushi) people (Inventaire National du Patrimoine Naturel, 2017).

14.2.15 Salomon Islands

The Mauritius Herbarium houses a flowering specimen of parthenium weed collected by P.O. Wiehe in 1939 from Île Boddam in the Salomon Islands or Salomon Atoll, within the Chagos Archipelago, which are British Indian Ocean Territory but previously Mauritian (Fig. 14.2). The Chagos Archipelago is a group of seven atolls, with more than 60 islands of different coralline rock structures, about 500 km south of the Maldives archipelago. Île Boddam is about 1.08 km² in area, and one of the main islands in the Salomon Islands group, housing the former main settlement but now uninhabited. In the latter half of the 18th century, coconut plantation workers from Mauritius settled on Île Boddam but inhabitants of the Salomon Islands were later evicted between 1967 and 1973 and resettled in Mauritius and the Seychelles (https://en.wikipedia.org/wiki/Salomon_Islands).

14.3 Management Methods Utilized in Southern Africa

Small-scale and subsistence farmers in Southern Africa are among those most affected by parthenium weed, as the cost of chemical control is prohibitive for many landowners. Commercial farming fairly rigorously manages weeds within cultivated land and along margins, using frequent herbicide applications.

South Africa proactively recognized the need for, and implemented, a management programme for parthenium weed using a chemical and biological control approach (Olckers, 2004; Strathie et al., 2011; Terblanche, 2014). This management programme is largely due to the vision and actions of the Working for Water Programme, administered previously through the Department of Water Affairs and Forestry and now by the Department of Environmental Affairs. This initiative commenced in 1995, combining the clearing of invasive plant species, with job creation and the training and poverty-relief of disadvantaged communities. The programme partners with local communities and national and provincial Departments of Agriculture, Environmental Affairs, Trade and Industry, research foundations and private companies (https://www.environment.gov.za/projectsprogrammes/wfw). It is currently recognized that broader scale intervention is required against parthenium weed, supported by increased resources and with national coordination (Terblanche et al., 2016).

South Africa additionally has a long and favourable history of research and implementation of the biological control of invasive plants using selective introduced natural enemies, dating back to 1913 (Moran et al., 2013). These two platforms have facilitated management of parthenium weed. Other African countries have benefited subsequently as technology on the biological control of parthenium weed has been transferred from the south to the east of Africa, particularly to Ethiopia and Tanzania, with others to follow.

Parthenium weed has been well managed in Australia through a combination of control methods, including mechanical, chemical and biological control, as well as containment strategies, competitive plant species and other cultural control methods (Navie et al., 1996; Adkins et al., 2005; Khan et al., 2010; Dhileepan and McFadyen, 2012). In Africa, disjointed and poorly integrated combinations of a few of these methods have been employed to manage parthenium weed as follows, with varying levels of success.

14.3.1 Mechanical control

A widely used, albeit labour-intensive, weed management practice in Africa is that of hand-weeding (Akobundu, 1991). On smallholder farms, this approach is the predominant method used for weed control (Vissoh *et al.*, 2004). In some countries, hand-weeding is supplemented with the use of a short hoe (International Fund for Agricultural Development, 1998). In Southern Africa, the management of parthenium weed in subsistence agriculture has relied largely on hand-weeding and mechanical clearing (Fig. 14.4). The immediate surrounds of homesteads are often cleared mechanically or manually. Consequently, those tasked with weeding (mostly women and children) have repeated, prolonged exposure to the weed, with risks of developing characteristic symptoms associated with parthenium weed allergy (Contact dermatitis and respiratory reactions; Allan *et al.*, Chapter 6, this volume). Limited or no resources for protective equipment such as gloves or respirators, access to water to wash off skin and clothing after contact with the

plant, and restricted resources or access to medication, result in a high probability of health risks developing. Additional to the physiological symptoms, the physical consequences of hand-weeding for prolonged periods include permanent spinal deformation (Oyedemi and Olajide, 2002). In an agricultural context, while hand-weeding may provide temporary relief from parthenium weed, the soil disturbance stimulates growth from parthenium weed seed banks.

14.3.2 Chemical control

The use of herbicides for crop weeds in Africa dates back to the 1950s (Idris, 1970). Herbicides that are registered for use on parthenium weed in South Africa include Access™ 240 SL with the active ingredient picloram

Fig. 14.4. Manual clearing of *Parthenium hysterophorus*, a common practice in Southern Africa, along roadsides in densely invaded rural communities such as KwaJobe, KwaZulu-Natal province in South Africa. (L. Strathie, Agricultural Research Council - Plant Health and Protection.)

240 g/l, and Plenum™ 160 ME with active ingredients fluroxypyr 80 g/l and picloram 80 g/l. This chemical management approach has been used sporadically, although intensively in selected areas, to control parthenium weed along roadsides in South Africa as a preventative measure to reduce further spread (Goodall *et al.*, 2010) and contain outlying populations (Terblanche *et al.*, 2016). After initial treatment, repeated applications of herbicide are required every growth season until native vegetation recovers, due to the significant regenerative ability of parthenium weed. Although chemical control is effective and useful, particularly for use along roadsides or in new areas of invasion, due to herbicide costs, this management approach will in most cases not be adopted broadly by smallholder farmers in Southern Africa. It is estimated that <5% of African smallholder or subsistence farmers do not use herbicides for this reason (Mavudzi *et al.*, 2001; Overfield *et al.*, 2001; Muoni *et al.*, 2013). Nevertheless, chemical control integrated within other management options remains an important tool to manage infestations of parthenium weed.

14.3.3 Biological control

Of the countries discussed earlier in this chapter, only South Africa has actively implemented a biological control programme on parthenium weed (Strathie *et al.*, 2011). This endeavour was initiated under the auspices of an 'Emerging Weeds Programme' of the Working for Water Programme, in which five promising candidate plant species were targeted for biological control during the early stages of their invasion (Olckers, 2004). In fact, by then, parthenium weed was already fairly widespread in South Africa but had not yet reached its full extent, nor had it reached the proportions that other major plant invaders had been achieved, although its capacity for severe invasion was recognized. The South African biological control programme against parthenium weed, which has evaluated six agents and so far introduced four

that target the leaves, stems and seeds (Dhileepan *et al.*, Chapter 7, this volume; Strathie *et al.*, 2011), was based on the successful management programme developed in Australia which utilized multiple natural enemies (McFadyen, 1992; Dhileepan and McFadyen, 2012). Additional agents are under consideration and further integration of biological control into management efforts is necessary (Terblanche *et al.*, 2016). While the practice of biological control of invasive alien plants in South Africa began more than 100 years ago, with considerable successes achieved, there is still scope for the scaling up of operations to maximize benefits (Zachariades *et al.*, 2017). Distribution mapping of parthenium weed populations (McConnachie *et al.*, 2011), the assessment of soil seedbanks (L. Strathie and A. McConnachie, unpublished data), the suitability of several biological control agents for release (Strathie *et al.*, 2011; Retief *et al.*, 2013; McConnachie, 2015a, 2015b), and the field release and evaluation of these biological control agents, have been undertaken in South Africa. This technology has also been transferred into East Africa through international cooperative projects (McConnachie and Witt, Chapter 15, this volume).

The winter rust fungus (*Puccinia abrupta* Diet. & Holw. var. *partheniicola* (Jackson) Parmelee) that is present on parthenium weed in South Africa (Wood and Scholler, 2002) is not documented anywhere else within the regions under discussion here, although its presence is likely as it has been discovered without deliberate introduction in several other countries.

There is a desperate need for sustainable, cost-effective management strategies, such as those using biological control, to manage parthenium weed on the African continent and the western Indian Ocean islands, where resources for other management interventions are limited. Greater regional and international collaboration and broader adoption of biological control methods are required. Utilization or expansion of existing networks such as the Weed Identification and Knowledge in the western Indian Ocean portal (WIKWIO) (Le Bourgeois *et al.*,

2015) could be useful. WIKWIO is developing a science and technology network to consolidate existing knowledge and facilitate sharing of knowledge on weeds of food and cash crops of the western Indian Ocean islands, and effective management practices.

14.3.4 Other management methods

Competitive plants

Several studies elsewhere have investigated the feasibility of growing competitive pasture and crop species as a management approach for controlling parthenium weed infestations and remediating degraded areas (O'Donnell and Adkins, 2005; Khan et al., 2010). This approach takes into account native or other crop or pasture species that are not only good competitors in relation to parthenium weed, but are also tolerant to the allelochemicals released by it. In Southern Africa, only one such study has investigated the competitive ability of native grasses in relation to parthenium weed and their tolerance of its allelochemicals. Van der Laan et al. (2008) demonstrated in a field trial that of the three native perennial grass species tested against parthenium weed – African lovegrass (Eragrostis curvula (Schrad.)), Guinea grass (Panicum maximum Jacq.) and woolly finger grass (Digitaria eriantha Steud.) – Guinea grass displayed the best overall growth performance and suppressed parthenium weed growth over time. However, in a seed germination laboratory study, Guinea grass seed was most sensitive to one of the allelochemicals (parthenin) of parthenium weed, while African lovegrass seed was least affected. Van der Laan et al. (2008) concluded that if Guinea grass was to be used in a control or rehabilitation programme, it would be difficult to establish it from seed in areas already infested with parthenium weed and therefore planting of seedlings would give the grass a head-start. The feasibility of this approach in an African context is questioned from two perspectives; the first is that parthenium weed in Southern Africa occurs in many areas which receive erratic rainfall, and establishment of grass seedlings could be challenging. The second obstacle centres on the fact that the most competitive grasses in the van der Laan et al. (2008) study are also highly palatable, which in turn would encourage grazing if grazers could not be excluded, resulting in disturbance that would exacerbate weed growth.

Distribution surveys

In 2005, a project entitled 'Management of the Weed Parthenium (Parthenium hysterophorus L.) in Eastern and Southern Africa using Integrated Cultural and Biological Measures' was initiated under the auspices of the Integrated Pest Management Collaborative Research Support Program (IPM CRSP), funded by the United States Agency for International Development (USAID). This multi-faceted, international collaborative project involved research and implementation of management options for the control of parthenium weed in parts of Africa. Due to the paucity of detailed information on the distribution and abundance of parthenium weed in affected African countries, objectives of one component of the study included: (i) prediction of the most suitable areas of the continent for invasion by parthenium weed based on eco-climatic conditions; (ii) consolidation of known distribution data for parthenium weed in Eastern and Southern Africa; (iii) expansion of this dataset by conducting distribution surveys to determine the extent of the weed in selected countries; and (iv) generation of a database of distribution data. This research revealed that much of sub-Saharan Africa was eco-climatically suitable for the growth of parthenium weed, and that the extent of invasion by the weed was far greater than what current databases at the time reflected (McConnachie et al., 2011). For Southern Africa, the distribution of parthenium weed was considered to have been fully measured in South Africa and Swaziland, although there has been subsequent spread. However, surveys are required in other Southern African countries and western Indian Ocean islands as the plant has not yet reached its maximum potential distribution. Focused efforts are required to increase knowledge of the existing distribution of parthenium weed. Networks such as WIKWIO are useful as they are building databases of weed species observations.

Containment

In a first for Africa, a national strategy was developed in 2014 to manage parthenium weed in South Africa (Terblanche *et al.*, 2016). The national strategy was developed to set policy and to monitor progress towards goals at a national level, and an implementation plan set goals and time-frames for their achievement at local levels. The strategy formulated goals for: (i) the prevention of spread of parthenium weed to new areas; (ii) local eradication of isolated populations; (iii) containment in areas where eradication was not possible; and (iv) asset protection where containment was no longer an option (Terblanche *et al.*, 2016).

Containment of parthenium weed infestations is a critical component in an integrated strategy to curtail the impacts and spread of this weed. This was high-lighted in Australia, where significant investment was allocated to containing the spread of parthenium weed to Queensland (Chamberlain and Gittens, 2004). Wash-down facilities operated within Queensland were able to reduce the potential spread of parthenium weed seed through contaminated vehicles and machinery from the core parthenium weed-infested areas in Central Queensland to New South Wales. In addition, The New South Wales Parthenium Weed Strategy (Blackmore and Charlton, 2011) put in place strategies of eradication, awareness and coordinated management, an entry minimization strategy.

In the National Strategy for the Management of Parthenium Weed in South Africa, much emphasis was placed on containment procedures, including: (i) placing all of the invaded areas into containment zones; (ii) the identification of the main spread pathways; (iii) implementation of active surveillance (and enforcement of legislation where necessary); (iv) promotion of

awareness; (v) and the development and implementation of management plans (and the subsequent monitoring and assessment thereof) (Terblanche, 2014). The success of containment in South Africa, let alone Southern Africa, will face many challenges arising from ecological features of the target plant, social and cultural practices that will influence management, inadequate levels of funding, and multiple political considerations (Terblanche *et al.*, 2016).

14.4 Management Recommendations

Management of parthenium weed within Southern Africa and the western Indian Ocean islands remains a challenge, as it does for most developing countries, for a number of reasons. These include one or more of the following: (i) lack of inventory of invasive species and knowledge of their extent and impacts; (ii) lack of broad awareness of the impacts or potential impacts of invasive alien species such as parthenium weed; (iii) lack of value placed on conservation of bio-diversity; (iv) lack of legislation on the management of invasive alien species; (v) lack of national long-term commitment and coordinated efforts to manage invasive species; (vi) limited resources for large-scale management programmes; and (vii) lack of expertise and/or facilities.

14.4.1 Land use and land management practices

Studies have clearly demonstrated that improved grass cover, achieved by means of reduced grazing pressure by lowering stocking rates, leads to improved management of parthenium weed. High rates of disturbance, particularly due to livestock overgrazing, are clearly evident in parts of Southern Africa where parthenium weed is rampant. Livestock stocking rates are a culturally sensitive issue in Southern Africa due to the link with wealth status. Attempts to address this issue are therefore unlikely to succeed dramatically in the near future. Manual control of weeds in rural subsistence farming systems, with limited use of herbicides, most likely exacerbates invasion levels by creating regular, high levels of disturbance. Improved land management practices among commercial and subsistence farmers are critically required through government-led extension programmes.

14.4.2 Awareness and strategic planning

A national or regional strategy provides high-level guidelines for the management of parthenium weed, providing a framework by the setting of goals and objectives, outlining a system for prioritization of areas requiring management interventions, recommending the type of interventions to be utilized and allocating resources (Terblanche et al., 2016). The National Strategy for the Management of Parthenium Weed in South Africa aligns with the legislation of the National Environmental Management: Biodiversity Act No. 10 of 2004 that was amended in 2014, constituting an invasive species control and eradication strategy (Terblanche et al., 2016). The guidelines are given effect in the National Implementation Plan for the Management of Famine Weed (parthenium weed) in South Africa, which outlines roles, responsibilities, resources and activities within given timeframes (Terblanche et al., 2016). A strategy requires a vision (for example, to minimize the impact of the weed on society, the economy and the environment). The division of the invaded or potential area for invasion into different management zones enables sets of management actions and targets to be defined for each zone. A strategy should have an objective (such as ensuring collaboration, coordination and prioritization of resources) to achieve defined goals that are specific, measureable, assignable in terms of responsibility, realistic and with a defined timeframe. Management activities must include planning, implementation, monitoring and evaluation (Terblanche et al., 2016).

Binggeli (2003) outlined the need for the development of a national strategy to manage invasive species in Madagascar, to halt introductions of potentially invasive species and contain ornamental and harmful species already present. He noted that international cooperation with Southern Africa and neighbouring Mascarene Islands is essential due to the commonality of invasive species in these regions. As Binggeli (2003) noted, political will, public support and financial resources are often lacking, particularly for active response to early detection of an invasive species. Wider use of biological control in Madagascar was recommended (Binggeli, 2003), but is yet to be implemented.

Ideally, countries should develop national strategies to deal with parthenium weed and foster international collaboration through programmes such as the Global Environment Facility-sponsored Removing Barriers to Invasive Plant Management in Africa and the Integrated Pest Management Innovation Lab project in East Africa.

14.5 Future Requirements

The earliest record of parthenium weed in the Southern African and western Indian Ocean island regions appears to be from Mauritius in the 1700s, with records for other countries reported in the following centuries. Whether this was indeed the earliest presence in the region, or it was due to better documentation of vegetation of that area at that time or that the plant was more apparent there, is not known. However, by many accounts, since then parthenium weed appears to have been recognized as problematic in only some countries, and mostly only in recent decades, although the earliest reports in Mauritius indicated that it was widespread on the island at the time. The historic patterns of introduction and movement of the plant in Southern Africa and the western Indian Ocean islands are not clear. Molecular techniques help to elucidate the genetic relationship and origin of various plant populations in the Southern African and western Indian Ocean island, but recent studies of parthenium weed have scant representation from these regions (Y. Geng, Yunnan University, 2014, personal communication). Distribution, abundance and impact of parthenium weed is clearly variable, but there is a lack of documented knowledge here. Studies of the impacts of parthenium weed on biodiversity, agricultural production and the health of humans and animals in these two regions are lacking, but needed.

Four main barriers (weak policy and institutional environment, unavailable critical information, inadequate implementation of prevention and control, and lack of capacity) preventing the effective management of invasive species that were identified in East Africa are applicable in much of Southern Africa and the western Indian Ocean islands.

Legislative policy regarding the management of invasive alien plants is required for countries affected by parthenium weed, followed by commitment from national governments to detect and reduce the spread of the weed within their own countries and to cooperate with countries within their region. This is required to maintain continuity and sustainability as external donor programmes have defined and limited resources. Wider awareness of the identification and impacts of parthenium weed would improve intervention response times. In this regard, suitable awareness materials should be shared among affected countries.

In much of the Southern African and western Indian Ocean island regions, comprehensive distribution records of parthenium weed are lacking, making management decisions difficult. Distribution surveys with online databases would benefit intervention efforts. No doubt, distribution is likely to be more widespread than is currently recorded and will inevitably expand with time. Databases such as the Southern African Plant Invaders Atlas, Swaziland Alien Plants Database, Catalogue of the Plants of Madagascar, Weed Identification and Knowledge in the Western Indian Ocean portal, Comisión Nacional para el Conocimiento y Uso de la Biodiversidad (CONABIO), Global Biodiversity Information Facility (GBIF) and Tropicos® are all useful for a broad-scale understanding of distribution, knowledge of which may be scant due to limited herbaria records or focused surveys. The development of regional databases could highlight priority areas for management interventions at a regional scale.

International networks facilitate the sharing of knowledge and resources (e.g. biological control agents, best management practices, awareness materials) among affected countries and could be further developed within Southern Africa and the western Indian Ocean islands, and Africa as a whole.

In order to implement prevention and control, the driving forces for parthenium weed invasion must be understood. Local land-use practices and pathways of spread should be understood to determine key factors influencing invasion. Planned management strategies at a regional or national level are beneficial, provided adequate resources are allocated. Systems that promote early detection and rapid response during the early stages of invasion, at country and regional level, should be promoted. Ultimately, best management practices adapted to local needs, and training programmes to enable implementation of these activities, are desirable. Technology transfer on chemical control is required, with acceptance of effective herbicides as a viable management option as part of an integrated control approach. Registration of suitable herbicides may be necessary in some affected countries, with compliance to accepted local and international norms and standards.

Broader acceptance of biological control as the most sustainable, cost-effective long-term solution is required from countries within Southern Africa and the western Indian Ocean islands. This should be followed by technology transfer to countries lacking appropriate expertise. Broader acceptance of biological control research outcomes is required within each region to accept introduction of agents into new countries in the region with only limited additional testing, as introduced natural enemies are likely to disperse beyond country borders regardless of legislation.

The commitment of national agricultural extension programmes to advise on improved land management practices (grazing, crop production, soil conservation), including improved communal grazing systems, would be beneficial to reduce the invasion risk of parthenium weed by decreasing soil disturbance. There is scope for the development and use of indigenous competitive plant species to compete with parthenium weed.

In order to achieve the above aspects, there needs to be sufficient capacity and technology transfer, and training should be undertaken to develop in-country capacity for the implementation of management activities such as biological and chemical control. Increased capacity is also required to improve risk assessment and biosecurity processes to prevent new introductions into countries without parthenium weed at this stage. Herbicide assistance programmes (where government provides herbicides and the landowner provides labour and undertakes spraying on his/her own land, based on approved management plans and using best management practices) have been useful in parts of South Africa.

There are many basic challenges in Southern Africa and the western Indian Ocean islands, such as access to water and sanitation, food, housing and employment. Without future commitment by governments, the detrimental impacts of parthenium weed on food security, human and animal health, and the conservation of biodiversity and natural resources will continue to increase, further impoverishing human populations. Coordinated efforts at local and regional levels are required to manage parthenium weed within Southern Africa and the western Indian Ocean islands.

Acknowledgements

We thank Lesley Henderson (South Africa), Mark Hyde (Zimbabwe), Cláudia Baider (Mauritius), Azaad Gaungoo (Mauritius), Thomas Le Bourgeois (CIRAD), Bruno Senterre (Seychelles), Charles Morel (Seychelles), Dominique Strasberg (La Réunion), Chris Birkenshaw (Missouri Botanical Garden), Sheila Gregory (United Kingdom) and Tomas Chiconela (Mozambique) for distribution records. We thank Ivan Riggs, Colette Terblanche, Ian Macdonald, Ian Rushworth and Ntombifuthi Shabalala for information on the South African distribution. Costas Zachariades (South Africa), Arne Witt (CABI), Christine Griffiths (Mauritius), Vincent Florens (Mauritius), Thierry Pailler (La Réunion) and José Martin (CIRAD) are also thanked. The Department of Environmental Affairs Natural Resources Management Programs, South Africa in particular, and the USAID-funded Integrated Pest Management Innovation Lab at Virginia Tech, together with Virginia State University, USA are gratefully acknowledged for financial support. Asad Shabbir is thanked for developing the maps for this chapter.

References

Adkins, S.W., Navie, S.C. and Dhileepan, K. (2005) Parthenium weed in Australia: Research progress and prospects. In: Ramachandra Prasad, T.V., Nanjappa, H.V., Devendra, R., Manjunath, A., Subramanya, S.C., Chandrashekar, V.K., Kiran Kumar, K.A., Jayaram, K.A., and Prabhakara Setty, T.K. (eds) *Proceedings of the Second International Conference on Parthenium Management.* University of Agricultural Sciences, Bangalore, India, pp. 11–27.

Akobundu, O. (1991) Weeds in human affairs in sub-Saharan Africa: Implications for sustainable food production. *Weed Technology* 5, 680–690.

Autran, J.C. (2010) Se soigner par les plantes des ILES OI – INRA, Ile Reunion 974.

Baker, J.G. (1877) *Flora of Mauritius and the Seychelles: A Description of the Flowering Plants and Ferns of those Islands.* L. Reeve & Co., London.

Bandeira, S., Massingue Manjate, A., Filipe, O. and Boana, E. (2006) An ecological assessment of the health of the Chibuto-Missavene wetland in the dry season, Mozambique – emphasis on resources assessment, utilization and sustainability analysis. In: *Wetlands-based Livelihoods in the Limpopo Basin: Balancing Social Welfare and Environmental Security.* Project report. CGIAR Challenge Program on Water and Food, Colombo, Sri Lanka.

Binggeli, P. (2003) Introduced and invasive plants. In: Goodman, S.M. and Benstead, J.P. (eds) *The Natural History of Madagascar.* University of Chicago Press, Chicago, Illinois, pp. 257–268.

Blackmore, P. and Charlton, S. (2011) *New South Wales Parthenium Strategy: 2010–2015.* New South Wales Department of Primary Industries, Orange, Australia.

Bojer, W. (1837) *Hortus Mauritianus: Énumération des plantes, exotiques et indigènes, qui croissent a l'Ile Maurice, disposees d'apres la methode naturelle.* Imprimerie d'Aimé Mamarot et Compagnie, Maurice.

Bory de Saint-Vincent, J.B. (1804) *Voyage dans les quatre principales îles des mers d'Afrique, fait par ordre du gouvernement, pendant les années neuf et dix de la République (1801 et 1802), avec l'histoire de la traversée du capitaine Baudin jusqu'au Port-Louis de l'Île Maurice.* Volume 2. F. Buisson, Paris.

Bouton, L. (1864) *Plantes médicinales de Maurice,* 2nd edn. Port Louis, Ile Maurice. Typography: E. Dupuy et P. Dubois.

Bowen, J. (2001) A study of the extent of alien species invasion into the Lubombo Conservancy, Swaziland. Unpublished report. Quest Overseas, UK. 6 pp.

Braun, K.P., Dlamini, S.D. and Dlamini, T.S. (2004) *Swaziland's Alien Plants Database.* Available at: http://www.sntc.org.sz/alienplants/ (accessed 22 June 2016).

Callmander, M.W., Phillipson, P.B., Schatz, G.E., Andriambololonera, S., Rabarimanarivo, M., et al. (2011) The endemic and non-endemic vascular flora of Madagascar updated. *Plant Ecology and Evolution* 144, 121–125.

Catalogue of the Plants of Madagascar (2017) Tropicos.org, Missouri Botanical Garden. Available at http://www.tropicos.org/Name/2701101 (accessed 8 June 2017).

Chamberlain, J. and Gittens, A. (2004) *Parthenium Weed Management: Challenges, Opportunities and Strategies.* The Department of Natural Resources, Mines and Energy, Brisbane, Australia.

Daruty, C. (1886) *Plantes Médicinales de l'Île Maurice et des Pays Intertropicaux.* General Steam Printing Company, Mauritius.

De Cordemoy, E.J. (1895) *Flore de l'Île de la Réunion (phanérogames, cryptogames, vasculaires, muscinées) avec l'indication des propriétés économiques & industrielles des plantes.* Paul Klinksieck, Paris.

Dhileepan, K. and McFadyen, R.E. (2012) *Parthenium hysterophorus* L. – parthenium. In: Julien, M., McFadyen, R.E. and Cullen, J. (eds) *Biological Control of Weeds in Australia: 1960 to 2010.* CSIRO Publishing, Melbourne, Australia, pp. 448–462.

Fosberg, F.R. and Renvoize, S.A. (1970) 3. Plants of Farquhar Atoll. In: Stoddart, D.R. (ed.) *Coral Islands of the Western Indian Ocean.* Atoll Research Bulletin No. 136. The Smithsonian Institution, Washington, DC, pp. 27–33.

Friedmann, F. (2011) *Flore des Seychelles: Dicotylédones.* Faune et Flore tropicales Vol. 44. Muséum National d'Histoire Naturelle, Paris.

Fusée-Aublet, J.B. (1775) *Histoire des plantes de la Guiane Françoise: rangées suivant la méthode sexuelle, avec plusieurs mémoires sur différens objets intéressans, relatifs à la culture & au commerce de la Guiane Françoise, & une notice des plantes de l'Île-de-France.* P.F. Didot jeune, Paris.

Goodall, J., Braack, M., De Klerk, J. and Keen, C. (2010) Study on the early effects of several weed-control methods on *Parthenium hysterophorus* L. *African Journal of Range and Forage Science* 27, 95–99.

Guillemin (1828) *Parthenium hysterophorus* L. In: Audouin, Bourdon, I., Brongniart, A., De Candolle, Delafosse, G., et al. *Dictionnaire Classique d'Histoire Naturelle,* Volume 13. PANPIV, Rey & Gravier, Paris, pp. 83–84.

Henderson, L. (2001) *Alien Weeds and Invasive Plants: A Complete Guide to Declared Weeds and Invaders in South Africa.* Handbook No. 12. Agricultural Research Council – Plant Protection Research Institute, Roodeplaat, South Africa.

Henderson, L. (2017) *Southern African Plant Invaders Atlas (SAPIA).* Agricultural Research Council, Pretoria, South Africa.

Henderson, L. and Wilson, J.R. (2017) Changes in the composition and distribution of alien plants in South Africa: An update from the Southern African Plant Invaders Atlas. *Bothalia – African Biodiversity and Conservation* 47, 1–26.

Hilliard, O.M. (1977) *Compositae in Natal.* University of Natal Press, Natal, South Africa.

Hind, D.J.N., Jeffrey, C. and Scott, A.J. (1993) 109. Composées. In: Bosser, J., Antoine, R., Ferguson, I.K., *Flore des Mascareignes: La Réunion, Maurice, Rodrigues.* Royal Botanic Gardens, Kew, London/Office de la Recherche Scientifique et Technique Outre-Mer/Mauritius Sugar Industry Research Institute/The Sugar Industry Research Institute, Mauritius.

Humbert, H. (1963) Flore de Madagascar et des Comores (Plantes Vasculaires). Issue 189. Famille – Composées. Tome III: p. 636. Publiée sous les auspices du gouvernement de la République Malgache et sous la direction de H. Humbert. Muséum National d'Histoire Naturelle, Laboratoire de Phanérogamie, Paris.

Husson, O., Charpentier, H., Chabaud, F.-X., Naudin, K., Rakotondramanana and Seguy, L. (2010) *Les principales plantes des jachères et adventices des cultures à Madagascar (Annexe 1, Manuel pratique du semis direct à Madagascar)*. GSDM/CIRAD, Montpellier/Antananarivo.

Hyde, M.A., Wursten, B.T., Ballings, P. and Coates Palgrave, M. (2016) Flora of Zimbabwe: Species information: Records of: *Parthenium hysterophorus*. Available at: http://www.zimbabweflora.co.zw/speciesdata/species-display.php?species_id=160290 (accessed 24 April 2017).

Idris, H. (1970) Chemical control of weeds in cotton in the Sudan Gezira. *Pest Articles and News Summaries* 16, 96–105.

International Fund for Agricultural Development, Japan Official Development Assistance, and Food and Agriculture Organization (1998) *Agricultural Implements used by Women Farmers in Africa*. FAO, Rome.

International Union for Conservation of Nature (1992) Kingdom of Swaziland. In: *Protected Areas of the World: A Review of National Systems. Volume 3: Afrotropical*. IUCN, Gland, Switzerland, pp. 291–294.

Inventaire National du Patrimoine Naturel (National Inventory of Natural Heritage) 2017. Muséum national d'Histoire naturelle (ed.),2003–2017. Available at: https://inpn.mnhn.fr (accessed 8 June 2017).

Kariuki, P. and Leigh, F. (2015) Swaziland. In: *Regional Development and Spatial Inclusion: African Economic Outlook 2015*. OECD, Paris, pp. 202–203.

Keatimilwe, K. and Mlangeni, J. (2003) Swaziland. In: *Southern African Institute for Environmental Assessment. Environmental Impact Assessment in Southern Africa*. Southern African Institute for Environmental Assessment, Windhoek, South Africa.

Khan, N., O'Donnell, C., Shabbir, A. and Adkins, S.W. (2010) Competitive displacement of parthenium weed with beneficial native and introduced pasture plants in central Queensland, Australia. In: Zydenbos, S.M. (ed.) *Proceedings of the Seventeenth Australasian Weed Conference*. Christchurch, New Zealand, pp. 131–134.

Kueffer, C. (2010) Invasive alien plants in the Seychelles: some lessons learnt. *Zwazo* January–June, 26–27.

Kull, C.A., Tassin, J., Moreau, S., Rakoto Ramiarantosa, H., Blanc-Pamard, C. and Carrière, S.M. (2012) The introduced flora of Madagascar. *Biological Invasions* 14, 875–888.

Kull, C.A., Tassin, J. and Carrière, S.M. (2014) Approaching invasive species in Madagascar.

Madagascar Conservation & Development 9, 60–70.

Le Bourgeois, T., Carrara, A., Dodet, M., Dogley, W., Gaungoo, A., *et al.* (2008). Advent-OI: Principales adventices des îles du sud-ouest de l'Océan Indien. Cirad ed. Montpellier, France, cdrom.

Le Bourgeois, T., Grard, P., Andrianaivo, A.P., Gaungoo, A., Ibrahim, Y., *et al.* (2015) WIKWIO – Weed Identification and Knowledge in the Western Indian Ocean – Web 2.0 participatory portal. European Union programme ACP S&T II, Cirad, IFP, MCIA/MSIRI, FOFIFA, CNDRS (eds). Available at: http://portal.wikwio.org (accessed 8 June 2017).

Mainali, K.P., Warren, D.L., Dhileepan, K., McConnachie, A.J., Strathie, L., *et al.* (2015) Projecting future expansion of invasive species: Comparing and improving methodologies for species distribution modelling. *Global Change Biology* 21(12), 4464–4480.

Mapaura, A. and Timberlake, J. (2004) A checklist of Zimbabwean vascular plants. *Southern African Botanical Diversity Network Report* No. 33. Sabonet, Pretoria.

Mascarine Cadetiana (2016) *Conservatoire National Botanique de Mascarin*. Available at: http://mascarine.cbnm.org/mascarine (accessed 24 April 2017).

Mavudzi, Z., Mashingaidze, A.B., Chivinge, O.A., Ellis-Jones, J. and Riches, C. (2001) Improving weed management in a cotton–maize system in the Zambezi Valley Zimbabwe. *Brighton Crop Protection Conference – Weeds*. BCPC, Farnham, UK, pp. 169–174.

McConnachie, A.J. (2015a) Host range and risk assessment of *Zygogramma bicolorata*, a defoliating agent released in South Africa for the biological control of *Parthenium hysterophorus*. *Biocontrol Science and Technology* 25(9), 975–991.

McConnachie, A.J. (2015b) Host range tests cast doubt on the suitability of *Epiblema strenuana* as a biological control agent for *Parthenium hysterophorus* in Africa. *BioControl* 60, 715–723

McConnachie, A.J., Strathie, L.W., Mersie, W., Gebrehiwot, L., Zewdie, K., *et al.* (2011) Current and potential geographic distribution of the invasive plant *Parthenium hysterophorus* (Asteraceae) in eastern and southern Africa. *Weed Research* 51(1), 71–84.

McFadyen, R.C. (1992) Biological control against parthenium weed in Australia. *Crop Protection* 11, 400–407.

McIntyre, G. (1991) *Weeds of Sugar Cane in Mauritius: Their Description and Control*. Mauritius

Sugarcane Industry Research Institute, Réduit, Mauritius.

Moran, V.C., Hoffmann, J.H. and Zimmermann, H.G. (2013) 100 years of biological control of invasive alien plants in South Africa: History, practice and achievements. *South African Journal of Science* 109(9/10), Art. #a0022. http://dx.doi.org/10.1590/sajs.2013/a0022 (accessed 24 August 2018).

Muoni, T., Rusinamhodzi, L. and Thierfelder, C. (2013) Weed control in conservation agriculture systems of Zimbabwe: Identifying economical best strategies. *Crop Protection* 53, 23–28.

Nath, R. (1988) *Parthenium hysterophorus* L. – a general account. *Agricultural Review* 9, 171–179.

Navie, S.C., McFadyen, R.E., Panetta, F.D. and Adkins, S.W. (1996) The biology of Australian weeds 27. *Parthenium hysterophorus* L. *Plant Protection Quarterly* 11(2), 76–88.

O'Donnell, C. and Adkins, S. (2005) Management of parthenium weed through competitive displacement with beneficial plants. *Weed Biology and Management* 5, 77–79.

Olckers, T. (2004) Targeting emerging weeds for biological control in South Africa: The benefits of halting the spread of alien plants at an early stage of their invasion. *South African Journal of Science* 100, 64–68.

Overfield, D., Murithi, F.M., Muthamia, J.N., Ouma, J.O., Birungi, K.F., *et al.* (2001) Analysis of the constraints to adoption of herbicides by smallholder maize growers in Kenya and Uganda. *Brighton Crop Protection Conference – Weeds.* BCPC, Farnham, UK, pp. 907–912.

Oyedemi, T. and Olajide, A. (2002) Ergonomic evaluation of an indigenous tillage tool employed in Nigerian agriculture. *ASAE Annual Meeting*, Paper Number 028001. American Society of Agricultural and Biological Engineers, St Joseph, Michigan.

Perrier de la Bâthie, H. (1928) Les pestes végétales à Madagascar. *Revue de Botanique Appliquée et d'Agriculture Coloniale* 8(77), 36–43.

Perrier de la Bathie, H. (1931) Les plantes introduites à Madagascar (suite). *Revue de Botanique Appliquée et d'Agriculture Colonial* 11(124), 991–999.

Phillipson, P.B., Schatz, G.E., Lowry II, P.P. and Labat, J.-N. (2006) A catalogue of the vascular plants of Madagascar. In: Ghazanfar, S.A. and Beentje, H.J. (eds) *Taxonomy and Ecology of African Plants, Their Conservation and Sustainable Use.* Royal Botanic Gardens, Kew, pp. 613–727.

Pourchez, L. (2011) *Savoirs des Femmes: Médecine Traditionnelle et Nature: Maurice, Réunion,* Rodrigues. *Savoirs Locaux et Autochtones, 1.* UNESCO, Paris.

Retief, E., Ntushelo, K. and Wood, A.R. (2013) Host-specificity testing of *Puccinia xanthii* var. *parthenii-hysterophorae*, a potential biocontrol agent for *Parthenium hysterophorus* in South Africa. *South African Journal of Plant and Soil* 30, 7–12.

Robertson, S.A. (1989) *Flowering Plants of Seychelles (An Annotated Checklist of Angiosperms and Gymnosperms with Line Drawings).* Royal Botanic Gardens, Kew.

Rochecouste, E., Vaughan, R.E., Autrey, L.J.C. and McIntyre, G. (1981) *Weeds of Mauritius. Indexes and glossary (1969–1981).* Mauritius Sugarcane Industry Research Institute, Réduit, Mauritius.

Seychelles Plant Gallery (2013) Plant Conservation Action Group. National History Museum, Victoria, Seychelles. Available at: http://www.seychellesplantgallery.com/Image/SEY012013_0348.JPG (accessed 5 June 2017).

Stoddart, D.R. (1970) 1. Introduction. In: Stoddart, D.R. (ed.) *Coral Islands of the Western Indian Ocean.* Atoll Research Bulletin No. 136. The Smithsonian Institution, Washington, DC, pp. 1–5.

Stoddart, D.R. and Poore, M.E.D. (1970) 2. Geography and ecology of Farquhar Atoll. In: Stoddart, D.R. (ed.) *Coral Islands of the Western Indian Ocean.* Atoll Research Bulletin No. 136. The Smithsonian Institution, Washington, DC, pp. 7–26.

Strathie, L.W., McConnachie, A.J. and Retief, E. (2011) Initiation of biological control against *Parthenium hysterophorus* L. (Asteraceae) in South Africa. *African Entomology* 19, 378–392.

Terblanche, C. (2014) *National Strategy for the Management of Famine Weed (Parthenium hysterophorus) in South Africa: 2014–2018.* The Department of Environmental Affairs Environmental Programmes, Natural Resource Management Programs, South Africa.

Terblanche, C., Nanni, I., Kaplan, H., Strathie, L., McConnachie, A.J., Goodall, J. and Van Wilgen, B. (2016) An approach to the development of a national strategy for controlling invasive alien plant species: the case of *Parthenium hysterophorus* in South Africa. *Bothalia* 46(1), a2053. http://dx.doi.org/10.4102/abc.v46i1.2053 (accessed 24 August 2018).

UICN (Comité français de l'Union Internationale pour la Conservation de la Nature en France) (2017) Les espéces envahissantes en outre-mer. Available at: http://www.especes-envahissantes-outremer.fr/autoComplete/

detailsuicn.php?fiche=950 (accessed 8 June 2017).

Van Der Laan, M., Reinhardt, C.F., Belz, R.G., Truter, W.F., Foxcroft, L.C. and Hurle, K. (2008) Interference potential of the perennial grasses *Eragrostis curvula*, *Panicum maximum* and *Digitaria eriantha* with *Parthenium hysterophorus*. *Tropical Grasslands* 42, 88–95.

Vaughan, R.E. and Wiehe, P.O. (1937) Studies on the vegetation of Mauritius: I. A preliminary survey of the plant communities. *Journal of Ecology* 25, 289–343.

Vissoh, P.V., Gbehounou, G., Ahanchede, A., Kuyper, T.W. and Roling, N.G. (2004) Weeds as agricultural constraint to farmers in Benin: Results of a diagnostic study. *NJAS – Wageningen Journal of Life Sciences* 52(3/4), 305–329.

Wiehe, P.O. (1938) Report on visit to Rodrigues. Unpublished report to Director of Agriculture. Mauritius. (Typescript housed at The Mauritius Herbarium, Réduit, Mauritius).

Wild, H. (1967) The Compositae of the Flora Zambesiaca area. 1. Heliantheae. *Kirkia* 6, 1–62.

Wood, J.M. (1897) *Report on Natal Botanic Gardens for the Year 1896*. Durban Botanic Society, Durban, South Africa.

Wood, A.R. and Scholler, M. (2002) *Puccinia abrupta* var. *partheniicola* on *Parthenium hysterophorus* in Southern Africa. *Plant Disease* 86, 327.

World Food Program (2015) Swaziland. Available at: https://www.wfp.org/countries/swaziland (accessed 22 June 2016).

Zachariades, C. (2017) An initial assessment of *Chromolaena odorata* and other terrestrial invasive alien plant species in southern Mozambique: January 2017. Unpublished report. Agricultural Research Council, Pretoria, South Africa.

Zachariades, C., Paterson, I.D., Strathie, L.W., Hill, M.P. and van Wilgen, B.W. (2017) Assessing the status of biological control as a management tool for suppression of invasive alien plants in South Africa. *Bothalia* 47(2), a2142.

15 History and Management – East and North Africa, and the Middle East

Andrew McConnachie[1]* and Arne Witt[2]

[1]Department of Primary Industries, Biosecurity and Food Safety, Orange, New South Wales, Australia; [2]CABI Africa, Nairobi, Kenya

15.1 Introduction

While attention was being drawn to the explosion of parthenium weed (*Parthenium hysterophorus* L.) in India and Australia during the 1970s, East and North Africa and the Middle East were in the early stages of being exposed to this weed. Affecting areas where communities were largely dependent on subsistence farming and pastoralism, and containing important conservation areas, parthenium weed was set to become a major issue throughout the region, affecting livelihoods and biodiversity. Forty-five years later and parthenium weed is now a well-established invader in many parts of East Africa and emerging in others. In North Africa and the Middle East, parthenium weed seems to be in an equilibrium phase, either as a result of being on the edge of its eco-climatic range or confined to only a few isolated areas where conditions favour its growth and persistence, such as along the Nile River Delta.

Numerous studies have considered the traits of successful plant invaders (Baker, 1965; Rejmánek and Richardson, 1996; Kolar and Lodge, 2001; Catford *et al.*, 2009) and the characteristics of recipient environments that make them susceptible to invasion (Rejmánek, 1989; Levine and D'Antonio, 1999; Mack *et al.*, 2000; Pyšek and

Richardson, 2006; Catford *et al.*, 2009). According to Catford *et al.* (2009), invasion is essentially a function of propagule pressure from the invading species, the abiotic characteristics of the invaded ecosystem and the characteristics of the recipient community. Parthenium weed possesses many of the traits associated with successful invaders, but its distribution is limited by the characteristics of the invaded ecosystem. Parthenium weed achieves optimal growth, reproduction and spread within a defined set of eco-climatic requirements (McConnachie *et al.*, 2011; Kriticos *et al.*, 2015; Mainali *et al.*, 2015; Shabbir *et al.*, Chapter 3, this volume), with soil moisture and type being two of the more significant criteria (Dale, 1981; Navie *et al.*, 1996). In East Africa, where vast infestations of the weed occur, the importance of both these features becomes evident (Wabuyele *et al.*, 2015) with suitable soils and two fairly distinct wet seasons.

However, another critical component which drives most plant invasions, including that of parthenium weed, is natural and human-mediated disturbance. The majority of people, especially in East and North Africa, are subsistence or small-scale farmers, which means that vast tracks of land have been transformed for crop production, while semi-arid regions unsuitable for crops

* Andrew.mcconnachie@dpi.nsw.gov.au

are used extensively for livestock production. Overgrazing and poor land-use practices have contributed to desertification, with over 73% of some African countries' agricultural drylands already degraded (ECA, 2007). Over one-quarter of land in Ethiopia is already considered to be degraded. Approximately 30% of Kenya was already affected by very severe to severe land degradation in the earlier part of this century (UNEP, 2002). More recent studies have found that 20% of all cultivated areas, 30% of forests and 10% of grasslands in Kenya are degraded (Muchena, 2008). These denuded landscapes are especially prone to parthenium weed invasion. War-mediated dispersal and disturbance (Tareke, 2000; Tamado and Milberg, 2000) as well as famine and food aid distribution are also thought to have played important roles in the spread of parthenium weed in East Africa.

During the time that parthenium weed has occurred in East and North Africa, and the Middle East, there has been a distinct lack of a coordinated response to the invasion. The reasons for this are many and varied and are reviewed by Boy and Witt (2013). One of the main reasons they cite for the lack of action in controlling invasive alien plants is a lack of policy or legislation, or implementation thereof. In addition, there is little to no information regarding the distribution and impacts of invasive alien plants and little capacity with regard to their management. There is also little coordination between various sectors within a country to manage such plants, with no regional response to any pest incursion. Because of an aversion to biological control, probably due to a lack of information on its safety and efficacy, and a lack of resources, no emphasis has been placed on initiating a biological control programme against parthenium weed (all classical biological control agents introduced into East Africa at that stage had been for water weeds). As a result, the only control measures implemented have been at the localized scale and involve mainly physical control (hand-weeding) with some herbicide applications. The impacts of parthenium weed on human and animal health are also not well documented; had they been, there

might have been a stronger impetus to tackle this invasive plant sooner.

Fortuitously, in 2005, this lack of awareness surrounding the impact of parthenium weed (in East and North Africa, at least) was about to change forever. Two international projects were initiated that year. The first, entitled 'Removing Barriers to Invasive Plant Management in Africa' (RBIPMA), was conceptualized by the Centre for Agriculture and Bioscience International (CABI) in collaboration with the International Union for the Conservation of Nature (IUCN), working with country partners in Ethiopia, Uganda, Ghana and Zambia. The aim of this project was to literally remove the barriers surrounding the management of alien invasive alien plants in these four countries. The project was implemented by United Nations Environment Program (UNEP) and executed by CABI in collaboration with the IUCN and four national executing agencies (Boy and Witt, 2013). Invasive alien species (IAS) targeted for direct action included parthenium weed, mesquite (*Prosopis juliflora* (Sw.) DC.), lantana (*Lantana camara* L.), giant sensitive plant (*Mimosa pigra* L.), paper mulberry (*Broussonetia papyrifera* (L.) Vent.), white bark senna (*Senna spectabilis* DC.) and citronella grass (*Cymbopogon nardus* L.). The four main objectives of the project were: (i) to raise awareness; (ii) to strengthen or develop policies; (iii) to build capacity; and (iv) to develop and implement best management practices for the selected IASs. The second project was also a four-year endeavour, coordinated by Virginia State University, USA, under the auspices of the United States Agency for International Development (USAID) and under the Integrated Pest Management Collaborative Research Support Program (IPM CRSP). It was focused specifically on the integrated control of parthenium weed in East (Uganda and Ethiopia) and southern Africa (South Africa, Swaziland and Botswana) (Strathie *et al.*, 2011). The project broadly looked at the distribution and impact of parthenium weed and assessed the efficacy of various control methods. The USAID IPM CRSP project has since received two more rounds of funding and continues to work in Ethiopia, Uganda,

Kenya and Tanzania, focusing on the biological control of parthenium weed (W. Mersie, Virginia, 2016, personal communication).

15.2 History of Invasion and Spread in East Africa, North Africa and the Middle East

15.2.1 East and North Africa

The earliest published reference for the presence of parthenium weed in northern Africa dates back to 1960. It was around this date that parthenium weed was apparently introduced into the Nile Region, Egypt (Fig. 15.1), through a single sowing of impure contaminated grass seed imported from Texas (Boulos and El-Hadidi, 1984). According to Zaharan and Willis (2009), the region it was introduced to is extremely fertile, due to the alluvial deposits of the Nile River. The region extends from the south of the Nile Valley and northwards towards the Nile Delta. It also extends west into the Nile Fayium, which was originally a marshy depression of the Western Desert and has a long history of cultivation. The Nile region, therefore, was a perfect entry route for parthenium weed into North Africa: rich, fertile soils; agricultural disturbance; and adequate soil moisture, together with regular floods which contributed to its further spread.

It is unknown if parthenium weed moved from Egypt into Ethiopia and then further south. One theory is that it may have been dispersed by troops and their equipment from Egypt into Somalia and then Ethiopia during the Ogaden/Ethiopian–Somali War (July 1977–March 1978). Egypt sent millions of dollars in arms, established military training camps and sent experts to Somalia in support of Egypt's long-standing policy of securing the

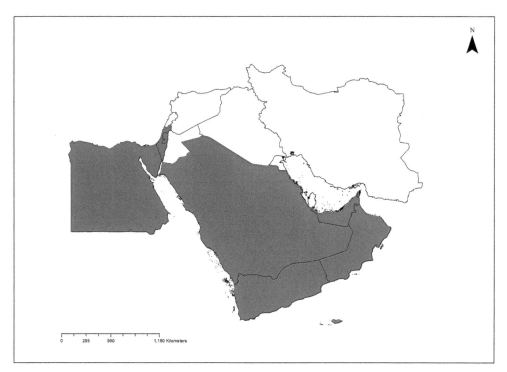

Fig. 15.1. Distribution of parthenium weed in Egypt, Israel, Palestine, Yemen, Oman United Arab Emirates and Saudi Arabia (populations discovered while chapter was in preparation) at a country scale. (A. Shabbir.)

Nile River (the Blue Nile has Lake Tana, Ethiopia, as its source) (Tareke, 2000). There are accounts of parthenium weed in Somalia during this time (Tamado and Milberg, 2000; Maal and Abdi, 2014). However, there are reports of parthenium weed in Ethiopia in 1968 (Tamado, 2001), confounding the theory that it may have been introduced during the 1977/78 conflict. Parthenium weed may also have found its way into Ethiopia as a seed contaminant in famine-relief supplies brought in during the 1980s (Boy and Witt, 2013). If this is correct, then it must have been a secondary introduction as parthenium weed was already detected in Ethiopia in 1968 (Tamado, 2001). Regardless of how parthenium weed entered Ethiopia, and whether it was from single or multiple introductions, the weed is found occupying a wide invaded area across most of the country (Fig. 15.2), invading both rangelands and cropping lands (Tamado and Milberg, 2000).

The first records for parthenium weed in Kenya were both from coffee (*Coffea arabica* L.) plantations; in 1973 from Ruiru (Beentje *et al.*, 2005) and in 1975 from the capital Nairobi (Njoroge, 1989). However, these populations are thought not to have spread much over the following 30 years. A possible second introduction (probably from Ethiopia) was thought to have established around 2008, and was considered to be a lot more aggressive, in terms of plant establishment and spread, especially when compared with the more benign first introduction (Fig. 15.2). Between 2008 and 2015, noticeable spread of parthenium weed was observed along three major roads leading out of Nairobi to Arusha (Tanzania), Mombasa and Nakuru, highlighting the important role of roads as primary pathways or disturbance corridors and vehicles as vectors. Road construction during this period, especially to the south of Nairobi to Namanga and then Arusha, largely contributed to the spread. Extensive road development throughout the region has played a key role in facilitating the spread of parthenium weed.

As part of the RBIPMA project, parthenium weed was recorded for the first time in Uganda (Fig. 15.2) on 16 November 2008 at Namalemba Primary School, near Busembatia village to the north of Lake Victoria (A. Witt and G. Howard, Nairobi, 2008, personal communication). At the same time, another infestation was documented near Bugembe. It is speculated that parthenium weed may have spread to these localities from populations in northern Uganda (possibly from Pader, an internally displaced people camp, where the World Food Program had supplied food aid).

In 2010, parthenium weed was recorded in Tanzania (Fig. 15.2) for the first time near Arusha International Airport (Wabuyele *et al.*, 2014). In 2011, two additional sites were documented in Karatu and Ngorongoro conservation area (Wabuyele *et al.*, 2014). The presence of parthenium weed in the world famous Serengeti-Masai Mara ecosystem (northern Tanzania and southwest Kenya) is of great concern. If its relentless advance cannot be checked, the survival of entire populations of wild herbivores may soon be at risk (Boy and Witt, 2013).

In 2013, parthenium weed was observed for the first time in Rwanda (Fig. 15.2). The small roadside infestation was found on the edge of Matimba, just south of the Uganda border. With favourable soils and rainfall in Rwanda, along with significant land disturbance, it is anticipated that parthenium weed will flourish in this country.

15.2.2 The Middle East

The Middle East has a significant link with North Africa, in particular Egypt (Fig. 15.1). Foreign policy in Egypt operates along a non-aligned level. Factors such as population size, historical events, military strength, diplomatic expertise and strategic geographical position give Egypt extensive political influence in the Middle East. Cairo has been a crossroads of the Arab world's commerce and culture for centuries, and its intellectual and Islamic institutions are at the centre of the region's social and cultural landmarks. With the first populations of parthenium weed having being recorded in Egypt in

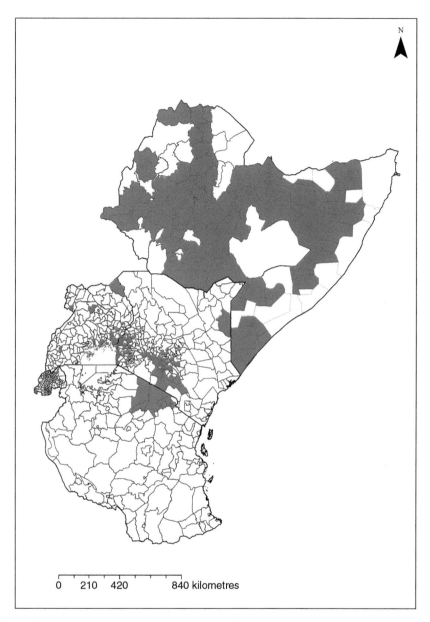

Fig. 15.2. Distribution of parthenium weed in Ethiopia, Kenya, Uganda, Rwanda and Tanzania at a district scale. (A. Shabbir.)

1960, it is a very distinct possibility that the plant was accidentally moved into the Middle East through trade. Alternatively, the weed may also have arrived in the Middle East through the mass movement of migrants and workers to the Gulf States from parthenium-infested areas of Asia, particularly India and Pakistan. Currently,

over 5 million Indians live in the Gulf States (Bollier and Haddad, 2013), by far the most populous immigrant nationality. Parthenium weed was first recorded in India in the 1950s (see Shabbir *et al.*, Chapter 12, this volume).

The first records for parthenium weed in Palestine and Israel (Fig. 15.1) are from Bet Shean Valley, Tirat Zvi, in 1979 (Dafni and Heller, 1982; Dafni and Heller, 1990; Danin, 2000). Areas invaded are listed as 'fields' and it is thought that it was introduced to both areas in contaminated cereal grain (Joel and Liston, 1986). This possible source of introduction is not surprising considering that Israel is almost completely dependent on grain imports, with the USA being one of the main exporters, but also including Argentina (Schachar, 2011).

Parthenium weed was found in Oman for the first time in 1998 (Fig. 15.1). It was growing along a track between Raysut and Qaftwawt (Dhofar) in open *Anogeissus–Commiphora* woodland (Kilian *et al.*, 2002; Alhammadi, 2010). In 2010, it was recorded in Yemen in Al-Mahra (just west of Jadib) on a track and along roadsides to Damquat (Kilian *et al.*, 2002; Fig. 15.1). Here, it is speculated that parthenium weed was spread from Dhofar, Oman, by a new road between both countries. Three years later (2013) it was detected for the first time in the United Arab Emirates (UAE) (Mahmoud *et al.*, 2015), this time in a garden at Hamryah coast, Sharjah (Fig. 15.1). The introduction was thought to be accidental, possibly as a result of the importation of

contaminated seed or within potting medium of imported plants. Oman, Yemen and the UAE all engage in high volumes of trade with the Middle East, East and North Africa, India and Pakistan, so the initial population in Oman could have originated in any one of the listed areas.

Several eco-climatic models have been developed for parthenium weed over the last 4 years in an attempt to determine its potential distribution and in so doing facilitate the development of management strategies (McConnachie *et al.*, 2011; Mainali *et al.*, 2015; Kriticos *et al.*, 2015). All of these models show areas that are suitable for the growth of parthenium weed in East and North Africa and the Middle East. Using the model of Mainali *et al.* (2015), East Africa is observed to have large areas that are eco-climatically suitable for parthenium weed, while in North Africa, the Nile Valley and Delta contain suitable habitat (Fig. 15.3). In Israel and Palestine, small pockets of fragmented suitable habitat exist along the Mediterranean coast. In the south-west corner of Yemen there are several small areas which appear to be suitable habitat for the weed. In most cases, the model predictions concur quite well with the current distribution of the weed.

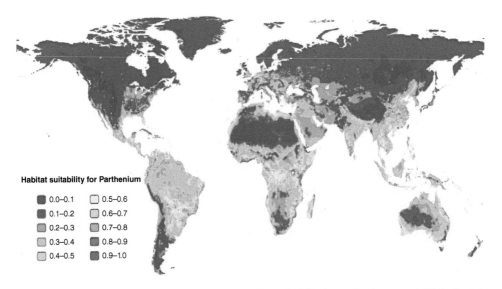

Habitat suitability for Parthenium

- 0.0–0.1
- 0.1–0.2
- 0.2–0.3
- 0.3–0.4
- 0.4–0.5
- 0.5–0.6
- 0.6–0.7
- 0.7–0.8
- 0.8–0.9
- 0.9–1.0

Fig. 15.3. Eco-climatic model depicting the global habitat suitability for parthenium weed. (Mainali *et al.*, 2015.)

Countries in East and North Africa and the Middle East for which there are no current records and for which the models show eco-climatic suitability (Fig. 15.3) for the growth of parthenium weed include: Burundi; Eritrea; the northern coast of Libya, Tunisia, and some parts of Algeria and Morocco; and a small area in north-eastern Iraq.

15.3 Land Use and Invasion

Parthenium weed as a species is particularly well adapted to land disturbance. In the absence of competition, it is able to dominate and rapidly colonize disturbed ground, such as roadsides, wastelands, cultivated lands and overgrazed pastures, and causes significant losses to agricultural production (Chippendale and Panetta, 1994; Adamson and Bray, 1999; Tamado *et al.*, 2002). This is particularly evident throughout East Africa,

with parthenium weed a serious invader of rangelands and cropping areas (Table 15.1). Roadside invasions are also particularly prominent, especially when locality data is plotted and all the main road networks are outlined. Disturbance-facilitated parthenium weed invasion is not always linked to anthropogenic activities, however; in conservation areas for example where parthenium weed has established, disturbance caused by events like fire, animal migrations or even relatively small disturbance events like those caused by termites may be all that's required to allow parthenium weed to establish in a new area.

15.4 Invasion Costs

15.4.1 Impacts on human health

Globally, the impact of parthenium weed on human health has been documented to

Table 15.1. Land use types invaded by parthenium weed in North and East Africa, and the Middle East.

Region	Country	Land use	Reference
North Africa	Egypt	Cultivated lands	Boulos and El-Hadidi (1984)
East Africa	Eritrea	Unknown	Tadesse (2004)
	Ethiopia and Somalia	Crop and rangelands; roadsides, conservation areas; fallow land, drainage ditches, woodlands (provided there is sufficient light)	Tamado and Milberg (2000) McConnachie *et al.* (2011)
	Kenya	Coffee plantations; croplands, conservation areas; roadsides, fallow land, drainage ditches, woodlands (provided there is sufficient light)	Njoroge (1986) Boy and Witt (2013)
	Tanzania	Abandoned or fallow fields, railway lines, conservation areas, urban open space, roadsides, disturbed areas, wasteland, pasture, cropland, gardens, drainage lines, culverts, waterways, floodplains, savannah, open woodland	Wabuyele *et al.* (2014)
	Rwanda	Roadside	A. Witt (pers. observations)
	Uganda	School, roadsides	Wabuyele *et al.* (2014)
Middle East	Oman	Track in open *Anogeissus–Commiphora* woodland, roadsides	Kilian *et al.* (2002) Alhammadi (2010)
	Palestine and Israel	Fallow fields	Joel and Liston (1986)
	Saudi Arabia	Found in agriculture lands, stream-banks, roadsides, wastelands, dry slopes.	(Thomas *et al.*, 2015)
	UAE	Residential garden	Mahmoud *et al.* (2015)
	Yemen	Track, roadside	Kilian *et al.* (2002)

range from allergic eczematous dermatitis (after prolonged skin contact) to respiratory problems (allergic rhinitis, bronchitis or asthma) from exposure to the pollen (Towers and Rao, 1992; McFadyen, 1995; Evans, 1997; see Allen *et al.*, Chapter 6, this volume).

Despite the scale of the problem in East Africa, especially Ethiopia, and its effect on the livelihoods of millions of people, there is a lack of information quantifying the impacts of this weed on human and animal health. In a recent survey, Dinwiddie (2014) noted that very few Ethiopian farmers, when interviewed, said that they experienced health problems associated with parthenium weed. This is in contrast to an earlier study which noted that most farmers reported problems due to parthenium weed (Wiesner *et al.*, 2007). In the latter study, the following symptoms were noted: general illness (80%), allergic reactions (90%), asthma (62%), irritation of skin and pustules on hands (30%), stretching and cracking of skin (21%), and stomach pains (22%). Rao *et al.* (1977) reported only a 4% incidence of contact dermatitis, which was similar to the 7% recorded in the study of Dinwiddie (2014). In Kenya, the only study documenting any impacts of parthenium weed on health was for livestock (Mutua *et al.*, 2014). With most respondents (>78%) of this study being unaware of any impacts of parthenium weed on animal health, the remaining respondents said that it caused coughing (10%), diarrhoea (5%), mouth ulcers (3%), emaciation (3%) and death (1%). Indirect effects of parthenium weed on human health have also been documented.

In a recent study by Nyasembe *et al.* (2015) in East Africa, the survival and fitness of the malaria vector, *Anopheles gambiae* (Diptera: Culicidae), was found to be enhanced when it utilized parthenium weed as a food source, and when compared with other weed species. These results highlight the potential epidemiological implications of higher disease transmission, especially where there is a high degree of overlap between malaria endemic and parthenium weed-infested areas.

15.4.2 Socio-economic impacts

Studies like those conducted by Beyene and Tessema (2015) have shown that parthenium weed has very real impacts on communities that practice subsistence agriculture. One of the more noticeable socio-economic costs picked up by the study was the significant number of respondents who spent time conducting early weeding of parthenium weed (c.92% of families). Failure to invest time in this practice would more than likely be the difference between crop success and failure.

15.4.3 Impacts on crop production

For some of the East African countries covered in this chapter, parthenium weed has a documented impact on crop productivity. In Ethiopia, studies showed that if parthenium weed was left uncontrolled in sorghum (*Sorghum bicolor* (L.) Conrad Moench) fields throughout the growing season (Fig. 15.4), grain yield losses could vary between 40% and 97% (Tamado *et al.*, 2002) and 69% even at low weed densities (Reda *et al.*, 2008). Another study from Ethiopia highlighted the allelopathic effect that parthenium weed has on seed germination of the endemic cereal crop, tef (*Eragrostis tef* Zucc.). This crop is a staple of the Ethiopian people, occupying 28% of cultivated land. In Kenya, parthenium weed has been reported as a weed of coffee plantations for over 30 years, requiring chemical intervention (Njoroge, 1986, 1991).

15.4.4 Impacts on livestock production

In Ethiopian grazing lands (Fig. 15.5), the expansion of weeds like parthenium weed is having a detrimental effect through displacing better quality and more palatable species of grasses (Gurmessa *et al.*, 2015). In addition, surveys conducted with Ethiopian farmers show that parthenium weed is considered to affect animal health (18% of respondents) and the quality of milk (97% of respondents) (Beyene and Tessema, 2015).

Fig. 15.4. Parthenium weed invading a sorghum (*Sorghum bicolor*) crop in Ethiopia. (A. McConnachie.)

15.4.5 Impacts on ecosystem services

The threat of parthenium weed to biodiversity management and conservation has been investigated in studies from India (Kohli *et al.*, 2006) and Australia (Grice, 2006). However, little research has focused on environmental impacts in East Africa and the Middle East. One study from Awash National Park, Ethiopia showed that after 3 years of manual control, parthenium weed infestations were reduced by as much as 75% (Boy and Witt, 2013). The resulting knock-on effects for the environment were not measured in this study; however, they would have been significant.

15.5 Management Approaches

15.5.1 Uses

In Ethiopia, parthenium weed has been recorded to have several uses, including its inclusion in floral bouquets (K. Dhileepan, Addis Ababa, 2014, personal communication), as a packing material for boxed tomatoes (*Solanum lycopersicum* L.) and as a compost (Dinwiddie, 2014). In such uses, the potential for increasing dispersal of parthenium weed seeds is high. Where it was used in tomato packaging, seeds contained within inflorescences were being moved from the fields to markets, and then on to new areas. Composting also presented a problem as Dinwiddie (2014) found that heat generated during plant decomposition was insufficient to kill all the seeds, with the result that viable seeds would be sold in compost, further facilitating the spread of the weed. In addition, allelochemicals from the parthenium weed used to make the compost would be contained in the mix, suppressing germination of crop seeds or affecting crop growth. During the time of the Ethiopian–Somali War, it is reported that the weed was used as a coagulant to

Fig. 15.5. Communal grazing area in Ethiopia invaded with parthenium weed. (A. McConnachie.)

stop bleeding in injured soldiers (Maal and Abdi, 2014).

In the town of Arusha, Tanzania, parthenium weed is also being used in floral bouquets. Many roadside florists use the inflorescences as a substitute for baby's breath (*Gypsophila paniculata* L.). In Uganda, parthenium weed inflorescences are used in bridal bouquets and in posies. In fact, some people were observed growing the weed specifically for this purpose. This often resulted in new invasions of parthenium weed around churches, where the bouquets/posies had been discarded. A second use recorded from Uganda was as a treatment for stomach ailments.

15.5.2 Control techniques: physical

In Ethiopia, as part of the RBIPMA project, parthenium weed was physically cleared from hundreds of hectares of farmland and pasture

on and around the Welenchiti and Amibara project sites (Boy and Witt, 2013). It was also physically removed from areas within and adjacent to the Awash National Park, which after 3 years succeeded in reducing the invasions by as much as 75% (however, it must be noted that it has since re-established in these areas). Much-improved crop yields were also recorded in farming areas cleared of parthenium weed (Boy and Witt, 2013). In 2011, parthenium weed was manually removed from the Masai Mara National Reserve (A. Witt, Nairobi, 2011, personal communication) by hand-pulling and subsequently burnt. Field staff in the Mara Triangle of the National Reserve still regularly undertake chemical control of the weed.

15.5.3 Control techniques: chemical

Herbicidal control has been investigated in Ethiopia to control parthenium weed in sorghum (Kassahun and Tefera, 1999; Tamado *et al.*, 2002; Tamado and Milberg, 2004). In Kenya, the efficacy of glyphosate and gramoxone has been studied for controlling parthenium weed in coffee (Njoroge, 1989, 1991). However, due to the fact that much of the invaded range of parthenium weed in East Africa is dominated by subsistence agriculture, chemical control is often going to be regarded as a cost-prohibitive control strategy.

15.5.4 Control techniques: biological

The first classical biological control agent recorded on parthenium weed in the region was the rust *Puccinia abrupta* var. *partheniicola* (Jackson) Parmelee, from Ethiopia in 2002 (Tessema *et al.*, 2002). The rust was found to have a significant impact on the morphology of parthenium weed, as well as dry matter and seed production under ideal conditions (Bekeko *et al.*, 2012). The distribution and impact of the rust throughout the invaded range in Ethiopia, however, is sporadic – probably as a result of climatic conditions (Bekeko *et al.*, 2012). Parthenium weed is also attacked by another disease in Ethiopia, a phytoplasma (Tessema *et al.* 2004). The disease causes a witches-broom effect, but its mode of transmission and impact is unknown.

In 2014, as part of the USAID IPM-CRSP project, the biological control agent *Zygogramma bicolorata* Pallister (Coleoptera: Chrysomelidae) was released from quarantine in Ethiopia. This was the first arthropod biological control agent for a weed to be officially tested and released in Ethiopia. A mass-rearing centre was set up in Wollenchiti for beetles to be released throughout the invaded range in Ethiopia (W. Mersie, Virginia, 2016, personal communication). The stem-boring weevil *Listronotus setosepennis* Hustache (Coleoptera: Curculionidae) is also being mass-reared at the centre and has been released into the surrounding area. The stem-galling moth *Epiblema strenuana* Walker (Lepidoptera: Tortricidae) was recently reported not to be suitable for

release in Ethiopia (McConnachie, 2015). The research conducted in South Africa showed that the East African native, niger (*Guizotia abyssinica* (L.f.) Cass.) was susceptible to attack by *E. strenuana*. Future agents to be considered for Ethiopia include the seed-feeding weevil *Smicronyx lutulentus* Dietz (Coleoptera: Curculionidae) and the root crown-feeding moth *Carmenta* sp. nr *ithacae* (Beutenmüller) (Lepidoptera: Sesiidae).

The biological control agent *Z. bicolorata* was first released in Arusha, Tanzania, in 2013 and there have since been an additional two releases. Plans are in place to release *L. setosepennis* there in the near future. As part of the current USAID IPM CRSP project, an application for the introduction of *Z. bicolorata* to Uganda for the biological control of parthenium weed has been successful.

In Israel, *E. strenuana* was recorded on ragweed (*Ambrosia* spp.) in 2008 (Yaacoby and Seplyarsky, 2011). Due to the effective dispersal ability of *E. strenuana*, it will only be a matter of time before it is found on parthenium weed populations in Israel.

15.5.5 Control techniques: cultural

Of all the countries covered in this region, parthenium weed has only been declared noxious in two, Ethiopia and Kenya. In Ethiopia, legislation was passed in 2001 by the Ethiopian Parliament (Dinwiddie, 2014) and in Kenya, it was declared noxious in 2010 under their Suppression of Noxious Weeds Act (Gazette Notice no. 4423).

Through the RBIPMA project, there was significant community-awareness creation around the identification, impacts and management of parthenium weed in Ethiopia and Uganda. Considerable amounts of publicity material was developed and disseminated, especially in Ethiopia and to a lesser extent Uganda. Many workshops and field days were held in Ethiopia to further raise awareness among communities. Communities became actively involved in many of the management activities.

In the Masai Mara National Reserve, funding from the Australian High Commission allowed for the employment and training of local communities in managing parthenium weed. Awareness material was also developed and distributed among communities living in and adjacent to the reserve. Some awareness material developed for use in the Masai Mara National Reserve was also disseminated in and around Arusha (Tanzania), and in Uganda.

Training workshops hosted by CABI, to highlight the impacts of invasive plants including parthenium weed, have been held in all countries across the region, including Rwanda. Socio-economic surveys to determine the impacts of parthenium weed have also been undertaken in and around Arusha (Tanzania), an activity that created significant awareness of the threats posed by this noxious weed. A short video was also produced detailing the impacts of four invasive plants in East Africa, including parthenium weed.

Since 2005, the USAID-funded IPM CRSP project has funded parthenium weed research and community extension activities in both southern and East African countries. The project has produced posters in multiple languages (English, Amharic, Oromiffa, Tigrigan and Somali) to raise awareness around the health impacts of parthenium weed (W. Mersie, Virginia, 2012, personal communication). Almost every year since inception, research meetings and field days have been held (primarily in Ethiopia).

In 2010, an 'International Workshop on the Biological Control and Management of *Parthenium hysterophorus*' was held in tandem with an 'International Workshop on the Biological Control and Management of *Chromolaena odorata* and other Eupatoriae' in Nairobi, Kenya. These workshops brought together African and International parthenium weed researchers. Besides the research value, the workshops generated much media interest, with a resulting increase of community awareness in Kenya.

15.6 Discussion

Having first been reported in the Australian literature in 1955, parthenium weed was only proclaimed a noxious plant some 20 years later (Auld *et al.*, 1982). It was first recorded in South Africa in 1880 (Hilliard, 1977), but was only considered to have become invasive a century later (Strathie and McConnachie, Chapter 14, this volume). In East Africa, it was first recorded in the 1970s; 30 years on, it was declared a noxious weed in Ethiopia. For reasons not quite clear, parthenium weed appears to have a significant lag phase between the time it is first introduced and when it becomes problematic. This phenomenon could be explained by two theories. The first proposes that different biotypes of parthenium weed exist, with some being more invasive than others (Navie *et al.*, 1996). The second hypothesis deals with invasion lag times and proposes that species introduced to novel environments require a certain period of adaptation before they can fully exploit their new habitat, i.e. become invasive (Aikio *et al.*, 2010).

The management of parthenium weed in East and North Africa and the Middle East has either been totally lacking, or at best haphazard. Ethiopia has had the most coordinated management programme out of any country in the region, thanks to the internationally funded donor projects of USAID IPM-CRSP and RBIPMA. Mechanical clearing, although labour intensive and potentially deleterious to human health, has achieved effective short-term results on a localized scale in some invaded areas in Ethiopia (Boy and Witt, 2013). However, without consistent and regular follow-up efforts, these populations will merely be replenished from the seedbank. This was documented after initial manual clearing of parthenium weed in the Masai Mara National Reserve in Kenya, where there was no follow-up in the area managed by the Narok County Council due to a lack of resources, and populations have largely recovered. Biological control efforts in

Ethiopia are still in their early stages; *Z. bicolorata* and *L. setosepennis* have yet to show signs of establishment (W. Mersie, Virginia, 2016 personal communication). However, it is still early days and there are several other agents that are currently being tested (*S. lutulentus*) or will be tested in the future (*Carmenta* sp. nr *ithacae*). *Zygogramma bicolorata* was also released in Tanzania for the first time in 2013, around the town of Arusha. As of yet, there is no formal confirmation that the leaf-feeding beetle has established. The stem-boring weevil, *L. setosepennis*, will hopefully be released in Tanzania in the near future.

Boy and Witt (2013) summarized the lessons that were learned from the RBIPMA project, and many of these lessons hold true for the region under consideration in this chapter in terms of the challenges to be faced in effectively tackling the ongoing parthenium weed invasion.

1. It takes time to establish and institute effective invasive plant management programmes in developing countries that have little to no prior awareness.
2. First-time projects undertaken in developing countries may have limited baseline information to work from in terms of a country's invasive species and their distribution and impacts.
3. Eliciting the commitment, early on, of politicians and policy makers at the highest level is usually essential to the genesis and long-term success of national IAS management programmes.
4. The need for preventive IAS management is particularly hard to 'sell' in developing countries, where in addition the implications of long-term biodiversity impacts, as opposed to immediate socio-economic impacts, are often relatively poorly understood by communities and governments.
5. In view of the immense scale of the problem and the continuing lack of available resources in most developing countries, much more emphasis needs to be placed on cost-effective management practices, such as biological control.

6. The enormity and the growing severity of the IAS problem is such that nothing short of a concerted global campaign – eliciting massive long-term commitment and funding from all major donor nations, international development and aid agencies and NGOs – is needed, if this scourge is to be brought under control.

The future of the parthenium weed invasion in East and North Africa and the Middle East is currently at a crossroads. While there is presently heightened awareness throughout most of the region, thanks to international programmes such as the RBIPMA and USAID IPM-CRSP projects, there need to be permanent, coordinated management resolutions made for not only parthenium weed but all invasive plants. If this does not happen, the regional impacts of invasive plant species will continue to increase. Now at the height of its invasion in Australia, parthenium weed affects *c*.600,000 km^2 of pasture land and is predicted to cost Queensland beef producers AUD$110 million by 2050 (Adamson and Bray, 1999). If the invasion reaches similar proportions in East and North Africa in particular, this would spell certain disaster for food security in the regions, due to the largely subsistence approach to agriculture. Lastly, let us not forget the impact of parthenium weed on the natural environment; many people throughout the region are heavily dependent on natural resources and the knock-on effects that this weed has on ecosystem services would be decisive in their survival.

Acknowledgements

The authors wish to acknowledge contributions made to the district-scale infestation data by: Hassan Ali (Benadir University, Mogadishu, Somalia), Jennifer Bisikwa (Makere University, Kampala, Uganda), Krissie Klarke (PAMS Foundation, Arusha, Tanzania) and Emily Wabuyele (National Museums of Kenya, Nairobi, Kenya), and Asad Shabbir for preparation of the maps.

References

Adamson, D. and Bray, S. (1999) The economic benefit from investing in insect biological control of parthenium weed (*Parthenium hysterophorus*). School of Natural and Rural System Management, University of Queensland, Brisbane, Australia.

Aikio, S., Duncan, R.P. and Hulme, P.E. (2010) Lag-phases in alien plant invasions: Separating the facts from the artefacts. *Oikos* 119, 370–378.

Alhammadi, A.S.A. (2010) Preliminary survey of exotic invasive plants in some western and high plateau mountains in Yemen. *Assiut University Bulletin for Environmental Research* 13, 1–11.

Auld, B.A., Hosking, J. and McFadyen, R. (1982) Analysis of the spread of tiger pear and parthenium weed in Australia. *Australian Weeds* 2, 56–60.

Baker, H.G. (1965) Characteristics and modes of origin of weeds. In: Baker, H.G. and Stebbins, G.L. (eds) *The genetics of colonizing species*. pp. 147–168. Academic Press, New York.

Beentje HJ, Jeffrey C, Hind DJN (2005) Compositae (Part 3). In: Beentje HJ, Ghazanfar SA (eds.) *Flora of Tropical East Africa*. Royal Botanic Gardens, Kew, London, UK, pp. 1–332.

Bekeko, Z., Temam, H. and Tessema, T. (2012) Distribution, incidence, severity and effect of the rust *Puccinia abrupta* var. *Partheniicola* on *Parthenium hysterophorus* L. in Western Hararghe Zone, Ethiopia. *African Journal of Plant Science* 6, 337–345.

Beyene, H. and Tessema, T. (2015) Distribution, abundance and socio-economic impacts of parthenium (*Parthenium hysterophorus*) in southern zone of Tigray, Ethiopia. *Journal of Poverty, Investment and Development* 19, 22–29.

Bollier, S. and Haddad, M. (2013) Interactive: Powering the Gulf, *Al Jazeera* 1 May 2013. https://www.aljazeera.com/indepth/interactive/2013/04/201342914169120172.html (accessed 29 May 2018).

Boulos, L. and el-Hadidi, M.N. (1984) *The Weed Flora of Egypt*. American University in Cairo Press, Cairo.

Boy, G. and Witt, A. (2013) Chapter 8. Taking stock: Lessons learned and recommendations arising from the experiences of the project. In: *Invasive Alien Plants and their Management in Africa*. UNEP/GEF Removing Barriers to Invasive Plant Management. CABI Africa, Nairobi.

Catford, J,A,, Jansson, R, and Nilsson, C. (2009) Reducing redundancy in invasion ecology by integrating hypotheses into a single theoretical framework. *Diversity and Distributions* 15, 22–40.

Chippendale, J.F. and Panetta, F.D. (1994) The cost of parthenium weed to the Queensland cattle industry. *Plant Protection Quarterly* 9, 73–76.

Dafni, A. and Heller, D. (1982) Adventive flora of Israel – phytogeographical, ecological and agricultural aspects. *Plant Systematics and Evolution* 140, 1–18.

Dafni, A. and Heller, D. (1990) Invasions of adventive plants in Israel. In: de Castri, E., Hansen, A.J. and Debussche, M. (eds) *Biological Invasions in Europe and the Mediterranean Basin*. Kluwer Academic Publishers, Dordrecht, Netherlands, pp. 135–160.

Dale, I.J. (1981) Parthenium weed in the Americas. *Australian Weeds* 1, 8–14.

Danin, A. (2000) The nomenclature news of flora Palaestina. *Flora Mediterranea* 10, 109–172.

Dinwiddie, R. (2014) Composting of an invasive weed species *Parthenium hysterophorus* L. – an agroecological perspective in the case of *Alamata woreda* in Tigray, Ethiopia. MSc thesis, University of Agricultural Science, Alnarp, Sweden.

Gurmessa, K., Tolemariam, T., Tolera, A., Beyene, F. and Demeke, S. (2015) Feed resources and livestock production situation in the highland and mid-altitude areas of Horro and Guduru districts of Oromia regional state, western Ethiopia. *Science, Technology and Arts Research Journal* 4, 111–116.

ECA (2007) *Economic Report on Africa 2007: Accelerating Africa's Development through Diversification*. UN Economic Commission for Africa, Addis Ababa.

Evans, H.C. (1997) *Parthenium hysterophorus*: a review of its weed status and the possibilities for biological control. *Biocontrol News and Information* 18, 89N–98N.

Grice, A.C. (2006) The impacts of invasive plant species on the biodiversity of Australian rangelands. *The Rangeland Journal* 28, 27–35.

Joel, D.M. and Liston, A. (1986) New adventive weeds in Israel. *Israel Journal of Botany* 35, 215–223.

Kassahun, Z.N.O and Tefera, O.A. (1999) Evaluation of pre-emergence herbicide for the control of *Parthenium hysterophorus* in sorghum. *Arem* 5, 130–137.

Kilian, N., Hein, P. and Hubaishan, M.A. (2002) New and noteworthy records for the flora of Yemen, chiefly of Hadhramout and Al-Mahra. *Willdenowia* 32, 239–269.

Kohli, R., Batish, D.R., Singh, H.P. and Dogra, K.S. (2006) Status, invasiveness and environmental threats of three tropical American invasive weeds (*Parthenium hysterophorus* L., *Ageratum conyzoides* L., *Lantana camara* L.) in India. *Biological Invasions* 8, 1501–1510.

Kolar, C.S. and Lodge, D.M. (2001) Progress in invasion biology: predicting invaders. *Trends in Ecology and Evolution* 16, 199–204.

Kriticos, D.J., Brunel, S., Ota, N., Fried, G., Lasink, A.G.J.M.O., *et al.* (2015) Downscaling pest risk analyses: Identifying current and future potentially suitable habitats for *Parthenium hysterophorus* with particular reference to Europe and North Africa. *PLOS ONE* 10(9), e0132807. DOI:10.1371/journal.pone.0132807.

Levine, J.M. and D'Antonio, C.M. (1999) Elton revisited: a review linking diversity and invisibility. *Oikos* 87, 15–26.

Maal, A.A. and Abdi, D.M. (2014) Impact of *Parthenium hysterophorus* L. on farm lands in Wajaale, Somaliland. BSc thesis, Gollis University, Hargiesa, Somaliland.

Mack, R.N., Simberloff, D., Lonsdale, W.M., Evans, H., Clout., *et al.* (2000) Biotic invasions: causes, epidemiology, global consequences, and control. *Ecological Applications* 10, 689–710.

Mahmoud, T., Gairola, S. and El-Keblawy, A. (2015) *Parthenium hysterophorus* and *Bidens pilosa*, two new records to the invasive weed flora of the United Arab Emirates. *Journal on New Biological Control Reports* 4, 26–32.

Mainali, K.P., Warren, D.L., Dhileepan, K., McConnachie, A., Strathie, L., *et al.* (2015) Projecting future expansion of invasive species: comparing and improving methodologies for species distribution modelling. *Global Change Biology* 21, 4464–4480.

McConnachie, A.J., Strathie, L.W., Mersie, W., Gebrehiwot, L., Zwedie, K., *et al.* (2011) Current and potential geographic distribution of the invasive plant *Parthenium hysterophorus* (Asteraceae) in eastern and southern Africa. *Weed Research* 51, 71–84.

McFadyen, R.E. (1992) Biological control against parthenium weed in Australia. *Crop Protection* 24, 400–407.

McFadyen, R.E. (1995) Parthenium weed and human health in Queensland. *Australian Family Physician* 24, 1455–1459.

Muchena, F.N. (2008) Indicators for Sustainable Land Management in Kenya's Context. GEF Land Degradation Focal Area Indicators, ETC-East Africa. Nairobi, Kenya.

Mutua B.M., Muriithi J.K. and Omwoyob, O. (2014) Farmers' awareness level on the effect of parthenium weed (*Parthenium hysterophorus* L.) on agricultural production and its control in Nyando division, Kenya. *International Journal of Agriculture*. Photon 125, 305–310.

Navie, S.C., McFadyen, R.E., Panetta, F.D. and Adkins, S.W. (1996) The biology of Australian weeds 27. *Parthenium hysterophorus* L. *Plant Protection Quarterly* 11, 76–88.

Njoroge, J.M. (1986) New weeds in Kenya coffee. *Kenya Coffee* 51, 333–335.

Njoroge, J.M. (1989) Glyphosate (Round-up 36% ai) low rate on annual weeds in Kenya coffee. *Kenya Coffee* 54, 713–716.

Njoroge, J.M. (1991) Tolerance of *Bidens pilosa* and *Parthenium hysterophorus* L. to paraquat (Gramaxone) in Kenya coffee. *Kenya Coffee* 54, 713–716.

Nyasembe, V.O., Cheseto, X., Kaplan, F., Foster, W.A., Teal, P.E.A., *et al.* (2015) The invasive American weed *Parthenium hysterophorus* can negatively impact malaria control in Africa. *PLoS ONE* 10(9): e0137836.

Pyšek, P. and Richardson, D.M. (2006) The biogeography of naturalization in alien plants. *Journal of Biogeography* 33, 2040–2050.

Rao, P.V.S., Mangala, A., Rao, B.S.S. and Prakash, K.M. (1977) Clinical and immunological studies on persons exposed to *Parthenium hysterophorus* L. *Cellular and Molecular Life Sciences* 33, 1387–1388.

Reda, F., Mekuria, M., Meles, K., Zewdie, K., Fessehaie, R., *et al.* (2008) Weed research in sorghum and maize. In: Tadesse, A. (ed.) *Increasing Crop Production through Improved Plant Protection – Volume I*. Plant Protection Society of Ethiopia (PPSE), 19–22 December 2006, Addis Ababa, pp. 303–324.

Rejmánek, M. and Richardson, D.M. (1996) What attributes make some plant species more invasive? *Ecology* 77, 1655–1661.

Strathie, L.W., McConnachie, A.J. and Retief, E. (2011) Initiation of biological control against Parthenium hysterophorus L. (Asteraceae) in South Africa. *African Entomology* 19, 378–392.

Schachar, G. (2011) *Israel – Grain and Feed Annual Report*. GAIN report number IS1102. USDA Foreign Agricultural Service, Washington, DC.

Tadesse, M. (2004) Asteraceae (Compositae). In: Hedberg, I., Friis, I. and Edwards, S. (eds) *Flora of Ethiopia and Eritrea, Vol. 4, Part 2, 408*. The National Herbarium, Addis Ababa.

Tamado, T. (2001) Performance and yield stability of medium duration ground-nut genotypes under erratic rainfall conditions in eastern Ethiopia. *Tropical Science* 41, 192–198.

Tamado, T. and Milberg, P. (2000) Weed flora in arable lands of Eastern Ethiopia with emphasis on the occurrence of *Parthenium hysterophorus*. *Weed Research* 40, 507–521.

Tamado, T. and Milberg, P. (2004) Control of parthenium (*Parthenium hysterophorus*) in grain sorghum (*Sorghum bicolor*) in the smallholder farming system in eastern Ethiopia. *Weed Technology* 18, 100–105.

Tamado, T., Ohlander, L. and Milberg, P. (2002) Interference by the weed *Parthenium hysterophorus* L. with grain sorghum: Influence of weed density and duration of competition. *International Journal of Pest Management* 48, 183–188.

Tareke, G. (2000) The Ethiopia–Somalia war of 1977 revisited. *International Journal of African Historical Studies* 33, 635–667.

Taye, T., Obermeier, C., Einhorn, G., Seemüller, E. and Büttner, C. (2004) Phyllody disease of parthenium weed in Ethiopia. *Pest Management Journal of Ethiopia* 8, 39–50.

Tessema, T., Gossmann, M., Einhorn, G., Buttner, C., Metz, R. *et al.* (2002) The potential of pathogens as biological control of parthenium weed (*Parthenium hysterophorus* L.) in Ethiopia. *Mededelingen – Faculteit Landbouwkundige en Toegepaste Biologische Wetenschappen, Universiteit Gent* 67, 409–420.

Tessema, T., Einhorn, G., Gossmann, M., Büttner, C. and Metz, R. (2004) The potential of parthenium rust as biological control of parthenium weed in Ethiopia. *Pest Management Journal of Ethiopia* 8, 83–95.

Thomas, J., Basahi, R., Al-Ansari, A. E., *et al.* (2015) Additions to the Flora of Saudi Arabia: two new generic records from the Southern

Tihama of Saudi Arabia. *National Academy Science Letters* 38(6): 513–516.

Towers, G.H.N. and Subba Rao, P.V. (1992) Impact of the pan-tropical weed, *Parthenium hysterophorus* L. on human affairs. In: Richardson, R.G. (ed.) *Proceedings of the First International Weed Control Congress*, Melbourne, Australia. Weed Science Society of Victoria, Melbourne, pp. 134–138.

UNEP (2002) African Environment Outlook: GEO-4. United Nations Environment Programme, Nairobi.

Wabuyele, E., Lusweti, A., Bisikwa, J., Kyenune, G., Clark, K., *et al.* (2014) A roadside survey of the invasive weed *Parthenium hysterophorus* (Asteraceae) in East Africa. *Journal of East African Natural History* 103, 49–57.

Wiesner, M., Tessema, T., Hoffmann, A., Wilfried, P., Buettner, C., *et al.* (2007) Impact of the pan-tropical weed *Parthenium hysterophorus* L. on human health in Ethiopia. Institute of Horticultural Science, Urban Horticulture, Berlin.

Yaacoby, T. and Seplyarsky, V. (2011) *Epiblema strenuana* (Walker, 1863) (Lepidoptera: Tortricidae), a new species in Israel. *EPPO Bulletin* 41, 421.

Zaharan, M.A. and Willis, A.J. (2009) Chapter 6. The Nile Region. In: *The Vegetation of Egypt*. Springer Sciences + Business Media B.V., Dordrecht, the Netherlands, pp. 251–303.

16 Conclusions

Steve W. Adkins,[1]* Asad Shabbir[2] and Kunjithapatham Dhileepan[3]

[1]The University of Queensland, Gatton, Queensland, Australia;
[2]University of the Punjab, Lahore, Pakistan; current affiliation:
the University of Sydney, Narrabri, New South Wales, Australia;
[3]Biosecurity Queensland, Department of Agriculture and
Fisheries, Brisbane, Queensland, Australia

16.1 Introduction

We are currently going through a period of significant global environmental change, the so-called Anthropocene. The interconnected concerns of climate change, habitat and biodiversity loss, and the anthropological influences steering these changes, are familiar to most. However, one major mediator of environmental change, invasive species, lingers unseen in most considerations of this global environmental change. Invasive species are being dispersed around the globe at an ever-increasing rate, largely unobstructed and with their movement expedited by human activities.

One of the most destructive invasive weeds of the present time, whose movement is unintentionally facilitated and has benefited by an upsurge in international trade and travel, is parthenium weed. The destruction that this species alone can cause is unquantifiable. In certain parts of the world it can significantly reduce crop and pasture production on a regional basis, in other areas it is destructive to the natural environment and can severely affect the health of humans, domesticated animals and also native animals. The weed can come to dominate certain landscapes, so much so that local communities have, over the course of time, come to accept its presence and believe it is a plant for which they should find a use and not manage.

Throughout this book we have analysed the available information on the biology and ecology of the weed, both in its native and introduced ranges. We have examined the modes of spread and the weed's impact on agricultural production, the environment and human health, and its potential uses. We have looked at the various forms of management adopted in different parts of the world and their effectiveness. The following discussion reiterates the main findings presented and we conceptualize why this plant has become such a 'superior weed', worthy of ranking in the top-five weeds worldwide, and a weed that has achieved this in just a few decades. We examine integrated management options and the way forward in the management of the weed under a changing environment. Finally, we look at the gaps in our knowledge and how we might close them with the use of collaborative research programmes around the globe.

16.2 Weed Problem and Status

16.2.1 Origin and biology

Parthenium weed is now recognized as one of the most invasive weeds worldwide. It is

* s.adkins@uq.edu.au

© CAB International 2019. *Parthenium Weed: Biology, Ecology and Management*
(eds S. Adkins, A. Shabbir and K. Dhileepan)

an annual, herbaceous plant from the family Asteraceae that has a deep tap root and an erect stem system, reaching a height of 2.0 m under ideal growing conditions. The plant is capable of flowering even when still in the rosette stage of growth and a single plant in the field can produce more than 150,000 seeds. These seeds can be distributed easily by vehicles, agricultural and road construction machinery, within animal fodder, pasture seed lots and stock feed, and naturally by wind and water. Thus, spread occurs either accidently or due to unchecked trade and transportation, which takes it across national and international borders. Unlike most invasive weeds, it has not been deliberately introduced to any location. The extraordinary and resourceful attributes of this weed have been sufficient for it to 'spread by itself', and it is still spreading.

The centre of diversity of the genus *Parthenium* is in northern Mexico and it seems most likely that this is the area of origin of the species, although a second centre of diversity also exists in central South America. The 'North American' race has white flowers and is the race that has been introduced into all other parts of the world. The 'South American' race has cream to yellow flowers and differs from the North American race in leaf morphology, pollen colour, capitula size, development of axillary branches, and size of disc florets and ray corollas. For some, it should be considered a subspecies. Hymenin is often the dominant sesquiterpene lactone found in the plants of the South American race, whereas parthenin is the dominant sesquiterpene lactone in the North American race. In the native range the most vigorous growth occurs in the warm, wet summer period with a smaller amount of growth occurring over the winter months, depending on winter rainfall.

Apart from being present in its native range, the weed has now invaded vast areas of land in over 45 countries and continues to spread into other countries at a rapid rate. In the invaded range it is responsible for the degradation of grasslands, peri-urban landscapes and wastelands as well as infesting numerous cropping systems in different countries. In addition, it is an environmental weed negatively affecting biodiversity and species richness in natural communities. The suppression of forage and grain crops due to competitive and allelopathic interference has reduced livestock production and challenged the sustainability of many farming systems. The direct hazards to human and livestock health further exacerbate the situation.

16.2.2 Ecology

In terms of its ecology, parthenium weed can germinate, grow and flower over a wide range of temperatures and photoperiods, hence in many countries it can be present in the field at any time of the year. Flowering occurs best at warm 27/22°C day/night temperatures and can set seed banks as large as 200,000 seeds/m^2. These seeds can remain viable in the soil seed bank for many years, with a half-life of c.5–7 years; on the soil surface, seeds die within 6 months. Maximum germination is best achieved under a diurnal temperature regime of c.21/16°C, but populations have a wide range of temperatures over which they can germinate (9–36°C).

Average seedling recruitment is about 110/m^2, of which only about 14 plants/m^2 attain maturity. Several cohorts of seedlings may be produced each year. The preferred habitats for parthenium weed include: soils that are predominantly dark cracking clay; riparian sites; cultivated areas, including field edges; abandoned and vacant land; roadsides; railway tracks; and many sunny and open areas.

There are conflicting reports as to whether parthenium weed is self- or cross-pollinated, and what the actual mechanism of pollination is. In the native range it is considered to undergo a high degree of self-pollination. However, in the invaded range the weed appears to be insect cross-pollinated or at least with pollen dispersed mainly by insects. A recent study has raised the very interesting idea that in certain parts of the invaded range, where the plants are extremely invasive, their mating system is one of cross-pollination, while less invasive communities are self-pollinated.

16.2.3 Climate change

The changes occurring in global climate are likely to have a significant impact upon future parthenium weed invasion. Despite having C4 photosynthesis in its rosette leaves, the main vegetative part of the plant operates C3 photosynthesis, and shows significant improvement in growth and biomass production under a higher atmospheric CO_2 concentration (480 ppm) as compared with the ambient concentration (360 ppm). This not only indicates a superior invasive potential under the present climatic condition but also warns about the growing severity of the issue in the future. In addition, warm temperatures of 25–35°C coupled with a high CO_2 concentration (700 ppm) have been shown to improve photosynthetic rate, water use efficiency and consequently the growth of the weed. In a further study, parthenium weed biomass was significantly increased at a high CO_2 concentration (550 ppm) when it was grown in combination with other grass and legume species. So, parthenium weed is likely to become more aggressive and grow more rapidly under elevated temperature and CO_2 conditions as well as reduced soil moisture in the future. Predictive modelling run under a climate change scenario indicates the potential of the weed to spread into new regions, including sub-Saharan African, the Asian-Pacific region and some European countries, such as Portugal, Italy, Spain and France.

16.2.4 Impacts

Parthenium weed is not just responsible for the reduction in biodiversity and species richness in natural ecosystems. It is also responsible for the degradation of grasslands, peri-urban landscapes and wastelands, and reductions in yield in over 40 crops in different countries. In Ethiopia, parthenium weed was the second most frequent weed found among 102 weed species in 240 crop fields and up to 90% of farmers ranked it as the most important weed in their region. In a study in Pakistan, it was found that parthenium weed had become the second most frequent weed out of 31 local weed species within just 2 years of its introduction. In India, parthenium weed has been reported to heavily infest upland rice, causing severe yield losses, and the weed has been shown to interfere strongly with crop growth and development. In other studies it has been reported to cause up to 40% yield loss in wheat (*Triticum aestivum* L.), maize (*Zea mays* L.), teff (*Eragrostis tef* Zucc. Trotter), blackgram (*Phaseolus mungo* L.), sunflower (*Helianthus annuus* L.) and sorghum (*Sorghum bicolor* L.). The negative impacts of parthenium weed on crop growth and yield depend on its density and its competition duration. In a weed–crop competition study in Ethiopia, it was found that parthenium weed caused complete failure (97% yield losses) of a grain sorghum crop at a lowland site. Only 3 plants/m² of parthenium weed caused a c.70% reduction in sorghum grain yield. In Pakistan, parthenium weed caused substantial yield losses when competing with a maize crop at varying densities (5–20 plants/m²)

16.2.5 Health

It has been long recognized that parthenium weed has detrimental effects on human and domesticated animal health. These detrimental effects are attributed predominantly to the occurrence of a range of sesquiterpene lactones in the plant, especially parthenin, which has been shown to cause adverse responses in both humans and domesticated animals. Parthenin is the major sesquiterpene lactone in the majority of global parthenium weed populations, especially the northern American populations from which all introduced populations have been derived. High levels of parthenin are found in the essential oil fraction extracted from the trichome hairs present on the leaf, stem and fruit, or in the coating materials of the pollen grains. The vast land areas of parthenium weed infestation, and the ability of the plant to produce considerable numbers of flowers, and hence pollen grains, have been a major causal factor in the severity of the human health problems. The largest impact on human health has been seen in India, where parthenium weed has been determined to be the major cause of contact dermatitis and has led to chronic disease and indirectly to several deaths.

Parthenium-induced allergenic diseases occur when a sensitized individual is in contact with or inhales airborne particles of parthenium weed, such as pollen, detached hairs or other fine plant debris. These diseases include an array of dermatitis forms, most commonly airborne contact dermatitis, caused by a Type IV, delayed hypersensitive reaction, and internal conditions such as rhinitis, asthma and atopic dermatitis, because of a Type I immediate hypersensitive reaction. Similar diseases are reported in domesticated animals that have either ingested or simply encountered parthenium weed.

The long-term solution to prevent the detrimental effects attributed to parthenium weed is the reduction of plant numbers in the invaded landscape. The eradication of the plant may not be possible once it is fully established, but management methods to reduce plant numbers and to reduce flowering, and therefore pollen production, can be implemented. These include physical methods such as plant removal or cultivation prior to flowering, the use of chemicals, biological control or the use of competitive plants to suppress parthenium weed growth.

16.2.6 Invasive traits

Several aspects of the plant's biology and ecology contribute to its invasiveness. The main inferences presented are that the plant has: (i) beneficial morphological characteristics, a unique reproductive biology, a strong competitive ability, the ability to escape from natural enemies in the non-native regions and a C3/C4 photosynthesis. In addition, (ii) well-developed tolerances to abiotic stresses and an ability to grow in a wide range of edaphic conditions are thought to be additional invasion tools on a physiological front. (iii) An allelopathic potential of the weed against crop, other weed and pasture species, with multiple modes of allelochemical expression, may also be responsible for its invasion success. Moreover, the release of novel allelo-chemicals in non-native environments might have a pivotal role in its invasion process. (iv) Genetic diversity found among different populations and biotypes of parthenium weed, based on geographic, edaphic, climatic and ecological ranges, might also be a strong contributor towards its invasion success. (v) Rising temperatures and atmospheric CO_2 concentrations and changing rainfall patterns, all within the present-day climate change prediction range, are favourable for parthenium weed growth and its reproductive output, and therefore driving its more recent and future spread and infestation. Such an understanding of the core phenomena regulating the invasion success has practical implications for its management. A better understanding of the interaction of physiological processes, ecological functions and genetic makeup within a range of environments may help to devise appropriate management strategies for parthenium weed.

16.3 Management

16.3.1 Introduction

Despite the development of many control approaches and the formulation of these into management strategies for parthenium weed, effective, strategic management is still a challenge in most situations. In many areas where parthenium weed has been introduced it is still spreading fast, and causing severe economic losses to agriculture and negative effects on native plant biodiversity and human and animal health. The resources required to manage this weed across such diverse ecosystems are significant and often unavailable.

16.3.2 Coordination

To maximize the benefits of the available management strategies and help contain parthenium weed and slow its further spread, coordination is a key step. For coordination to be effective it should be fashioned around a national strategy that has clear, measurable and time-focused objectives. Unfortunately, in most developing countries where the weed is now present, such national strategies are yet to be developed. It is therefore highly recommended that invaded countries should develop, implement and then periodically review their parthenium weed management strategies to contain it, slow its further spread and abate economic losses.

16.3.3 Legislation and early detection

Only a few countries have developed effective legislation that prevents movement of parthenium weed into that country or into uninfested areas within that country. Examples of those countries that have developed legislation include Australia, South Africa, Sri Lanka and Kenya, with the legislation largely focused on preventing its movement into uninfested areas. In India, the state of Karnataka categorised parthenium weed as an agricultural pest in 1969, but this legislation could not be enforced and therefore was unable to prevent its spread to other states. In other countries where parthenium weed has become a problem, legislative measures and their implementation should now be considered a top priority to help contain the spread of parthenium weed and to contain it to the regions it is presently found.

Early detection, followed by rapid, effective action is fundamental to eradication of the weed in new areas, especially those that are considered ideal for parthenium weed growth. Further, compliance-based detection and eradication of the weed in such areas is vital if the weed is to be contained and prevented from spreading further.

16.3.4 Community awareness and empowerment

Just as important as early detection is awareness and networking if long-term management of parthenium weed is to be achieved. Awareness is often very poor (Fig. 16.1) and increasing awareness often depends on the engagement of all stakeholders, including within the government, the private sector and the community, to get involved and to start campaigns against the weed at all levels. Awareness and the implementation of best management practice is an important component of the overall management strategy for parthenium weed. Effective coordination between working and research groups in different countries, and between regional groups, is important to learn from each other's experiences and to create awareness about parthenium weed. The International Parthenium Weed Network, a network founded in Australia, currently enables the exchange of information on parthenium weed management between its more than 350 members from 30 countries.

16.3.5 Physical and chemical control

Manual removal is generally only feasible for treating small and isolated areas of parthenium weed, such as in residential zones, along roadsides or small agricultural fields. It is not considered economical or practical for large areas (e.g. grazing areas, national parks and protected forests), particularly in countries where the cost of labour is high. Pulling out plants, including the root system, when the soil is sufficiently moist, or hoeing to remove a section of the root system are the most effective manual control options. Treatments that only cut off the above-ground part and leave the root system can result in the plant reshooting. Where possible, control should be undertaken before plants produce seed and start replenishing the seed bank.

Mechanical treatments can be effective in managing parthenium weed but they often create favourable seed beds and promote large-scale seedling regrowth, which may exacerbate the problem if follow-up control is not undertaken. Mowing or slashing can also result in the rapid regrowth of plants from the lateral parts of the shoot that are close to the ground. Ploughing and

Fig. 16.1. Awareness of parthenium weed and its health implications is often very poor. People sitting in a field of parthenium weed in South Africa during a pension pay-out day. (Ezemvelo KZN Wildlife and Department of Environmental Affairs, South Africa.)

mulching are the most appropriate methods for controlling parthenium weed while preparing land for planting with a crop or improved pasture, particularly if it can be done at the vegetative stage before flowering occurs.

The impact of fire on parthenium weed populations is restricted to a combination of some anecdotal evidence, case studies and the findings from a single trial undertaken in northern Australia. In areas where parthenium weed is dense, there is unlikely to be sufficient fuel to carry a fire. However, in such areas that have been burnt, a greater parthenium weed incidence has been observed and measured following burning when compared to nearby unburnt areas. Based on such findings, burning of parthenium-infested areas has tended to be discouraged.

Synthetic herbicides can be expensive to purchase and apply over large areas, which has tended to restrict their use mainly to eradication programmes, control in urban areas and along roadsides and for minimizing the impacts of parthenium weed in high-value crops. They are much less frequently used in grazing areas, public and uncultivated areas and forests. Some people are also hesitant to use them due to either perceived or actual risks that they pose to beneficial and native plants, the environment and human health.

16.3.6 Biological control

The classical biological control approach should be implemented as an essential fundamental tool in the management of parthenium weed. In countries where management is most effective, the management strategy for parthenium weed is built around a classical biological control programme to which other management tools are added. In Australia, 11 biological control agents have been released, with some of these agents becoming widely established, having a significant negative impact upon weed populations and no reported off-target consequences. Following Australia's example, several other countries have now released biological control agents against parthenium weed and several others have plans to release agents in the future. Countries seeking biological control agents can benefit greatly from the expertise and experiences of the Australian and South African biological control programmes through cost savings on surveys for natural enemies in the native range, sharing host-specificity testing data for common plants, seeking help on the selection of the most suitable agents and providing ready access to starter colonies from introduced field-established populations.

16.3.7 Ecological suppression

The use of beneficial plants that can suppress the growth of parthenium weed has been successfully tested in various locations. These suppressive plants, acting through competition or their allelopathic nature, not only inhibit the growth of parthenium weed but also provide a good source of fodder for livestock production. Further studies have suggested that the management of parthenium weed may be enhanced by supplementing the existing management strategies involving biological control with the sowing of suppressive plants. Moreover, the suppressive plants can act early before parthenium weed becomes established, and this is different to classical biological control agents, which generally emerge after parthenium weed has established in the field.

16.3.8 Best management practice

Best management practice guides specific to situations and locations (e.g. diversified cropping systems, rangelands, native environments) would not only help contain the spread of parthenium weed but also greatly enhance the effectiveness of integrated management. Roadside management of the

weed is a key step to stop the further local spread of parthenium weed. In most places where parthenium weed has become established along the roadsides, it then moves from these locations into the cropping and rangeland agro-ecosystems. Chemical control in such situations may be a financially feasible option. Where possible, well-constructed vehicle wash-down facilities along roadsides should be used, especially at state and national boundaries.

16.3.9 Uses

Despite the negative impacts of parthenium weed, numerous reports have claimed potential benefits and therefore uses of the weed. However, a critical appraisal of the actual uses that the plant has been put to has shown few demonstrable and therefore meaningful uses. Although compelling evidence is provided for some uses, most studies remain exploratory, descriptive and speculative. It is to be concluded that a precautionary approach would be sensible in promoting parthenium weed utilization as a management tool.

16.4 Research Gaps and Future Research

16.4.1 Introduction

Parthenium weed has been regarded as a significant problem for over six decades in both India and Australia, and consequently most of the research so far has come from these two countries. However, with the emergence of parthenium weed problems in many other countries, especially in Africa, the research effort on parthenium weed and its management has grown in recent years, but more is needed.

16.4.2 Distribution

There have been considerable improvements in the collection of information on the distribution of parthenium weed in several of the affected countries. One exceptional example is Queensland, Australia, where systematic surveys are undertaken every year. It would be good if such a standardized approach could be used in more countries, if not annually, at least once in 5 years. This will enable an understanding of the distribution and the rate of spread of the weed within those countries and this information can be used to help direct management and its future planning. Future research should therefore focus on simplifying the presently labour-intensive monitoring protocols, for example by using emerging tools such as drone technology or by placing more emphasis on stakeholder involvement in gaining additional field data.

16.4.3 Phylogenetics

Because of the lack of a modern phylogenetic analysis, the relationships of parthenium weed with other members of the genus remain unclear. The relationship of the white-flowered invasive North American race with that of the yellow-flowered non-invasive South American race is of great interest. This would help provide a better understanding of the invasion process, as would an analysis of the spread pathways around the world.

16.4.4 Economics

Despite parthenium weed becoming a major weed in many countries over many decades, affecting a range of industries, especially those of crop and animal production, forestry and protected landscapes, only limited information has become available on the economic impact of the weed. The exceptions are the losses to the Australian beef cattle industry and recent studies on cereal crop yield losses in Ethiopia and Pakistan. Future research should therefore focus on determining the economic losses in other crop production systems, forestry plantations, cattle production systems (especially those in developing countries), national parks and wastelands. Such information is needed to better understand the impact of the weed on the overall economy of the affected countries. This will require the involvement of economists and social scientists in future research programmes. A better effort to factor in the impacts of parthenium weed on the ecological cost to the environment should be included in future economic studies, so that a more complete measure of impact can be formulated.

16.4.5 Health

Although parthenium weed is known to cause human and animal health problems across its invaded range, no systematic studies have been undertaken at the individual or at the population level. The only exceptions are a couple of localized epidemiological/monitoring studies that were undertaken on human susceptibilities to parthenium weed in countries like Australia and India. However, with the spread of parthenium weed into more areas, and into areas with higher human and animal densities, this aspect of research should be a priority for most countries in the years to come. It would be beneficial to know if parthenium weed is still having a devastating impact on human health, especially in India, whether cultural methods and medical treatments are being successfully implemented and whether the human immune system is developing tolerance or is becoming less responsive to parthenium weed.

There are also limited data on the impact of parthenium weed on human health in the countries in the plant's native range. It is reported that the impact on health is significantly less in these countries than in invaded countries. This may reflect tolerance developed by the immune system of native peoples over time, or it may reflect different biochemical profiles of the native compared to the invasive populations of parthenium weed. Further work is warranted to address these suppositions.

As parthenium weed continues to spread through Africa and Asia, information needs to be collected from those countries that have recently been invaded and those that are proximal to infestations. There is a real opportunity to collect baseline data so that the true impact of parthenium weed, on the environment, agricultural production systems and human and animal health, can be ascertained. The collection of data and the monitoring of parthenium invasion may ultimately be used to manage both the spread and impact of parthenium weed.

More research is required to better understand the issues of tainted meat and milk coming from domesticated animals feeding on the weed. This is especially important if these products are to be consumed by sensitive individuals, including babies and infants. Also, the potential health impacts of parthenium weed on wild animal health is an unexplored area that needs greater research focus in future.

16.4.6 Biodiversity

With parthenium weed now invading many national parks, conservation areas and wildlife sanctuaries, the impact of the weed on plant community biodiversity (e.g. native pasture and palatable food plants) and its follow-on effects on large herbivores (e.g. antelope) and their predators (e.g. lion, tiger, leopard) needs to be looked at more systematically. It is also important to determine the recovery rate of different landscapes once parthenium weed has been removed.

16.4.7 Uses

With parthenium weed becoming more and more abundant, there have been some attempts to look at the ways the weed could be used for beneficial purposes. Most of the research on this topic so far has been laboratory based and undertaken in India. Although several interesting uses of the weed have reported, no follow-up research

on the practicality of such uses has been attempted. Hence, any future research on the uses of parthenium weed needs to focus on its feasibility at the commercial scale. This may include making use of the unique chemicals it contains as bioherbicides and biopesticides.

16.4.8 Management through biological control

As parthenium weed infestation levels and their nature vary across countries and continents, the approaches and research needs for managing parthenium weed will be different for different countries. Among all the management methods available, classical biological control is the most cost effective and efficient, and should form the core management method for parthenium weed in all countries, and under all situations (e.g. in nature reserves, natural pastures, cropping areas, roadsides and wasteland). Other management methods can then be added to form a highly effective integrated management strategy. This strategy will vary from country to country, and possibly within country, based on the nature and area of infestation.

Classical biological control research in Australia over four decades has resulted in the testing, release and establishment of 11 agents originally collected from the native range. Not all potential agents have been properly investigated, and there are still possibilities for further introductions from these potential agents. Detailed surveys to be undertaken in lowland Bolivia might also lead to the discovery of additional agents.

In view of the ongoing success in managing parthenium weed in the core infestation areas in central Queensland, Australia, future research on exploiting these released agents in other areas of Queensland where they currently do not occur, as well in other parthenium weed affected countries, is now a priority.

In Australia, parthenium weed is still spreading and entering southern Queensland and northern New South Wales, where many of the effective and widespread biological control agents in central

Queensland have not yet extended. Hence, future research should focus on redistributing the best established and effective biological control agents into these new areas where the biological control agents are currently not available. This will need to be in partnership with local stakeholder communities in the affected areas.

Future research will also need to focus on developing new ways for producing artificial diets, and for more efficiently mass-rearing and storing parthenium weed biological control agents without exposing the participating researchers to the adverse health effects of handling parthenium weed.

There has been increasing interest in the biological control of parthenium weed in several locations outside of Australia. South Africa, Ethiopia and India have already commenced the testing and release of some biological control agents known to be effective in Australia. Though there has been no research interest on exploiting classical biological control in other countries, countries like China and Pakistan are now showing interest in testing and releasing agents, in partnership with Australia and CABI, Pakistan. Countries like Malaysia and Nepal have the expertise and scientific manpower to take up this task.

Future research on biological control will also need to focus on using the various system modelling approaches (e.g. STELLA) for non-target risk assessment before introducing new biological control agents, and use climate matching models (e.g. CLIMEX) to predict the climatically most suitable areas for field releases. This would be relevant to countries like South Africa, India and Ethiopia, where testing and release of biological control agents are already in progress. With continued climate change, further research on how the parthenium biological control agents will perform under elevated atmospheric CO_2 levels, variable rainfall and increased temperatures is needed.

16.4.9 Other forms of management

Mycoherbicides are fungal-based bioherbicides. Information on the mycoflora associated with parthenium weed is now available, but further research is needed for their exploitation as active agents in the formulation of a mycoherbicide. Mycoherbicides have a potential to either replace or supplement the use of herbicides in managing parthenium weed in high-value crops and in urban areas for parthenium weed-related health-risk minimization.

In its native range, parthenium weed has developed resistance to a small number of herbicides used to control it in different agricultural situations in South, Central and North America. However, there is little or no report of herbicide resistance evolution within the parthenium weed populations in its introduced range, despite many chemicals being used repeatedly to control this weed in different situations. However, in the future, it will be necessary to monitor resistance to herbicides in parthenium weed populations that continue to be subjected to chemical control using herbicides with a similar mode of action, especially those populations along roadsides where herbicides are frequently used.

16.4.10 Rapid evolution – changing phenology, ecology and chemistry

Parthenium weed has quickly become one of the most invasive plants spreading and dominating around the globe. Its invasiveness is attributed to numerous unique eco-biological traits, some of which are thought to be rapidly evolving. The 'evolution of increased competitive ability hypothesis' suggests that exotic plants, when released from their natural herbivory in the newly invaded environment, can reallocate more resources towards their growth and fecundity, rather than to 'the now not needed' defensive mechanisms. With the passage of time, populations then become more vigorous and fertile and enhance their competitive ability as selection for these more advantageous traits occurs. Around the globe, introduced populations of parthenium weed have been reported to have extended their life cycle, so they now flower over much longer periods of time than seen in the native range. In addition, populations are establishing at much higher altitudes, in colder and dryer locations than seen in the native range. The invasion history of parthenium weed in different parts of the world varies from just a few years to more than a century. It would therefore be interesting to undertake a comparative, common garden study to look more closely at this possibility of rapid evolutionary change, and to do this for different geographic populations that have invaded at different periods in time. These studies should be replicated in both the native and introduced ranges of parthenium weed.

Apart from these evolving characters, allelopathy and production of novel chemicals in the introduced parthenium weed populations could also be considered as another driver of invasion success. However, recent work undertaken in Australia suggests that this is not necessarily the case, but detailed studies are still required to ascertain the chemical constituents and the amounts found within the different populations in the native and introduced ranges of parthenium weed. The knowledge of population-specific diversity and prevalence of chemical constituents found in the native and introduced populations of parthenium weed will have practical implications in understanding the invasion mechanism, the impact of the plant on human health and in the development of effective management strategies.

16.4.11 Role of donors and politicians for financial support of management

In its introduced range, parthenium weed has invaded many countries around the world. With few exceptions (e.g. Australia), most of the invaded countries are less developed and many considered to be under developed. Indeed, many of the countries that are worst affected by parthenium weed are among the poorest nations in the world, whose economies are largely dependent upon international aid. Generally, such countries do not have the capacity, knowledge and resources to manage the large-scale infestations of parthenium weed. The lack of financial resources for research on parthenium weed and its management are the main reasons for the rapid spread of weed in these countries, and to neighbouring countries. International donor agencies should generously support the required research and undertake capacity building of these less developed nations to combat the problem of invasive weeds. The support of politicians in highlighting the issue of parthenium weed to the policy makers and in convincing donors is therefore highly desired.

16.4.12 International collaboration and knowledge sharing

Parthenium weed is now a weed of global significance, affecting over 45 countries, mainly within Africa, Asia, the Pacific and Australia. However, introduction histories and infestation levels do vary widely between these countries – from the 1950s in the case of India and Australia to more recent introduction histories in countries like Malaysia, Thailand, Tanzania and now the Kingdom of Saudi Arabia. As a result, considerable amounts of research on parthenium weed and its management have already been undertaken in Australia and India, while in other countries research on parthenium has only just started or is yet to commence. Hence, it would be valuable to extend the network of international collaboration under the recently formed International Parthenium Weed Network (IPAWN). Membership should be encouraged for all nations that have the weed, or are under threat from invasion by the weed. Another example of recent knowledge sharing is the biological control expertise in Australia and South Africa. Australia has been involved in biological control research (native range surveys, agent prioritization, host specificity, rearing and release methods, and evaluation and impact assessment) for over four decades (since the mid-1970s). This work has resulted in the introduction and establishment of 11 biological control agents for the successful biological control of parthenium weed in the core infested areas in Central Queensland, where biological control has been used since the mid-1980s. Founded on Australian experience and knowledge, several effective biological control agents have now been tested and released in other countries, including India, South Africa, Ethiopia (through South Africa), Sri Lanka and Tanzania (through South Africa). Through further effective international collaboration, sharing knowledge, expertise and biological control agents, countries including Nepal, Pakistan and China could benefit from Australian experience and resources. The testing of agents for any specific countries could also be done either fully or partially in Australia.

Other areas that will benefit from international collaboration and knowledge sharing include: (i) the recent developments in molecular tools for the assessment of parthenium weed introduction and spread histories; (ii) the development of climatic and systems modelling approaches; (iii) allelopathic research; and (iv) the refining of rearing methods for biological control agents. These are all feasible through scientist-to-scientist contacts via the IPAWN and its newsletter. The dedicated Parthenium Weed Research Workshops that have been held at regular intervals (approximately every 3 years) at international weed conferences should be continued to enhance multi-agency collaborations and knowledge sharing.

16.4.13 Financial support for research

For chemical control research, funding from agrochemical companies is the most likely source. However, to enable management-directed research to take place, a long-term financial and resource commitment is required from reliable sources (i.e. local or national governments). One example where this has taken place is in Australia. The Queensland government has invested heavily in parthenium weed management research for four decades (over AUS$11 million), with an emphasis on biological control, and its commitment is continuing. This effort has resulted in the development of one of the state's most recent and successful biocontrol programmes. At the same time, the University of Queensland has been involved in parthenium weed research for over 25 years. Through postgraduate and postdoctoral research, it has undertaken numerous studies on the biology, ecology and management of parthenium weed. It has also set up the IPAWN, and runs the newsletter and regular workshops that have taken place in conjunction with major international weeds conferences.

Likewise, in India, the Indian Council for Agricultural Research has for 40 years been committed to parthenium weed research through its various research agencies and by supporting (financially) research in various agricultural universities across the country. In South Africa, the research on parthenium weed is primarily by the Agricultural Research Council's Plant Protection Research Institutes, and various universities, with financial support from the government's Working for Water Program. This has ensured that long-term research on parthenium weed has been possible in South Africa. It is essential that similar financial support and/or commitments are also made in other countries. However, for many developing countries, financial support for research is often provided by international donor agencies and delivered on an ad hoc basis with no long-term follow-up commitments, or by universities as one-off postgraduate research studies. For example, for any viable biological control research to be undertaken, a minimum commitment of 10 years of funding is needed. Hence, to continue the research initially commenced with financial support from donor agencies or local governments, long-term commitment and financial support is required from the recipient countries if this work is to be effective.

Index